T0281556

Advanced Mathematical
Modeling with Technology

Advances in Applied Mathematics

Series Editors:
Daniel Zwillinger, H. T. Banks

https://www.routledge.com/Advances-in-Applied-Mathematics/book-series/CRCADVAPPMTH?pd=published,forthcoming&pg=1&pp=12&so=pub&view=list

Advanced Mathematical Modeling with Technology

William P. Fox

College of William and Mary

Robert E. Burks

Naval Postgraduate School

CRC Press
Taylor & Francis Group

A CHAPMAN & HALL BOOK

Library of Congress CataloginginPublication Data
Names: Fox, William P., 1949– author. | Burks, Robert E., author.
Title: Advanced mathematical modeling with technology/
William P. Fox, Robert E. Burks.
Description: First edition. | Boca Raton : C&H\CRC Press, 2021. |
Series: Advances in applied mathematics |
Includes bibliographical references and index.
Identifiers: LCCN 2021000940 (print) | LCCN 2021000941 (ebook) |
ISBN 9780367494421 (hardback) | ISBN 9781003046196 (ebook)
Subjects: LCSH: Mathematical models. |
Decision making–Mathematical models. | Mathematics–Data processing.
Classification: LCC QA401 .F692 2021 (print) |
LCC QA401 (ebook) | DDC 511/.8–dc23
LC record available at https://lccn.loc.gov/2021000940
LC ebook record available at https://lccn.loc.gov/2021000941

ISBN: 9780367494421 (hbk)
ISBN: 9781032001814 (pbk)
ISBN: 9781003046196 (ebk)

Typeset in Palatino
by Newgen Publishing UK

Contents

Preface

Mathematical modeling is both a skill and an art; as such, you need to practice it in order to maintain and enhance your ability to use those skills and develop the craft of modeling. Essentially, the more practice that you get the better you will become in the craft of modeling. Topics covered in this book are the typical topics of the mathematical modeling courses that we teach or have taught over the past thirty years. We begin each chapter with a problem to motivate the reader into mathematical modeling. The problem tells "what" the issue is or problem that needs to be solved. In the chapter, we apply the principles of mathematical modeling to that problem and present the steps in obtaining a model. The key focus is the mathematical model and technology is presented as a method to solve that model or perform sensitivity analysis. We utilize technology to build, compute, or implement the model, and then analyze the model. Over the past ten years the authors have taught a three-course mathematical modeling sequence to students where Excel was chosen as the software of choice because of its availability, ease of use, familiarity with students, and access in their future jobs and careers.

Audience

This book is not intended as an introduction to mathematical modeling and would be best used for individuals or groups who have already taken an introductory mathematical modeling course and have been introduced to technology at some level. *Applied Advanced Mathematical Modeling* would be of interest to mathematics departments that offer mathematical modeling courses focused on discrete modeling or modeling for decision-making.

The following groups would benefit from using this book:

- Undergraduate students in quantitative methods courses in business, operations research, industrial engineering, management sciences, industrial engineering, or applied mathematics.
- Graduate students in discrete mathematical modeling courses covering topics from business, operations research, industrial engineering, management sciences, industrial engineering, or applied mathematics.

Objectives

The objective of *Advanced Mathematical Modeling with Technology* is to present and illustrate advanced applied mathematical modeling techniques that are accessible to students from many disciplines. The goal is to support the development of a desire for lifelong mathematical learning, habits of mind and competent and confident problem solvers for the twenty-first century. The goal is to be illustrative in nature. Chapter 1: Perfect Partners helps set the tone for the incorporation of mathematical modeling and technology. This chapter provides a process for thinking about the problem and illustrates many scenarios and examples. We establish the solution process and will provide solutions to these examples in later chapters.

Technology

We have selected Maple©, Excel ©, and R© to support the mathematical modeling process, where applicable to the content, over the many other technologies because of their wide accessibility. It is important that individuals have a basic fundamental understanding of one of these three technologies to get the most out of this book. We provide the appropriate code for the relevant technologies that we think are applicable to that chapter. In this way, the chapter can be covered in general discussion and the technology chosen from those provided to illustrate the models. However, we do not provide a discussion covering the basic setup of the technology.

Technology, such as Excel, Maple, and R, is really just a means to an end. Model building and interpretation are two essential elements and technology is merely a good tool to perform necessary calculations needed in modeling. Besides taking advantage of the many features in Excel and its add-ins, the Solver and Data Analysis Tool Pack, we present features that anyone can create using Excel. Many of the figures in this book are generated with technology. It is important to note that one limitation of Excel is its limited graphics capabilities, especially in three dimensions. Although we attempt to use all features available in Excel, occasionally we felt the need to create Excel templates, macros, and programs to solve the problem. This material will be made available upon request. Maple and R are both great compliments to Excel, and both have great graphics capabilities and many integrated commands to assist with mathematical modeling. Maple templates and example solution are also available, upon request, from the authors.

Organization

This book contains information that could easily be covered in an advanced semester course focused on mathematical modeling or a semester survey course of the topics. The book is designed to provide instructors the flexibility to pick and select material to support their course. Chapter 2 through Chapter 11 provide material and solution techniques to address the problems introduced in Chapter 1.

- Chapter 2 provides a review of both linear and nonlinear models of discrete dynamical systems. The chapter provides examples to determine and understand the long-term behavior of these systems. It also looks at the important concepts of equilibrium and stability in these models.

- Chapter 3 covers using ordinary differential equations (ODEs) to provide a more robust solution to problems. The chapter provides an understanding of analytical solution techniques for separable and linear ODEs. It also addresses numerical methods and the importance of slope fields.

- Chapter 4 builds upon the concepts introduced in Chapter 3 and extends them to address models of a system of ordinary differential equations. In addition, the chapter develops an understanding of direction fields and phase portraits in solving this models. The chapter also introduces the concept of applying qualitative assessment to autonomous systems.

- Chapter 5 addresses the use of regression tools for analyzing simple linear regression to advanced regression methods. The chapter provides an understanding of the concepts of correlation and linearity in solving regression models. It will also use several examples to build and interpret nonlinear, logistics, and Poisson regression models.

- Chapter 6 covers mathematical programming (linear, integer, and nonlinear) to solve problems in support of decision-making. We start with defining the mathematical programming methods and illustrate some formulations concepts. Technology is used to solve the formulated problems. This chapter also sets the foundation for future chapters covering data envelopment analysis and game theory.

- Chapter 7 is designed to build upon the concepts of Chapter 6 and introduces single variable nonlinear optimization problems. The chapter covers solution techniques that include numerical search techniques such as the Golden Section, Bisection method, and Newton's method to solve these problems.

- Chapter 8 continues to build the nonlinear capability of the reader by expanding from single variable to multivariable optimization problems and techniques. In addition, the chapter looks at constrained and inequality constrained optimization problems and expands the solution techniques to include Kuhn-Tucker.

- Chapter 9 discusses the power and limitations of simulations in developing solutions to problems. The chapter will develop an understanding of the concept of algorithms while building both deterministic and stochastic simulations to solve problems.

- Chapter 10 discusses the use of multi-attribute decision-making (MADM) in solving problems. Many problems in the real world have multiple criteria to consider in weighing alternatives and courses of actions. The chapter will cover multiple weighting schemes to include: entropy, rank order centroid, ratio, and pairwise comparison. In addition, the MADM methods of data envelopment analysis, simple of additive weights, and technique of order performance by similarity to ideal solutions are covered in the chapter.

- Chapter 11 covers the use of game theory in solving problems. The chapter covers both total and partial conflict games. The chapter uses multiple case studies to demonstrate the use of game theory to show the type of real decision problems and analysis.

The length of this book prevents us from addressing every potential nuance in modeling real world problems. We attempt to provide a useful set of models and potential appropriate techniques to obtain useful results for a common problems the reader may encounter as a guide. We do assume a basic or fundamental background in mathematical modeling and therefore only spend a little time establishing the procedure before we return to providing examples and solution techniques.

While the focus of this book is on utilization in a mathematical modeling course, it does have application for decision-makers in any discipline. The book provides an overview to the decision-maker of the wide range of applications of quantitative approaches to aid in the decision-making process. As we constantly remind our students, mathematical modeling is not designed to tell you what to do but it does provide insights and is designed to support critical thinking and the decision-making process. We view the mathematical modeling process as a framework for decision-makers. This framework consist of four key components: the formulation process, the solution process, interpretation of the solution in the context of the actual problem, and sensitivity analysis. The users of mathematical modeling should question the procedures and techniques and assumptions used in the analysis during every step in the process. At a minimum, you should always consider, "did I use an appropriate technique" to obtain a solution and why were other techniques discounted during the process? Another question could be "did I oversimplify the modeling process" so that any solution I develop will not really apply in this situation?

We thank all the mathematical modeling students that we have had over the last thirty years as well as all the colleagues who have taught mathematical modeling with us during this adventure. We particularly single out the following who helped in our three course mathematical modeling sequence at the Naval Postgraduate School over the years: Bard Mansger, Mike Jaye, Steve Horton, Patrick Driscoll, and Greg Mislick. We are especially appreciative of the mentorship of Frank R. Giordano over the past thirty – plus years.

College of William and Mary
William P. Fox

Naval Postgraduate School
Robert E. Burks

1

Perfect Partners: Mathematical Modeling and Technology

Objectives

1. Understand the mathematical modeling process.
2. Understand the process of decision modeling.
3. Understand that models have both strengths and limitations.

Consider the importance of decision-making in such areas as business (B), industry (I), and government (G). Decision-making under uncertainty is incredibly important in many areas of life but it is particularly important in business, industry, and government (BIG). BIG decision-making is essential to success at all levels and we do not encourage "shooting from the hip." However, we do recommend good analysis for the decision-maker to examine and question in order to find the best alternative course of action. So, why mathematical modeling?

A **mathematical model** may be defined as a description of a real-world system using mathematical concepts to facilitate the explanation of change in the system or to study the effects of different elements of the system and to understand changes in the patterns of behavior.

Mathematical models are used not only in the natural sciences (such as physics, biology, earth science, meteorology) and engineering disciplines (such as computer science, artificial intelligence), but also in the social sciences (such as business, economics, psychology, sociology and political science); physicists, engineers, statisticians, operations research analysts and economists use mathematical models most extensively. A model may help to explain a system and to study the effects of different components, and to make predictions about behavior.

Mathematical models can take many forms, including but not limited to dynamical systems, statistical models, differential equations, or game theoretic models. These and other types of models can overlap, with a given model involving a variety of abstract structures. In general, mathematical models may include logical models, as far as logic is taken as a part of mathematics. In many cases, the quality of a scientific field depends on how well the mathematical models developed on the theoretical side agree with results of repeatable experiments. Lack of agreement between theoretical mathematical models and experimental measurements often leads to important advances as better theories are developed.

1.1 Examples of Really Big Problems and the Process of Mathematical Modeling

Consider for a moment a basic real-world search situation. Two observation posts 5.43 miles apart pick up a brief radio signal. The sensing devices were oriented at 110° and 119° respectively when a signal was detected. The devices are accurate to within 2° (that is ±2° of their respective angle of orientation). According to intelligence, the reading of the signal came from a region of active terrorist exchange, and it is inferred that there is a boat waiting for someone to pick up the terrorists. It is dusk, the weather is calm, and there are no currents. A small helicopter leaves a pad from Post 1 and is able to fly accurately along the 110° angle direction. This helicopter has only one detection device, a searchlight. At 200 feet, it can just illuminate a circular region with a radius of 25 feet. The helicopter can fly 225 miles in support of this mission due to its fuel capacity. Some basic pre-launch decisions would include; where do we search for the boat? How many search helicopters should you use to have a "good" chance of finding the target?

Photochemical smog permeates the Los Angeles basin most days of the year. While this problem is not unique to the Los Angeles area, conditions in the basin are well suited to this phenomenon. The surrounding mountains and frequent inversion layers create the stagnant air that give rise to these conditions. Can we build a model to examine this? Can we use such a model to study or analyze the poor air quality in Los Angeles due to traffic pollution or measuring vehicle emissions?

In the sport of bridge jumping, a willing participant attaches one end of bungee cord to himself, attaches the other end to a bridge railing, and then drops off a bridge. Is this a safe sport? Can we describe the typical motion?

You are a new city manager in California. You are worried about earthquake survivability of your city's water tower. You need to analyze the effects of an earthquake on your water tower and see if any design improvements are necessary. You want to prevent catastrophic failure.

If you have flown lately, you may have noticed that most airplanes are full. As a matter of fact, many times an announcement is made that the plane is overbooked, and the airlines are looking for volunteers to take a later flight.

Why do airlines overbook? Should they overbook? What impact does this have on the passengers? What impact does it have on the airlines?

These events all share one common element – we can model them using mathematics to support making decisions. This textbook will help you understand what a mathematical modeler might do for you as a confident problem-solver using the techniques of mathematical modeling. As a decision-maker, understanding the possibilities and asking the key questions will enable better decision to be made.

1.2 The Modeling Process

In this chapter, we turn our attention to the process of modeling and examine many different scenarios in which mathematical modeling can play a role.

Mathematical modeling requires as much art as it does science. Thus, modeling is more of an art than a science. Modelers must be creative and willing to be more artistic or original in their approach to the problem. They must be inquisitive and question their assumptions, variables, and hypothesized relationships. Modelers must also think outside the box in order to analyze the models and its results and ensure their model and results pass the commonsense test. Science is very important and understanding science enables one to be more creative in viewing and modeling a problem. Creativity is extremely advantageous in problem-solving with mathematical modeling.

To gain insight we should consider one framework that will enable the modeler to address the largest number of problems. The key is that there is something changing for which we want to know the effects. We call this the system under analysis. The real world system can be very complicated or very simplistic. This requires a process that allows for both types of real world systems to be modeled within the same process.

Consider striking a golf ball with a golf club from a tee. Our first inclination is to use the equations about distance and velocity that we used in high school mathematics class. These equations are very simplistic and ignore many factors that could impact the fall of the ball such as wind speed, air resistance, mass of the ball, and other factors. As we add more factors, we can improve the precision of the model. Adding these additional factors makes the model more realistic and more complicated to produce. Understanding this model might be a first start in building a model for such situations or similar situation such as a bungee jumper or bridge swinger. These systems are similar for part of the model: the free fall portion has similar characteristics.

Figure 1.1 provides a closed loop process for modeling. Given a real-world situation like the one above, we collect data in order to formulate a mathematical model. This mathematical model can be one we derive or select from a

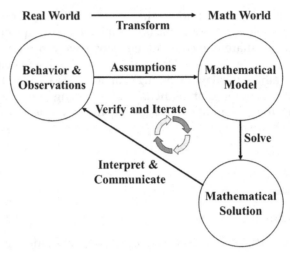

FIGURE 1.1
Modeling real world systems with mathematics.

collection of already built mathematical models depending on the required level of sophistication. We then analyze the model that we used and reach mathematical conclusions about it. Next, we interpret the model and either make predictions about what has occurred or offer explanation as to why something has occurred. Finally, we test our conclusion about the real-world system with new data. We may refine or improve the model to improve its ability to predict or explain the phenomena. We might even reformulate a new mathematical model.

Mathematical Modeling

We will build some mathematical models describing change in the real world. We will solve these models and analyze how good our resulting mathematical explanations and predictions are. The solution techniques that we employ in subsequent chapters take advantage of certain characteristics that the various models enjoy. Consequently, after building the models, we will classify the models based on their mathematical structure.

When we observe change, we are often interested in understanding why change occurs the way it does, perhaps to analyze the effects of different conditions, or perhaps to predict what will happen in the future. Often, a mathematical model can help us understand a behavior better, while allowing us to experiment mathematically with different conditions. For our purposes, we will consider a mathematical model to be a mathematical construct designed to study a particular real-world system or behavior. The model allows us to use mathematical operations to reach mathematical conclusions about the model as illustrated in Figure 1.1.

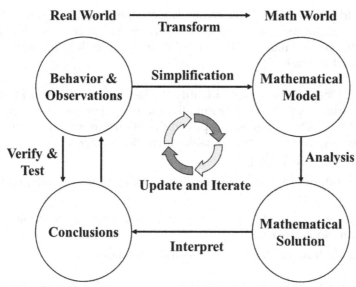

FIGURE 1.2

In reaching conclusions about a real-world behavior, the modeling process is a closed system (adapted from Giordano et al., 2014).

Models and Real-World Systems

A system is an assemblage of objects joined by some regular interaction or interdependence. Examples include sending a module to Mars, handling the United States debt, a fish population living in a lake, a TV-satellite orbiting the earth, delivering mail or locations of service facilities. The person modeling is interested in understanding not only how a system works but also what interactions cause change and how sensitive the system is to changes in these inputs. Perhaps the person modeling is also interested in predicting or explaining what changes will occur in the system as well as when these changes might occur.

Figure 1.2 suggests how we can obtain real-world conclusions from a mathematical model. First, observations identify the factors that seem to be involved in the behavior of interest. Often we cannot consider, or even identify, all the relevant factors, so we make simplifying assumptions excluding some of them. Next, we conjecture tentative relationships among the identified factors we have retained, thereby creating a rough "model" of the behavior. We then apply mathematical reasoning that leads to conclusions about the model. These conclusions apply only to the model and may or may not apply to the actual real-world system in question. Simplifications were made in constructing the model and the observations upon which the model is based invariably contain errors and limitations. Thus, we must carefully account for these anomalies and test the conclusions of the model against real-world observations. If the model is reasonably valid, we can then draw inferences about the real-world behavior from the conclusions drawn from the model. In summary, we have the following procedure for investigating real-world behavior:

Step 1. Observe the system, identify the key factors involved in the real-world behavior, simplify initially, and refine later as necessary.

Step 2. Conjecture or guess the possible relationships or inter-relationships among the factors and variables identified in Step 1.

Step 3. Solve the model.

Step 4. Interpret the mathematical conclusions in terms of the real-world system.

Step 5. Test the model conclusions against real-world observations– the commonsense rule.

Step 6. Perform model testing or sensitivity analysis.

There are various kinds of models that we will introduce as well as methods or techniques to solve these models in the subsequent chapters. An efficient process would be to build a library of models and then be able to recognize various real-world situations to which they apply. Another task is to formulate and analyze new models. Still another task is to learn to solve an equation or system in order to find more revealing or useful expressions relating to the variables. Through these activities we hope to develop a strong sense of the mathematical aspects of the problem, its physical underpinnings, and the powerful interplay between them.

Most models do simplify reality. Generally, models can only approximate real-world behavior. Next, let's summarize a process for formulating a model.

Model Construction

Let's focus our attention on the process of model construction. An outline is presented as a procedure to help construct mathematical models. In the next section, we will illustrate this procedure with a few examples.

These nine steps, a modification of the six-step approach presented by Giordano (Giordano et al., 2014) act as a guide for thinking about the problem and getting started in the modeling process.

Step 1. Understand the problem or the question asked.

Step 2. Make simplifying assumptions. Justify your assumptions.

Step 3. Define all variables and provide units.

Step 4. Construct a model.

Step 5. Solve and interpret the model.

Step 6. Verify the model.

Step 7. Identify the strengths and weaknesses of your model.

Step 8. Sensitivity Analysis or Model Testing of the model. Do the results pass the "common sense" test?

Step 9. Implement and maintain the model for future use.

Let's discuss each step in more depth.

Step 1. Understand the problem or the question asked:

Identifying the problem to study is usually difficult. In real life no one walks up to you and hands you an equation to be solved. Usually, it is a comment like, "we need to make more money," or "we need to improve our efficiency." We need to be precise in our formulation of the mathematics to describe the situation.

Step 2. Make simplifying assumptions:

Start by brainstorming the situation. Make a list of as many factors, or variables, as you can. Realize we usually cannot capture all these factors influencing a problem. The task is simplified by reducing the number of factors under consideration. We do this by making simplifying assumptions about the factors, such as holding certain factors as constants. We might then examine to see if relationships exist between the remaining factors (or variables). Assuming simple relationships might reduce the complexity of the problem. Once you have a shorter list of variables, classify them as independent variables, dependent variables, or neither.

Step 3. Define all variables:

It is critical to define all your variables and provide the mathematical notation to be used for each.

Step 4. Select the modeling approach and formulate the model:

Using the tools in this text and your own creativity build a model that describes the situation and whose solution helps to answer important questions.

Step 5. Solve and interpret the model:

We take the model we constructed in Steps 1–4 and solve it. Often this model might be too complex or unwieldy so we cannot solve it or interpret it. If this happens, we return to Steps 2–4 and simplify the model further.

Step 6. Verify the model:

Before we use the model, we should test it out. There are several questions we must ask. Does the model directly answer the question or does the model allow for the answer to the questions to be answered? Is the model useable in a practical sense (can we obtain data to use the model)? Does the model pass the commonsense test?

We like to say that we corroborate the reasonableness of our model rather than verify or validate the model.

Step 7. Strengths and weaknesses:

No model is complete with self-reflection of the modeling process. We need to consider not only what we did correctly, but what we did that might be suspect as well as what we could do better. This reflection also helps in refining models.

Step 8. Sensitivity analysis and model testing:

A modeler wants to know how the inputs affect the ultimate output for any system. Passing the common sense is essential. One of the authors once had a class model Hooke's law with springs and weights. He asked them all to use their model to see how far the spring would stretch using their weight. They all provided the numerical answers, but none said that the spring would break under their weight.

Step 9. Refine, implement, and maintain the model:

A model is pointless if we do not use it. The more user-friendly the model the more it will be used. Sometimes the ease of obtaining data for the model can dictate its success or failure. The model must also remain current. Often this entails updating parameters used in the model.

1.3 Illustrative Examples

We now demonstrate the modeling process that was presented in the previous section. Emphasis is placed on problem identification and choosing appropriate (useable) variables. We do not build the models as these modeling examples are repeated later in the book and the models are completed and discussed there.

Example 1.1 The Size of Prehistoric Creatures

Scenario. *Titanuswalleri* really lived about two million years ago on the oak and grass savannahs of what is now Florida. Dr. Bob Chandler, one of the world's experts on fossil birds, from Georgia College and State University, has dredged up a number of bits and pieces of this bizarre predatory bird from the Santa Fe River near Gainesville, Florida. Here are some facts about the terror bird from Chandler (1994) (adapted from http://discovermagazine. com/1997/jun/terrortaketwo1149):

- Giant, flightless predatory bird.
- Lived in South America 30 million years before the Interchange.
- Fossils have been found of the terror bird, *Titanuswalleri*, in Florida.
- Suspected to be a fierce hunter who would lie in wait and ambush its prey and attack from the tall grasslands.
- Suspected to pin down its prey with its beak.
- Suspected to use 4–5 inch inner toe claw with its beak to shred its prey.
- Another unique feature was this bird had arms, not wings (more powerful than the arms of the velociraptor.

We realize that bones provide a good indication of size (height) but not reliable body weight (weight is not easily fossilized). We want to model its weight based upon the information that can be determined from the fossils.

Understanding the Problem: What is the relationship between the bones found and the original "grown" size of the creature? This might be too general so let's restrict our problem. The fossil is a femur bone that is intact. We want to know if we can find a relationship between the weight of an animal and the size of its femur (measured as circumference). If we can find a relationship, then we can use the model to predict the size (in weight measures) of the terror bird as a function of the circumference of the femur.

Assumptions: We assume that the thickness (measured by circumference) of the femur is important to the size of the animal. We might conjecture that thicker bones support more weight. We know that the terror bird is an ancestor of modern birds, but should birds be used to help the build the model? We might assume that bird data is sufficient and use only that data to build a model. Or should we rely on species that lived about when the terror bird roamed the earth, like the dinosaurs and use their data to help build the model? We could assume that dinosaur data is more appropriate to use to build a model. We might want to try both approaches and compare results. If bird data is available, we might address the appropriateness of using modern bird data to build a model for a bird that lived two million years ago and did not even fly. If more terror bird fossils are discovered, we might attempt to model their size and get a sense of differing terror bird sizes.

Thus, the model that we build is affected by the assumptions we used and data available.

Example 1.2 Prescribed Drug Dosage

Scenario. Consider a patient that needs to take a newly marketed prescribed drug. To prescribe a safe and effective regimen for treating the disease, one must maintain a blood concentration above some effective level and below an unsafe level.

Understanding the Problem: Our goal is a mathematical model that relates dosage and time between dosages to the level of the drug in the bloodstream. What is the relationship between the amount of drug taken and the amount in the blood after time, t? By answering this question, we are empowered to examine other facets of the problem of taking a prescribed drug.

Assumptions: We should choose or know the disease in question and the type (name) of the drug that is to be taken. We will assume the drug is rythmol, a drug taken to control the heart rate and which is called an anti-arrhythmic. We need to know or to find decaying rate of rythmol in the blood stream. This might be found from data that has been previously collected. We need to find the safe and unsafe levels of rythmol based upon the drug's "effects" within the body. This will serve as bounds for our model. Initially, we might assume that the patient size and weight has no effect on the drug's decay rate. We might assume that all patients are about the same size and weight. All are in good health and no one takes other drugs that affect the prescribed drug. We assume all internal organs are functionally properly. We might assume that we can model this using a discrete time period even though the absorption rate is a continuous function. These assumptions help simplify the model.

Example 1.3 Determining Heart Weight

TABLE 1.1

Data for mammals' heart sizes (adapted from special projects in MA 381 at West Point)

Animal	Heart weight(g)	Diameter (mm)
Mouse	0.13	0.55
Rat	0.64	1.00
Rabbit	5.80	2.20
Dog	102.00	4.00
Sheep	210.00	6.50
Ox	2,030.00	12.00
Horse	3,900.00	16.00

Let's assume we are interested in building a model that models heart weight as a function of the size of the heart. We might have access to data that relates, for the following seven mammals, their heart weight in grams and their diameter of the left ventricle of the heart measured in millimeters (mm) shown in Table 1.1.

Problem Identification: Find a relationship between heart weight and the diameter of the left ventricle of the heart.

Assumptions: We will assume that the heart weight is typical for a healthy animal. We further assume that the diameter is measured the same way for each heart. We might assume that all mammals are scale models of other mammals. We assume that all mammals are in good health. We assume that the data was collected the same way for each mammal.

Example 1.4 Bridge Too Far

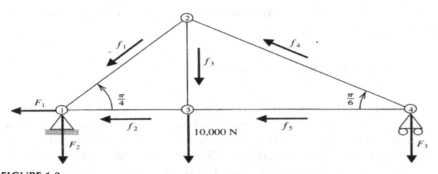

FIGURE 1.3
Bridge truss (from Burden and Faires, 1997, p. 423).

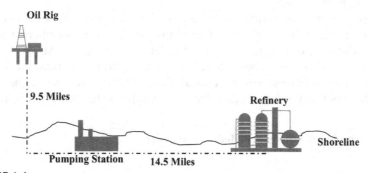

FIGURE 1.4
Oil rig, pumping station, and refinery.

Consider an engineering design for a truss bridge as shown in Figure 1.4. Trusses are lightweight structures capable of carrying heavy loads. In civil engineering bridge design, the individual members of the truss are connected with rotatable pin joints that permit forces to be transferred from one member of the truss to another. The accompanying Figure 1.4 shows a truss that is held stationary at the lower left endpoint #1, is permitted to move horizontally at the lower right endpoint #4, and has pin joints at #1, #2, #3, and #4 as shown. A load of 10 kilo newtons (kN) is placed at joint #3, and the forces on the members of the truss have magnitude given by f1, f2, f3, f4, and f5 as shown.

The stationary support member has both a horizontal force F1 and a vertical force F2, but the movable support member has only the vertical force F3.

If the truss is in static equilibrium, the forces at each joint must add to the zero vector, so the sum of the horizontal and vertical components of the forces at each joint must be zero. This produces the system of linear equations.

Problem Identification: Determine the forces on each joint as a function of the angles between joints.

Assumptions: We assume the bridge is sturdy and will not falter or collapse. We assume the motion of the bridge due to normal motion is negligible. We assume a design such as the one provided. We might assume the bridge is not in an earthquake zone.

Example 1.5 Oil Rig and Pumping Station Location

Consider an oil-drilling rig that is 8.5 miles offshore. The drilling rig is to be connected by underwater pipe to a pumping station. The pumping station is connected by land-based pipe to a refinery, which is 14.7 miles down the shoreline from the drilling rig (Figure 1.5). The underwater pipe costs $31,575 per mile, and land-based pipe costs $13,342 per mile. You are to determine where to place the pumping station to minimize the cost of the pipe.

Problem Identification: Build a model to minimize the cost of a pipe from the oil rig to the refinery.

Assumptions: We assume that the costs based upon solid estimates and will not fluctuate during the building of the pipeline. We assume no weather delays or any other natural or unnatural delays to building. All workers are competent in their jobs. The materials are not flawed in any way. We assume no natural disaster or storms that might disrupt work. We assume that when built the pipeline works successfully. A schematic is shown in Figure 1.4

Example 1.6 Emergency Medical Response

TABLE 1.2

Average travel times from Zone*i* to Zone *j* in perfect conditions

Zones	1	2	3	4	5	6
1	1	8	12	14	10	16
2	8	1	6	18	16	16
3	12	18	1.5	12	6	4
4	16	14	4	1	16	12
5	18	16	10	4	2	2
6	16	18	4	12	2	2

TABLE 1.3

Population in each zone

Zone	Population
1	50,000
2	80,000
3	30,000
4	55,000
5	35,000
6	20,000
Total	270,000

The Emergency Service Coordinator (ESC) for a county is interested in locating the county's three ambulances to maximize the residents that can be reached within eight minutes in emergency situations. The county is divided into six zones and the average time required to travel from one region to the next under semi-perfect conditions are summarized in Table 1.2.

The population in zones 1, 2, 3, 4, 5, and 6 are given in Table 1.3:

Problem Identification: Determine the location for placement of the ambulances to maximize coverage within the allotted time.

Assumptions: We assume that the population in the zones do not change in the short run. We also assume traffic patterns and time of day do not affect the times to travel from zone i to zone j.

Example 1.7 Bank Service Problem

The bank manager is trying to improve customer satisfaction by offering better service. The management wants the average customer to wait to be less than two minutes and the average length of the queue (length of the line waiting) to be two or fewer. The bank estimates about 150 customers per day. The existing arrival and service times are given in Tables 1.4 and 1.5.

TABLE 1.4

Arrival times

Time between arrival (min.)	Probability
0	0.10
1	0.15
2	0.10
3	0.35
4	0.25
5	0.05

TABLE 1.5

Service times

Service Time (min.)	Probability
1	0.25
2	0.20
3	0.40
4	0.15

Problem Identification: Build a mathematical model to determine whether the bank is meeting its goals. Determine if the current customer service is satisfactory according to the manager guidelines. If not, determine through modeling the minimal changes for servers required to accomplish the manager's goal.

Assumptions: We assume customer demands remains consistent over time.

1.4 Technology

The mathematical modeling necessary to address the examples above is enhanced with the use of technology. In addition, technology allows you to address problems that are much more robust and difficult that the example problems. This is one of the reasons this chapter is called "Perfect Partners." The partnering of technology with modeling is both key and essential to good modeling principles and practices. In this book, we illustrate three different technologies, Excel, Maple, and R as we look at and solve many real world related problems. The intent is to not make the reader an expert in any of the technologies but to demonstrate the capabilities of these technology platforms to address relevant problems.

Although Excel might not be the "go-to" technology for many mathematicians or academicians, it is a common tool used by modeling practitioners in the real world. The goal of illustrating Excel in this book is to demonstrate that it has the capability of solving many different types of problems and to empower students in math, science, and engineering to use Excel properly for solving future problems.

MAPLE is an excellent technology for mathematics and operations research majors. Its power and graphical interface in two and three dimensions make it an excellent tool.

R is a free software programming language and a software environment for statistical computing and graphics. The R language is widely used among statisticians and data miners for developing statistical software and data analysis. R is a language and environment for statistical computing and graphics. R provides a wide variety of statistical techniques (linear and nonlinear modeling, classical statistical tests, time-series analysis, classification, clustering, …) and graphical techniques, and is highly extensible. The S language is often the vehicle of choice for research in statistical methodology, and R provides an Open Source route to participation in that activity.

One of R's strengths is the ease with which well-designed publication-quality plots can be produced, including mathematical symbols and formulae where needed. Great care has been taken over the defaults for the minor design choices in graphics, but the user retains full control.

R is available as Free Software under the terms of the Free Software Foundation's GNU General Public License in source code form. It compiles and runs on a wide variety of UNIX platforms and similar systems (including FreeBSD and Linux), Windows, and MacOS.

The R Environment

R is an integrated suite of software facilities for data manipulation, calculation and graphical display. It includes:

- an effective data handling and storage facility;
- asuite of operators for calculations on arrays, in particular matrices;
- a large, coherent, integrated collection of intermediate tools for data analysis;
- graphical facilities for data analysis and display either on-screen or on hardcopy; and
- a well-developed, simple, and effective programming language which includes conditionals, loops, user-defined recursive functions, and input and output facilities.

The term "environment" is intended to characterize it as a fully planned and coherent system, rather than an incremental accretion of very specific and inflexible tools, as is frequently the case with other data analysis software.

Many users think of R as a statistics system. We prefer to think of it as an environment within which statistical techniques are implemented. R can be extended (easily) via packages. There are about eight packages supplied with the R distribution and many more are available through the CRAN family of Internet sites covering a very wide range of modern statistics.

1.5 Exercises

1. How would you approach a problem concerning a drug dosage? Do you always assume the doctor is right?

2. In modeling the size of any prehistoric creature, what information would you like to be able to obtain? What additional assumptions might be required?

3. In the oil rig problem (Example 1.5), what other factors might be critical in obtaining a "good" model that predicts reasonably well? What variables could be important that were not considered?

4. For the model in Example 1.3, are the assumptions about the data reasonable? How would you collect data to build the model? What other variables would you consider?

5. In Example 1.4, what is the impact of the location of the bridge?

6. In Example 1.7 on the bank queue problem, discuss the criticality of the assumptions. Do you feel that more training is as valuable as adding another server?

1.6 Projects

1. Is Michael Jordan the greatest basketball player of the century? What variables and factors need to be considered?

2. What kind of car should you buy when you graduate from college? What factors should be in your decision? Are car companies modeling your needs?

3. Consider domestic decaffeinated coffee brewing. Suggest some object-ives that could be used if you wanted to market your new brew. What variables and data would be useful?

4. Replacing a coaching legend at a school is a difficult task. How would you model this? What factors and data would you consider? Would you equally weigh all factors?

5. How would go about building a model for the "best pro football player of all time"?

6. Rumors abound in major league baseball about steroid use. How would you go about creating a model that could imply the use of steroids? Relate baseball's steroids rules to the Yankee's Alex Rodriquez case.

7. Since 2013, the America League has won 7 straight All-star games and 20 of the last 23 games dating back to 1997. Help the National League prepare to win by designing a model for players or a line-up that could help them change their outcome.

1.7 References and Suggested Further Reading

Chandler, R. M. (1994). The wing of Titaniswalleri (Aves: Phorusrhacidae) from the Late Blancan of Florida. *Bulletin of the Florida Museum of Natural History, Biological Sciences*, 36: 175–180.

Giordano, F., W. Fox, and S. Horton (2014). *A First Course in Mathematical Modeling*. 5th ed. Cengage Publishers, Boston, MA.

COMAP, Modeling Competition Sites found atwww.comap.com/contests

2

Review of Modeling with Discrete Dynamical Systems and Modeling Systems of DDS

Objectives

1. Review of linear and nonlinear discrete dynamical systems.
2. Review use of technology with Excel, Maple, or R.
3. Define, model, and solve systems of discrete dynamical systems.
4. Analyze the long-term behavior of systems of discrete dynamical systems.
5. Understand the concepts of equilibrium and stability in systems.
6. Model both linear and nonlinear systems of discrete dynamical systems.

2.1 Introduction and Review of Modeling with Discrete Dynamical Systems

Consider car options today: should we buy or lease a car? How do we go about financing a new car? Once you have looked at the makes and models to determine what type of car you like, it is time to consider the "costs" and finance packages that lure potential buyers into the car dealerships. This process can be modeled as a dynamical system. Payments are made typically at the end of each month. The amount owed is predetermined as we will see later in the chapter.

What if your company is faced with a decision as to buying or leasing new computers for the company use for the next short-term horizon? How could this decision be analyzed? We might employ discrete dynamical systems to model and analyze the possible decisions.

A *discrete dynamical system* (DDS) is a system that changes over time. In this case the time interval is discrete such a minutes, hours, days, weeks, or years. A DDS is easy to model but maybe difficult or even impossible to solve in closed form. In this chapter, although we present some easy closed form solutions, we concentrate more on building the models and obtaining numerical and graphical solutions.

Let's define the dynamical system $A(n)$ to be the amount of antibiotic in our blood stream after n *time* periods. The domain is non-negative integers representing the time periods from 0, 1, 2, ... that will be the inputs to the function. Since the domain is discrete then our function is a discrete function. The range is the values of $A(n)$ determined for each value of the domain. Thus, $A(n)$ also represents the dependent variable. For each input value of the domain from 0, 1, 2, ... the result is one and only one $A(n)$, thus $A(n)$ is a function.

There are three components to dynamical systems: an equation for sequence representing $A(n)$, the time period n is well defined, and at least one starting value. This starting value is called an **initial condition**. For example, if we start with no antibiotic in our system then $A(0) = 0$ mg is our initial condition, but if we started after we took an initial 200 mg tablet then $A(0) = 200$ mg would be our initial condition. An example of a discrete dynamical system with its initial condition would be:

$$A(n+1) = 0.5\, A(n), A(0) = 500$$

We are interested in modeling discrete *change*. Modeling with Discrete Dynamical Systems employs a method to explain certain discrete behaviors or make long-term predictions. A powerful paradigm that we use to model with discrete dynamical systems is:

future value = present value + change.

The dynamical systems that we will study with this paradigm may differ in appearance and composition, but we will be able to solve a large class of these "seemingly" different dynamical systems with similar methods. In this chapter we will use iteration and graphical methods to answer questions about the discrete dynamical systems.

We will use flow diagrams to help us see how the dependent variable changes. These flow diagrams help to see the paradigm and put it into mathematical terms. Let's consider financing a new Ford Mustang. The cost is $25,000 and you can put down $2,000, so you need to finance $23,000. The dealership offers you 2% financing over 72 months. Consider the flow diagram for financing the car depicted in Figure 2.1.

We use this flow diagram to help build the discrete dynamical model. Let $A(n)$ = the amount owed after n months. Notice that the arrow pointing into the circle is the interest to the unpaid balance. This increases your debt. The

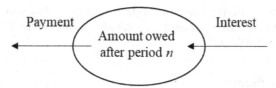

FIGURE 2.1
Flow diagram for financing a car.

arrow pointing out of the circle is your monthly payment that decreases your debt. We define the following variables:

$A(n+1)$ = the amount owed in the future;

$A(n)$ = amount currently owed;

Change as depicted in the flow diagram is $i\,A(n) - P$, so the model is

$$A(n+1) = A(n) + i\,A(n) - P,$$

where

i is the monthly interest rate and

P is the monthly payment.

We model dynamical systems that have only **constant coefficients**. A dynamical system with constant coefficients may be written in the form

$$a(n + 3) = b_2\,a(n + 2) + b_1\,a(n + 1) + b_0\,a(n),$$

where b_0, b_1, and b_2 are arbitrary constants.

Solutions to Discrete Dynamical Systems

Although some DDS have closed form analytical solution that we will discuss, our emphasis will be on iterative and graphical solution that we will illustrate on Excel. We will show that DDS are relatively easy to model. The solutions may always be obtained by iterative and graphical methods. Not all DDS have closed form analytical solutions.

Let's begin with a simple DDS model with a closed form solution.

Theorem 2.1 The solution to a linear discrete dynamical system $a(n + 1) = r\,a(n)$ for $r \neq 0$ is

$$a(k) = b^k\,a(0), \tag{2.1}$$

where $a(0)$ is the initial condition of the system at time period 0 and k is a generic time period.

Technology is an integral part of models with discrete dynamical systems. Every DDS has a numerical and graphical solution, which can easily be attained with technology.

A Drug Dosage Example

Suppose that a doctor prescribes that their patient takes a pill containing 100 mg of a certain drug every hour. Assume that the drug is immediately ingested into the bloodstream once taken. Also, assume that every hour the patient's body eliminates 25 per cent of the drug that is in his/her bloodstream. Suppose that the patient had 0 mg of the drug in his/her bloodstream prior to taking the first pill. How much of the drug will be in his/her bloodstream after 72 hours?

Problem Statement: Determine the relationship between the amount of drug in the bloodstream and time.

Assumptions: The system can be modeled by a discrete dynamical system. The patient is of normal size and health. There are no other drugs being taken that will affect the prescribed drug. We assume that there are no internal or external factors that will affect the drug absorption rate. The patient always takes the prescribed dosage at the correct time. Figure 2.2 shown the change diagram.

Variables:

Define $a(n)$ to be the amount of drug in the bloodstream after period n, $n = 0, 1, 2,\ldots$hours.

Change Diagram:

Model Construction:

Let's define the following variables:

$a(n+1)$ = amount of drug in the system in the future;

$a(n)$ = amount currently in system.

We define change as follows: change = dose – loss in system

$$\text{change} = 100 - 0.25\, a(n);$$

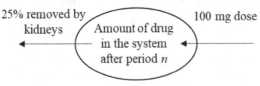

FIGURE 2.2

Change diagram for drugs in system.

so, *Future = Present + Change* is

$$a(n + 1) = a(n) - 0.25\, a(n) + 100$$

or

$$a(n + 1) = 0.75\, a(n) + 100.$$

We note that this is not in the form as Theorem 2.1 so we introduce Theorem 2.2.

Theorem 2.2 The solution to a linear discrete dynamical system $a(n+1) = r\, a(n) + d$ for $r \neq 0$ and $d \neq 0$ is

$$a(k) = b^k C + (d/(1-r)), \tag{2.2}$$

where $d/(1-r)$ is the equilibrium value of the system, C is the initial condition of the system at time period 0 and k is a generic time period that provides $a(0) = V$.

We can employ Theorem 2.2 and find

$$A(k) = 0.75\, k\, C + d/(1-r)$$

$$D = 100$$

$$1 - 7 = 0.25, \text{ so } d/(1-r) = 400.$$

Now,

$$A(0) = 0$$

So, $0 = C(0.75)^0 + 400$

$$C = -400.$$

The model, in general form, is

$$A(k) = 0.75^k(-400) + 400.$$

Let's iterate the system numerically. If we let $a(n)$ (we say "*a* at *n*" or "*a* of *n*") be the number of milligrams of drug in the bloodstream after *n* hours, and the initial amount in the bloodstream is 0 mg, then

$$a(0) = 0$$
$$a(1) = 0.75(0) + 100 = 100$$
$$a(2) = 0.75(100) + 100 = 175$$
$$a(3) = 0.75(175) + 100 = 231.25 \text{ mg.}$$

We could write these equations where we do not substitute the numerical values that we calculated:

$$a(0) = 0$$
$$a(1) = 0.75\, a(0) + 100$$
$$a(2) = 0.75\, a(1) + 100$$
$$a(3) = 0.75\, a(2) + 100.$$

Here, we see that the amount of drug in the bloodstream is related to the amount of drug in the bloodstream after the previous time period. Specifically, the amount of drug in the bloodstream after any of the first three time periods is 0.75 times the amount of drug in the bloodstream after the previous time period plus an additional 100 mg that is injected every hour.

We see that a pattern has developed that describes the amount of drug in the bloodstream. We are now prepared to conjecture (make an educated guess) about the amount of drug in the bloodstream after any hour. Mathematically, we say that the amount of drug in the bloodstream after n hours is

$$a(n + 1) = 0.75\, a(n) + 100.$$

The relationship above describes the amount of drug in the bloodstream after n hours. With this, we can see the change that occurs every hour within this "system" (amount of drug in the bloodstream), and state of the system, after any hour, is dependent on the state of the system after the previous hour. This is a *discrete dynamical system (DSS)*.

Analyzing the DDS: We want to find the value of $a(72)$. First, we can apply Theorems 2.2 and 2.3 as before.

$$a_e = 100/(0.25) = 400$$
$$a(k) = c(0.75)^k + 400$$

Since $a(0) = 0$

$$0 = c + 400$$
$$c = -400.$$

The solution is $a(k) = -400(0.75)^k + 400$ as illustrated with Theorem 2.2 before.

For $a(72) = -400\,(0.75)^{72} + 400 = 399.9999996$.

Interpretation of Results: The DDS shows that the drug reaches a value where change stops and eventually the concentration in the bloodstream levels at 400 mg as shown in Figure 2.3. If 400 mg is both a safe and effective dosage level, then this dosage schedule is acceptable. We discuss this concept of change stopping (equilibrium) later in this chapter.

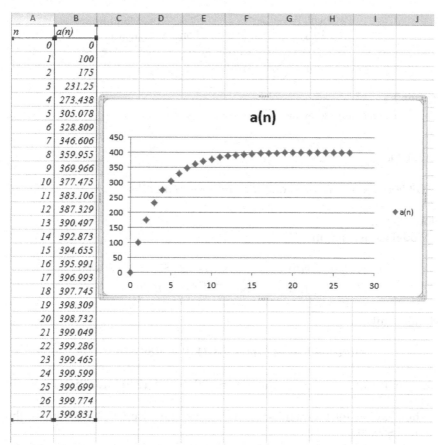

A	B	C	D	E	F	G	H	I	J
n	*a(n)*								
0	0								
1	100								
2	175								
3	231.25								
4	273.438								
5	305.078								
6	328.809								
7	346.606								
8	359.955								
9	369.966								
10	377.475								
11	383.106								
12	387.329								
13	390.497								
14	392.873								
15	394.655								
16	395.991								
17	396.993								
18	397.745								
19	398.309								
20	398.732								
21	399.049								
22	399.286								
23	399.465								
24	399.599								
25	399.699								
26	399.774								
27	399.831								

FIGURE 2.3
Behavior of drugs in our systems.

A Simple Mortgage Example

Five years ago your parents purchased a home by financing $80,000 for 20 years, paying monthly payments if $880.87 with a monthly interest of 1%. They have made 60 payments and wish to know what they actually owe on the house at this time. They can use this information to decide whether or not they should refinance their house at a lower interest rate for the next 15 or 20 years. The change in the amount owed each period increases by the amount of the interest and decreases by the amount of the payment. The flow diagram is presented in Figure 2.4.

Problem Identification: Build a model that relates the time with the amount owed on a mortgage for a home.

Assumptions: Initial interest was 12%. Payments are made on time each month.

FIGURE 2.4
Flow diagram for mortgage example.

Variables:

Let $b(n)$ = amount owed on the home after n months

Flow Diagram:
   Model Construction:

$$b(n + 1) = b(n) + 0.12/12\ b(n) - 880.87,\ b(0) = 80{,}000$$
$$b(n + 1) = 1.01\ b(n) - 880.87,\ b(0) = 80{,}000$$

Model Solution:

$$\text{Mortgage Owed } (n) = -8087(1.01)n + 88{,}087.$$

We can iterate this over the entire 20 years (240 months) in Excel (Figure 2.5):

After paying for 60 months, your parents still owe $73,395.37 out of the original $80,000.

In addition, they have paid in a total of $52,852.20 and only $6,605 went towards the principal payment of the home. The rest of the money went towards paying only the interest. If the family continues with this loan, then they will make 240 payments of $880.87 or $211,400.80 total in payments. This is $133,400.80 in interest. They have already paid $46,647.20 in interest. They would pay an additional $86,753.60 in interest over the next 15 years. What should they do? Perhaps an alternative scheme is available, such as refinance.

2.2 Equilibrium and Stability Values and Long-Term Behavior

We have previous mention Theorem 2.1 concerning equilibrium values to linear DDS. Let's further discuss these equilibriums.

n	b(n)			int
0	80000	37	76400.67	
1	79919.13	38	76283.8	
2	79837.45	39	76165.77	
3	79754.96	40	76046.56	
4	79671.64	41	75926.15	
5	79587.48	42	75804.55	
6	79502.49	43	75681.72	
7	79416.64	44	75557.67	
8	79329.94	45	75432.38	
9	79242.37	46	75305.83	
10	79153.92	47	75178.02	
11	79064.59	48	75048.93	
12	78974.37	49	74918.55	
13	78883.24	50	74786.86	
14	78791.2	51	74653.86	
15	78698.24	52	74519.53	
16	78604.36	53	74383.85	
17	78509.53	54	74246.82	
18	78413.76	55	74108.42	
19	78317.02	56	73968.64	
20	78219.32	57	73827.45	
21	78120.65	58	73684.86	
22	78020.98	59	73540.84	
23	77920.32	60	73395.37	
24	77818.66			
25	77715.97			
26	77612.26			
27	77507.51			
28	77401.72			
29	77294.87			
30	77186.95			
31	77077.95			
32	76967.85			
33	76856.66			
34	76744.36			
35	76630.93			
36	76516.37			

FIGURE 2.5
Screenshot of mortgage iterations in Excel.

Equilibrium Values

Let's go back to our original paradigm,

$$Future = Present + Change.$$

When change stops, the change equals zero and future equals the present. The value for which this happens, if any, is the equilibrium value. This gives us a context for the concept of the equilibrium value.

Models of the Form $a(n + 1) = r\,a(n) + b$, where r and b are Constants

Let's return to our drug dosage problem and consider adding the constant dosage each time period (time periods might be 4 hours). Our model is $a(n + 1) = 0.5\,a(n) + 16$ mg. We will also assume that there is an initial dosage applied prior to beginning the regime. We will let these initial values be as follows and graphic:

$a(0) = 20$

$a(0) = 10$

$a(0) = 40.$

Regardless of the starting value, the future terms of $a(n)$ approach 32 as depicted in Figure 2.6. Thus, 32 is the equilibrium value. We could have solved for this algebraically as well.

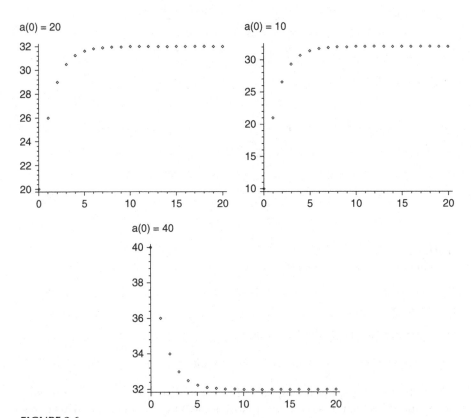

FIGURE 2.6
Plot of drugs in our systems with different initial starting conditions.

$$a(n + 1) = 0.5\,a(n) + 16$$
$$ev = 0.5ev + 16$$
$$0.5\,ev = 16$$
$$ev = 32.$$

Another method of finding the equilibrium values involves solving the equation $a = ra + b$ and solving for a (where a is ev) we find:

$$a = \frac{b}{1 - r}, \text{ if } r \neq 1.$$

Using this formula in our previous example, the equilibrium value is

$$a = 16/(1 - 0.5) = 32.$$

Stability and Long-Term Behavior

For a dynamical system, $a(n + 1)$ with a specific initial condition, $a(0) = a_0$, we have shown that we can compute $a(1)$, $a(2)$, and so forth. Often these particular values are not as important as the long-term behavior. By long-term behavior, we refer to what will eventually happen to $a(n)$ for larger values of n. There are many types of long-term behavior that can occur with DDS, we will only discuss a few here.

If the $a(n)$ values for a DDS eventually get close to the equilibrium value, ev, no matter the initial condition, then the equilibrium value is called a **stable equilibrium value** or **an attracting fixed point.**

Often, we characterize the long-term behavior of the system in terms of its stability. If a DDS has an equilibrium value and if the DDS tends to the equilibrium value from starting values near the equilibrium value, then the DDS is said to be stable.

Thus, for the dynamical system $a(n+1) = r\,a(n) + b$, where $b \neq 0$:

TABLE 2.1

Stability for linear functions

Value of r	DDS Form	Equilibrium	Stability of Solution	Long-term Behavior
$r = 0$	$a(n + 1) = b$	b	Stable	Stable equilibrium
$r = 1$	$a(n + 1) = a(n) + b$	*None*	Unstable	
$r < 0$	$a(n + 1) = r^*a(n) + b$	$b/(1 - r)$	Depends on $|r|$	Oscillations
$|r| < 1$	$a(n + 1) = r^*a(n) + b$	$b/(1 - r)$	Stable	Approaches $b/(1 - r)$
$|r| > 1$	$a(n +1) = r^*a(n) + b$	$b/(1 - r)$	Unstable	Unbounded

If $r \neq 1$, an equilibrium exists at $a = b/(1 - r)$.
If $r = 1$, no equilibrium value exists.

Relationship to Analytical Solutions

If a discrete dynamical system has an *ev* value, we can use the *ev* value to find the analytical solution.

Recall the mortgage example from Section 2.1,

$$B(n + 1) = 1.00541667\ B(n) - 639.34, B(0) = 73{,}395.$$

The *ev* value is found as 118,031.9274.

The analytical solution may be found using the following form:

$$B(k) = (1.00541667^k)C + D \text{ where } D \text{ is the } ev.$$
$$B(k) = (1.00541667^k)C + 118{,}031.9274, B(0) = 73{,}395.$$

Since $B(0) = 73{,}395 = 1.00541667^0(C) + 118{,}031.9274.$

$$C = -44{,}636.92736.$$

Thus,

$$B(k) = -44{,}636.92737\ (1.00541667^k) + 118{,}031.9274.$$

Let's assume we did not know the payment was 639.34 month. We could use the analytical solutions to help find the payment.

$$B(k) = (1.00541667^k)C + D.$$

We build a system of two equations and two unknowns.

$$B(K) = (1.00541667^k)C + D$$
$$B(0) = 73395 = C + D$$
$$B(180) = 0 = 1.00541667^{180}C + D$$
$$C = -44638.70, D = 118033.7$$
$$B(K) = -44638.70(1.00541667)^k + 118033.7.$$

D represents the equilibrium value and we accepted some round-off error. From our model form:

$$B(n + 1) = 1.00541667\ B(n) - P, \text{ we can find } P.$$

Solving analytically for the equilibrium value,

$$X - 1.00541667X = -P$$

$$X = P/0.00541667$$

$$X \text{ is } 118{,}033.70 \text{ so}$$

$$118{,}033.70 = P/0.00541667$$

$$P = 639.34.$$

Example 2.1 Growth of a Bacteria Population

We often model population growth by assuming that the change in population is directly proportional to the current size of the given population. This produces a simple, first-order DDS similar to those seen earlier. It might appear reasonable at first examination, but the long-term behavior of growth without bound is disturbing. Why would growth without bound of a yeast culture in a jar (or controlled space) be alarming?

There are certain factors that affect population growth. Things include resources (food, oxygen, space, etc.). These resources can support some maximum population. As this number is approached, the change (or growth rate) should decrease and the population should never exceed its resource supported amount.

Problem Identification: Predict the growth of yeast in a controlled environment as a function of the resources available and the current population.

Assumptions and Variables: We assume that the population size is best described by the weight of the biomass of the culture. We define $y(n)$ as the population size of the yeast culture after period n. There exists a maximum carrying capacity, M, that is sustainable by the resources available. The yeast culture is growing under the conditions established.

Model:

$$y(n+1) = y(n) + k\, y(n)(M - y(n)) \text{ where}$$

$y(n)$ is the population size after period n
n is the time period measured in hours
k is the constant of proportionality
M is the carrying capacity of our system.

We have data shown in Table 2.2 for the growth of bacteria in a Petri dish. The variable, $y(n)$, is the number of bacteria at the end of period n.

It is often convenient to think about the way the variables change between time periods. We compute $\Delta S(n) = y(n + 1) - y(n)$. The values are provided in Table 2.3 and used to find the proportionality constant.

We find the constant slope is 0.0008. We also find the *ev* of 621.

TABLE 2.2

Bacteria growth in a petri dish

N	0	1	2	3	4	5	6	7	8	9
Y	10.3	17.2	27	45.3	80.2	125.3	176.2	256.6	330.8	390.4
n	10	11	12	13	14	15	16	17	18	19
y	440	520.4	560.4	600.5	610.8	614.5	618.3	619.5	620	621

TABLE 2.3

Bacteria growth in a petri dish solution values

N	0	1	2	3	4	5	6	7	8
$y(n)$	10.3	17.2	27	45.3	80.2	125.3	176.2	256.6	330.8
$\Delta y(n)$	6.9	9.8	18.3	34.9	45.1	50.9	79.4	74.2	59.6
	9	10	11	12	13	14	15	16	17
$y(n)$	390.4	440	520.4	560.4	600.5	614.5	618.3	619.5	620
$\Delta y(n)$	46.4	79.4	40	40.1	14	3.9	1.2	0.5	1

In our experiment, we first plot $y(n)$ versus n and find a stable equilibrium value of approximately 621. Next, we plot $y(n + 1) - y(n)$ versus $y(n)$ $(621 - y(n))$ to find the slope, k, is approximately 0.0008, with $k = 0.0008$ and the carrying capacity in biomass is 621. This model is

$$y(n + 1) = y(n) + .0008\, y(n)(621 - y(n)).$$

Again, this is nonlinear because of the $y^2(n)$ term. There is no closed form analytical solution for this equation, however we may obtain a solution through iteration and graphing the iterated values in Excel.

The model shows stability in that the population (biomass) of the yeast culture approaches 621 as n gets large. Thus, the population is eventually stable at approximately 621 units as shown in Figure 2.7.

Example 2.2 Spread of a Contagious Disease

There are 5,000 students in college dormitories, and some students have been diagnosed with COVID-19, a highly contagious disease. The health center wants to build a model to determine how fast the disease will spread.

Problem Identification: Predict the number of students affected with COVID-19 as a function of time.

Assumptions and Variables: Let $m(n)$ be the number of students affected with COVID-19 after n days. We assume all students are susceptible to

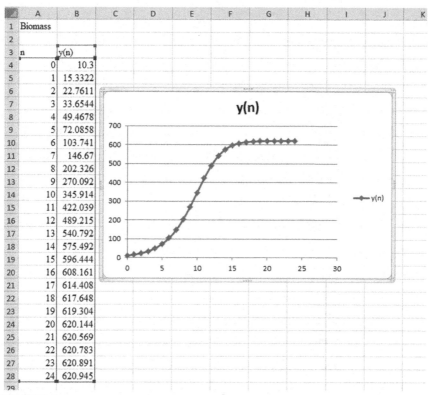

	A	B
1	Biomass	
2		
3	n	y(n)
4	0	10.3
5	1	15.3322
6	2	22.7611
7	3	33.6544
8	4	49.4678
9	5	72.0858
10	6	103.741
11	7	146.67
12	8	202.326
13	9	270.092
14	10	345.914
15	11	422.039
16	12	489.215
17	13	540.792
18	14	575.492
19	15	596.444
20	16	608.161
21	17	614.408
22	18	617.648
23	19	619.304
24	20	620.144
25	21	620.569
26	22	620.783
27	23	620.891
28	24	620.945

FIGURE 2.7
Screenshot of DDS for bacteria growth in a petri dish.

the disease. The possible interactions of infected and susceptible students are proportional to their product (as an interaction term).

The model is,

$$m(n + 1) - m(n) = k\,m(n)(5{,}000 - m(n)) \text{ or}$$

$$m(n + 1) = m(n) + k\,m(n)(5{,}000 - m(n)).$$

Two students returned from spring break with COVID-19, so $m(0) = 2$. The rate of spreading per day is characterized by $k = 0.00090$. It is assumed that there is no vaccine that can be introduced to slow the spread.

Interpretation: The results show that most students will be affected within 2 weeks. Since only about 10% will be affected within one week, every effort must be made to get the vaccination at the school and get the students vaccinated within one week. This is illustrated graphically in Figure 2.8.

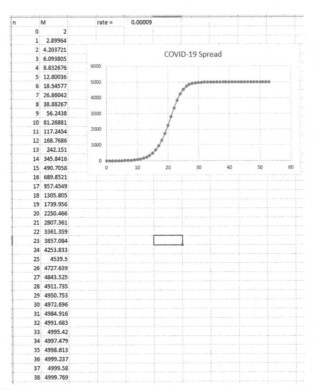

n	M	rate =	0.00009
0	2		
1	2.89964		
2	4.203721		
3	6.093805		
4	8.832676		
5	12.80036		
6	18.54577		
7	26.86042		
8	38.88267		
9	56.2438		
10	81.26881		
11	117.2454		
12	168.7686		
13	242.151		
14	345.8416		
15	490.7058		
16	689.8521		
17	957.4549		
18	1305.805		
19	1739.956		
20	2250.466		
21	2807.361		
22	3361.359		
23	3857.084		
24	4253.833		
25	4539.5		
26	4727.639		
27	4843.525		
28	4911.735		
29	4950.753		
30	4972.696		
31	4984.916		
32	4991.683		
33	4995.42		
34	4997.479		
35	4998.613		
36	4999.237		
37	4999.58		
38	4999.769		

FIGURE 2.8

Screenshot of plot and iteration for the spread of meningitis.

Example 2.3 Modeling for Number Theory

Introduction Square Integers

My friend was teaching his young granddaughter about squares of numbers. He recently went blind, so he was using mental mathematics to help. He conjectured a pattern. After 2^2, all perfect squares were the previous square plus the previous number plus the current number. Thus, he conjectured that $y^2 = x^2 + x + y$. He pondered, does this always work?

Let us solve the equation, $y^2 = x^2 + x + y$. Separating and factoring we obtain $y(y - 1) = x(x - 1)$. Solving for the relationship means we know to add a key assumption that $x \neq y$ and more specifically that x and y are consecutive numbers. In this case, let us state they are integers. Thus $x = y - 1$. The reader can see references Albright (2010) and Alfred (1967) for a more detailed and complete explanation.

How does this help students? Students generally recall the easy squares, maybe up to 10.

So, let us find 11^2. Thus, $11^2 = 100 + (10 + 11) = 121$.

What if you want to know the square of 10.1?

First, $10.1^2 = 102.01$. If we use other algorithms, then 10.1^2 should equal $100 + (10 + 10.1)$. However, $120.1 \neq 102.01$. So, let's find an algorithm that works not just for integers but for all real numbers.

We choose to use Discrete Dynamical Systems to develop this algorithm later.

We also noted a pattern that adding odd numbers from 5 on the squares from $2^2 = 4$ also provide perfect squares of numbers.

$$3^2 = 4 + 5 = 9,$$

$$11^2 = 4 + \text{sum}\ (5, 7, 9, 11, 13, 15, 17, 19, 21) = 121,$$

$$23^2 = 529.$$

Method 1: We know $20^2 = 400$.

So, $400 + 2*(20 + 21 + 22) + 3 = 529$.

Method 2:

With odd numbers 4+ sum (5, 7, 35) = 529.

Next, we examine if patterns for all real numbers exist. We choose discrete dynamical systems to model and look for patterns.

TABLE 2.4

Example data for pattern recognition

N	A(n)
2	4
3	9
4	16
5	25
6	36
7	49
8	64
9	81
10	100
11	121
12	144
13	169
14	196
15	225
16	256
17	289
18	324
19	361
20	400

For example, let $A(n)$ = square of n. We define our initial condition to be $A(2) = 4$. Now, $A(n1) = A(n) + \Delta$, where we define Δ to be the sum of $n - 1 + n$.

We illustrate in Excel to iterate from $n = 2$ to 20.

We note that in our modeling change there is a multiplier of 1 for $(n + (n - 1))$ for the change term. This is important as we shall see in the next section.

Extending to Squares of Real Numbers Other than Integers

Although dealing with integers is fun, we conjectured that there must be something similar for squares of decimals. Next, we built a discrete dynamical system to apply to decimals. Depending on the number of decimal places we find a multiplier must be used in the addition of the numbers to get them into the proper decimal equivalents. The multiplier that works is the size of the step used to iterate from n to $n + step\ size$, where n and the step size are not necessarily an integer.

$$A(n + 1) = A(n) + step\ size*(n - 1 + n).$$

We illustrate and iterate three examples.

$$(1)\ B(n + 1) = B(n) + 0.01(n + n - 1).$$

Initial Condition: $B(1) = 1$.

$$(2)\ C(n + 1) = C(n) + 0.2(n + n - 1).$$

Initial Condition: $C(10) = 100$.

$$(3)\ D(n + 1) = D(n) + 0.5(n + n - 1).$$

Initial Condition: $D(10) = 100$.

We iterate to show the values found, which as the corresponding squares. These are displayed in Figure 2.9.

Finding Patterns of Cubes with Discrete Dynamical Systems

After experimenting with squares and the discovering the use of DDS to model the merging patterns to obtain a formula for all real squares, I thought to explore cubes. Cubes are fascinating.

n	B(n)	n	C(n)	n	D(n)
	stepsize		stepsize		stepsize
	0.01		0.2		0.5
1	1	10	100	10	100
1.01	1.0201	10.2	104.04	10.5	110.25
1.02	1.0404	10.4	108.16	11	121
1.03	1.0609	10.6	112.36	11.5	132.25
1.04	1.0816	10.8	116.64	12	144
1.05	1.1025	11	121	12.5	156.25
1.06	1.1236	11.2	125.44	13	169
1.07	1.1449	11.4	129.96	13.5	182.25
1.08	1.1664	11.6	134.56	14	196
1.09	1.1881	11.8	139.24	14.5	210.25
1.1	1.21	12	144	15	225
		12.2	148.84	15.5	240.25
		12.4	153.76	16	256
		12.6	158.76	16.5	272.25
		12.8	163.84	17	289
		13	169	17.5	306.25
		13.2	174.24	18	324
		13.4	179.56	18.5	342.25
		13.6	184.96	19	361
				19.5	380.25
				20	400

FIGURE 2.9
Screenshot of DDS iterations for our three examples in Excel.

In most literature, cubic numbers are defined only by multiplication. The result is using a whole number in a multiplication three times.

Example: $3 \times 3 \times 3 = 27$, so 27 is a cube number.

Here are the first few cube numbers:

$$1 \quad (= 1 \times 1 \times 1)$$

$$8 \quad (= 2 \times 2 \times 2)$$

$$27 \quad (= 3 \times 3 \times 3)$$

$$64 \quad (= 4 \times 4 \times 4)$$

$$125 \quad (= 5 \times 5 \times 5)$$

... etc.

We begin by assuming that there has to be more as we search for a distinguishable pattern.

Since we were able to use DDS to explore squares of numbers it makes sense to start with DDS to explore cubes.

Cubes and DDS with Cubes

A cube, written with exponents as x^3, is also known as a product of $x*x*x$. For example, $6^3 = 6*6*6 = 216$.

So, we created a spreadsheet in Excel of the numbers 1, 2, 3, …n. The next column we cubed those numbers to get, 1, 8, 27, 64, …. We then created columns subtracting the differences in consecutive cubes obtaining, 7, 19, 37, 61, 91, …. T first these look like prime numbers but 91 is not prime but the product to 2 primes, 13 and 7. We took the differences of these odd numbers to obtain, 12, 18, 24, 30, 36, … We note a pattern that the next difference is always 6. Thus, we can work backwards to obtain a discrete dynamical system model.

We define $A(n)$ to be the cube of the nth number.

The conjectured pattern appears to be a second order DDS of the form:

$$A(n + 2) = (n - 1)*6 + A(n + 1) + [A(n + 1) - A(n)].$$

This simplifies to $A(n + 2) = (n - 1)*6 + 2*A(n + 1) - A(n)$ with initial conditions $A(1) = a_1$ and $A(0) = a_0$.

For example, given $A(1) = 8$, $A(0) = 1$, determine $A(7)$.

We iterate our DDS to obtain, $A(7) = 343 = 7^3$.

n	$A(n)$
1	1
2	8
3	27
4	64
5	125
6	216
7	343

Further, for any real consecutive cube, we have:

$$A(n + 2) = 2*A(n + 1) - A(n) + step\text{-}size\ (n - step\text{-}size)*(step\text{-}size*6).$$

TABLE 2.5

Calculating differences

n	A	1st diff	2nd diff	3rd diff
1	1			
2	8	7		
3	27	19	12	
4	64	37	18	6
5	125	61	24	6
6	216	91	30	6

TABLE 2.6

Iterated sequence

N	A(n)
1.00	1.000
1.10	1.331
1.20	1.728
1.30	2.197
1.40	2.744
1.50	3.375
1.60	4.096
1.70	4.913
1.80	5.832
1.90	6.859
2.00	8.000
2.10	9.261
2.20	10.648
2.30	12.167

For example, consider wanting the cube of 1.2 through 2.3 given $1^3 = 1$, $1.1^3 = 1.331$. We use our DDS and iterate to obtain the sequence:

For example, let us assume we want to start at 1, and get cubes by a step size of 2. Thus, we want cubes for 1, 3, 5, 7. We need two initial conditions so we know $A(0) = 1$ and $A(1) = 27$. We use our DDS formula,

$$A(n + 2) = 2*A(n + 1) - A(n) + step\ size(n - step\ size)*(step\ size*6),$$

and iterate to obtain the cubes for 5, 7, 9, etc.

n	Cubes
5	125
7	343
9	729

2.3 Introduction to Systems of Discrete Dynamical Systems

In the previous sections we reviewed linear and nonlinear DDS models. Now, we extend the discussion to systems of systems, but we still use our paradigm:

$$future = present + change.$$

Consider wanting to retire on a lake that you stock with bass and trout for endless fishing. Will it be endless? Can the species go exist in your lake? How often do you need to restock the lake? This is an example of a competitive hunter model where both species compete for the same resources.

Let's define a system of DDS.

$$A(n) = f(A(n), B(n)),$$

$$B(n) = g(A(n), B(n)).$$

As before, simple linear systems of DDS have closed form analytical solution however most systems do not. We will analyze all those DDS through iteration and graphs.

For selected set of initial conditions, we build numerical solutions to get a sense of long-term behavior for the system. For the systems that we will study, we will find their equilibrium values. We then explore starting values near the equilibrium values to see if by starting close to an equilibrium value, the system:

a.   will remain close;
b.   approaches the equilibrium value;
c.   does not remain close.

What happens near these values gives great insight concerning the long-term behavior of the system. We can study the resulting pattern of the numerical solutions and the resulting plots.

Simple Linear Systems and Analytical Solutions

Let's consider school vouchers. There are both students in the public school (PS) and the private magnet (PM) school. Suppose a pre-survey of families used as historical records determined that 75% of the magnet school remain while 25% preferred to transfer. We found the 65% of the public school want to remain but 35% preferred to transfer. Let's build a model to determine the long-term behavior of these students based upon this historical data. The change diagram is shown in Figure 2.10.

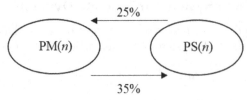

FIGURE 2.10
Change diagram for the school vouchers.

Problem Identification: Determine the school voucher over time.

Assumptions and Variables: Let n represent the number of student months. We define

$PS(n)$ = the number of students in public school at the end of n months;

$PM(n)$ = the number of students in the magnet school at the end of n months.

We assume that no other incentives are given to the students for either staying or moving.

The Model:

As build the dynamical systems. Mathematically, this is written as:

$$PM(n + 1) = 0.75\, PM\,(n) + 0.35\, PS(n),$$

$$PS(n + 1) = 0.25\, PM(n) + 0.65\, PS(n).$$

There are initially 1,500 students in the magnet school and 2,000 students in the public school. We seek to find the long-term behavior of this system.

We rewrite the model as a system of DDS:

$$PM(n + 1) = 0.75\, PM\,(n) + 0.35\, PS(n),$$

$$PS(n + 1) = 0.25\, PM(n) + 0.65\, PS(n).$$

$PM(0) = 1,500$ and $PS(0) = 1,000$, respectively.

This is a simple linear system. We may use the initial conditions to iterate this system of discrete dynamical systems as shown in Figure 2.11. We see from the solution plot that there are stable equilibria at about 2,042 and 1,458 in the magnet and public schools respectively.

Analytical Solutions

Analytical solutions assume knowledge of linear algebras through eigenvalues and eigenvectors.

We rewrite the DDS in matrix form:

$$X(n + 1) = MX(n),\ X(0) = B.$$

$$\begin{bmatrix} PM(n+1) \\ PS(n+1) \end{bmatrix} = \begin{bmatrix} 0.75 & 0.35 \\ 0.25 & 0.65 \end{bmatrix} \begin{bmatrix} PM(n) \\ PS(n) \end{bmatrix}, \begin{bmatrix} PM(0) \\ PS(0) \end{bmatrix} = \begin{bmatrix} 1500 \\ 2000 \end{bmatrix}$$

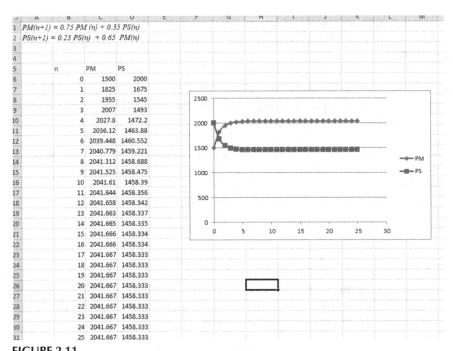

FIGURE 2.11
Screenshot iterative solution and plot for voucher students.

We define the analytical solution with two distinct real eigenvalues and eigenvectors. The general form of the solution is

$$X(k) = \lambda_1{}^k c_1 V_1 + \lambda_2{}^k c_2 V_2, \text{ where}$$

λ_1 & λ_2 are the two distinct eigenvalues,
V_1 & V_2 are the corresponding eigenvectors.
c_1 & c_2 are the constant.

The characteristic polynomial is $\lambda^2 - 1.4\lambda + 0.40 = 0$. This provides two eigenvalues: 1, 0.4. The corresponding eigenvalues can be found easily as vectors for $\lambda = 1$ of [0.35, 0.25] and for $\lambda = 0.40$ as [1, –1].

The general solution is

$$X(k) = c_1\left(1^k\right)\begin{bmatrix} 0.35 \\ 0.25 \end{bmatrix} + c_2\left(.4^k\right)\begin{bmatrix} 1 \\ -1 \end{bmatrix}$$

With our initial conditions we find c_1 and c_2 to be 416.6666 and –4.16666, so

$$X(k) = 5833.333\left(1^k\right)\begin{bmatrix} .35 \\ .25 \end{bmatrix} - 541.6666\left(.4^k\right)\begin{bmatrix} 1 \\ -1 \end{bmatrix}$$

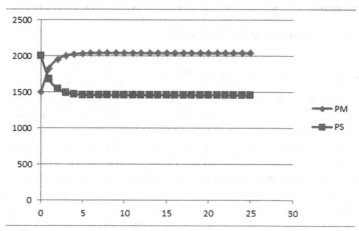

FIGURE 2.12
Plot of DDS for student's relocation example.

When $k = 10$, we find 2041.609 and 1458.31 for students, respectively. We see this again in Figure 2.12.

Analytically, we can solve for the equilibrium values. We let $X = D(n)$ and $Y = M(n)$. From the DDS, we obtain the equations:

$$X = 0.75\ X + 0.35\ Y,$$

$$Y = 0.25\ x + 0.65\ Y,$$

and both equations reduce to $X = 0.35/0.25\ Y$. There are two unknowns, so we need a second equation.

From the initial conditions, we know that $X + Y = 3500$. We can use the equations

$X + Y = 3,500$ and $X = 0.35/0.25\ Y$ to find the equilibrium values:
$X = 2041.6667$ and $Y = 1458.3333$.

We previously iterated the solution and now we start with initial conditions near to those equilibrium values and we find the sequences tend toward those values. We conclude the system has *stable* equilibrium values.

You should go back and change the initial conditions and see what behavior follows.

Interpretation: The long-term behavior shows that eventually (without other influences) of the 3,500 students about 1,458 remain in public school and 2,042 go to the magnet school. We might want to try to attract students with advertising and perhaps add incentives.

2.4 Iteration and Graphical Solution

Example 2.4 Competitive Hunter Models

Competitive hunter models involve species vying for the same resources (such as food or living space) in the habitat. The effect of the presence of a second species diminishes the growth rate of the first species. We now consider a specific example concerning trout and bass in a small pond. Hugh Ketum owns a small pond that he uses to stock fish and eventually allows fishing. He has decided to stock both bass and trout. The fish and game warden tell Hugh that after inspecting his pond for environmental conditions he has a solid pond for growth of his fish. In isolation, bass grow at a rate of 20% and trout at a rate of 30%. The warden tells Hugh that the interactions for the food affects trout more than bass. They estimate the interaction affecting bass is 0.0010 bass*trout and for trout is 0.0020 bass*trout. Assume no changes in the habitant occur.
Model:
　　Let's define the following variables:

$B(n)$ = the number of bass in the pond after period n.
$T(n)$ = the number of trout in the pond after period n.
$B(n) * T(n)$ = interaction of the two species.

$$B(n + 1) = 1.20\, B(n) - 0.0010\, B(n)\, {*}T(n).$$

$$T(n + 1) = 1.30\, T(n) - 0.0020\, B(n){*}\, T(n).$$

The equilibrium values can be found by allowing $X = B(n)$ and $Y = T(n)$ and solving for X and Y.

$$X = 1.2\, X - 0.001\, X{*}Y,$$

$$Y = 1.3\, Y - 0.0020\, X{*}Y.$$

We rewrite these equations as

$$0.2\, X - 0.001\, X{*}Y = 0,$$

$$0.3\, Y - 0.002\, X{*}Y = 0.$$

We can rewrite the two equations to obtain:

$$X\, (0.2 - 0.001\, Y) = 0,$$

$$Y(0.3 - 0.002\, X) = 0.$$

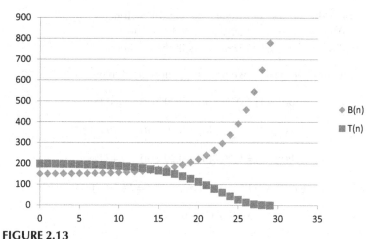

FIGURE 2.13
Bass and trout over time.

Solving we find $X = 0$ or $Y = 2,000$ and $Y = 0$ or $X = 1,500$.

We want to know the long-term behavior of the system and the stability of the equilibrium points.

Hugh initially considers 151 bass and 199 trout for his pond. The solution is left to the student as an exercise. From Hugh's initial conditions, bass will grow without bound and trout will eventually die out.

We iterated the system and obtained the plot, Figure 2.13, of bass and trout over time. Trout die out at about period 29.

This is certainly not what Hugh had in mind so he must find ways to improve the environment for the fish that alter the parameters from the model.

Example 2.5 Fast Food Tendencies

Consider that your student union center desires to have three fast food chains available to students serving: burgers, tacos, and pizza. These chains run a survey of students and find the following information concerning lunch: 75% that ate burgers will eat burgers again at the next lunch, 5% will eat tacos next, and 20% will eat pizza next. Of those who ate tacos last, 20% will eat burgers next, 60% will stay will tacos, and 35% will eat pizza next. Of those who ate pizza, 40% will eat burgers next, 20% tacos, and 40% pizza again.

We formulate the problem as follows:

Let n represent the nth day's lunch and so we define the following variables:

$B(n)$ = the number of burger eaters in the nth lunch.
$T(n)$ = the number of taco eaters in the nth lunch.
$P(n)$ = the number of pizza eaters in the nth lunch.

Formulating the system, we have the following dynamical system:

$$B(n + 1) = 0.75\ B(n) + 0.20\ T(n) + 0.40\ P(n),$$

$$T(n + 1) = 0.05\ B(n) + 0.60\ T(n) + 0.20\ P(n),$$

$$P(n + 1) = 0.20\ B(n) + 0.20\ T(n) + 0.40\ P(n).$$

Analytically, we let $X = B(n)$, $Y = T(n)$, and $Z = P(n)$ so that

$$X = 0.75\ X + 0.2\ Y + 0.4\ Z,$$

$$Y = 0.05\ X + 0.6\ Y + 0.2\ Z,$$

$$Z = 0.2\ X + 0.2\ Y + 0.4\ Z.$$

These equations reduce to

$$X = 20/9\ Z,$$

$$Y = 7/9\ Z,$$

$$Z = Z.$$

Since we have 14,000 students, then we assume that $X + Y + Z = 14,000$
We substitute and solve for Z first.

$$4\ Z = 14,000,$$

$$Z = 3,500,$$

$$X = 20/9\ Z = 20/9\ (3,500) = 7,777.77,$$

$$Y = 7/9\ Z = 7/9\ (3,500) = 2,722.222.$$

Suppose the campus has 14,000 students that eat lunch. The graphical results also show that an equilibrium value is reached at a value
of about 7,778 burger eaters, 2,722 taco eaters, and 3,500 pizza eaters.
This allows the fast food establishments to plan for a projected future.
We see this in the iterated table and Figure 2.14. By varying the initial
conditions for 14,000 students we find that these are stable equilibrium
values.

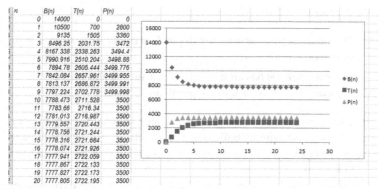

n	B(n)	T(n)	P(n)
0	14000	0	0
1	10500	700	2800
2	9135	1505	3360
3	8496.25	2031.75	3472
4	8167.338	2338.263	3494.4
5	7990.916	2510.204	3498.88
6	7894.78	2605.444	3499.776
7	7842.084	2657.961	3499.955
8	7813.137	2686.872	3499.991
9	7797.224	2702.778	3499.998
10	7788.473	2711.528	3500
11	7783.66	2716.34	3500
12	7781.013	2718.987	3500
13	7779.557	2720.443	3500
14	7778.756	2721.244	3500
15	7778.316	2721.684	3500
16	7778.074	2721.926	3500
17	7777.941	2722.059	3500
18	7777.867	2722.133	3500
19	7777.827	2722.173	3500
20	7777.805	2722.195	3500

FIGURE 2.14

Screenshot Excel iterated solution and plot for fast food on campus.

2.5 Modeling of Predator–Prey Model, SIR Model, and Military Models

Example 2.6 A Predator–Prey Model: Foxes and Rabbits

In the study of the dynamics of a single population, we typically take into consideration such factors as the "natural" growth rate and the "carrying capacity" of the environment. Mathematical ecology requires the study of populations that interact, thereby affecting each other's growth rates. In this module we study a very special case of such an interaction, in which there are exactly two species, one of which the predators eat the prey. Such pairs exist throughout nature such as lions and gazelles, birds and insects, pandas and eucalyptus trees, and Venus fly traps and flies.

To keep our model simple, we will make some assumptions that would be unrealistic in most of these predator–prey situations. Specifically, we will assume that

- the predator species is totally dependent on a single prey species as its only food supply;
- the prey species has an unlimited food supply; and
- there exist no other threats to the prey other than the specific predator.

In this modeling process, we will use the Lotka-Volterra model for predator–prey. Students can read more about the Lotka-Volterra models in the suggested readings. Here we simply present the model that we use.

We repeat our two key assumptions:

- The predator species is totally dependent on the prey species as its only food supply.
- The prey species has an unlimited food supply and no threat to its growth other than the specific predator.

If there were no predators, the second assumption would imply that the prey species grows exponentially without bound, i.e., if $x = x(n)$ is the size of the prey population after a discrete time period n, then we would have $x(n + 1) = a\, x(n)$.

But there *are* predators, which must account for a negative component in the prey growth rate. Suppose we write $y = y(n)$ for the size of the predator population at time t. Here are the crucial assumptions for completing the model:

- The rate at which predators encounter prey is jointly proportional to the sizes of the two populations.
- A fixed proportion of encounters lead to the death of the prey.

These assumptions lead to the conclusion that the negative component of the prey growth rate is proportional to the product xy of the population sizes, i.e.,

$$x(n+1) = x(n) + ax(n) - bx(n)y(n).$$

Now we consider the predator population. If there were no food supply, the population would die out at a rate proportional to its size, i.e. we would find $y(n+1) = -cy(n)$.

We assume that is the simple case that the "natural growth rate" is a composite of birth and death rates, both presumably proportional to population size. In the absence of food, there is no energy supply to support the birth rate. But there is a food supply: the prey. And what's bad for hares is good for lynx. That is, the energy to support growth of the predator population is proportional to deaths of prey, so

$$y(n+1) = y(n) - cy(n) + px(n)y(n)$$

This discussion leads to the discrete version of the Lotka–Volterra Predator–Prey Model:

$$x(n+1) = (1+a)x(n) - bx(n)y(n)$$
$$y(n+1) = (-c)y(n) + px(n)y(n)$$
$$n = 0,1,2,...$$

where a, b, c, and p are positive constants.

The Lotka–Volterra model consists of a system of linked dynamical systems equations that cannot be separated from each other and that cannot be solved in closed form. Nevertheless, they can be solved numerically and graphed in order to obtain insights about the scenario being studied.

Let us return to our foxes and hares' scenario. Let's assume this discrete model is as explained above. Further, data investigation yields the following estimates for the parameters that we require: {a,b,c,p} = {0.04,0.0004,0.09, 0.001}. Let's further assume that initially there are 600 rabbits and 125 foxes.

We iterate and plot the results for rabbits and foxes versus time and then plot rabbits versus foxes shown in Figures 2.15–2.16.

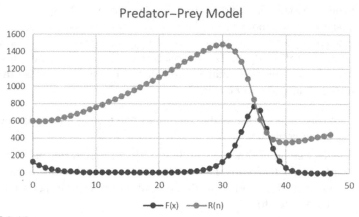

FIGURE 2.15
Foxes and rabbits over time.

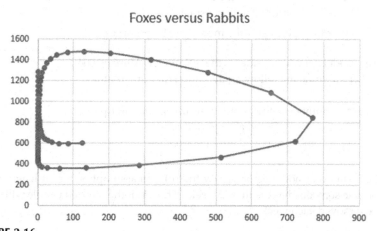

FIGURE 2.16
Foxes versus rabbits in a spiral motion.

If we ran this model for many more iterations, we would find the plot of foxes versus rabbits spiral in a similar fashion as above. We conclude that the model appears reasonable. We could find the equilibrium values for the system. There are is set of feasible equilibrium points for rabbits and foxes at (0,0) and at (2725, 960). The orbits of the spiral indicate that the system is moving away from both (0,0) and (2725, 960) so we conclude the system is not stable.

In models of predator–preys, it is often that "managers" of the ecological system must intervene in some way to keep both species flourishing.

Example 2.7 Discrete SIR Model of Epidemics

Consider a generic contagious disease, such as a new flu strand, that is spreading throughout the United States. The CDC is interested in experimenting with a model for this new disease prior to it actually becoming a "real" epidemic. Let us consider the population being divided into three categories: susceptible, infected, and removed. We make the following assumptions for our model:

- No one enters or leaves the community and there is no contact out-side the community.
- Each person is either susceptible, S (able to catch this new flu); infected, I (currently has the flu and can spread the flu); or removed, R (already had the flu and will not get it again that includes death).
- Initially every person is either S or I.
- Once someone gets the flu this year they cannot get it again.
- The average length of the disease is 2 weeks, after which the person is deemed infected and can spread the disease.
- Our time period for the model will be per week.

The model we will consider is the SIR model (Allman,2004).

Let's assume the following definition for our variables.

$S(n)$ = number in the population susceptible after period n.
$I(n)$ = number infected after period n.
$R(n)$ = number removed after period n.

Let's start our modeling process with $R(n)$. Our assumption for the length of time someone has the flu is 2 weeks. Thus, half the infected people will be removed each week:

$$R(n + 1) = R(n) + 0.5\,I(n).$$

The value, 0.5, is called the removal rate per week. It represents the proportion of the infected persons who are removed from infection each week. If real data is available, then we could do "data analysis" in order to obtain the removal rate.

$I(n)$ will have terms that both increase and decrease its amount over time. It is decreased by the number that are removed each week, $0.5*I(n)$.It is increased by the numbers of susceptible that come into contact with an infected person and catch the disease, $aS(n)I(n)$. We define the rate, a, as the rate in which the disease is spread or the transmission coefficient. We realize this is a probabilistic coefficient. We will assume, initially, that this rate is a constant value that can be found from initial conditions.

Let's illustrate as follows. Assume we have a population of 1,000 students in the dorms. Our nurse found 3 students reporting to the infirmary initially. The next week, 5 students came in to the infirmary with flu like symptoms. $I(0) = 3$, $S(0) = 997$. In week 1, the number of newly infected is 30.

$$5 = a\,I(n)S(n) = a\,(3)\,*(995),$$

$$a = 0.00167.$$

Let's now consider $S(n)$. This number is decreased only by the number that becomes infected. We may use the same rate, a, as before to obtain the model:

$$S(n + 1) = S(n) - aS(n)I(n).$$

Our coupled SIR model is:

$$R(n+1) = R(n) + 0.5I(n)$$
$$I(n+1) = I(n) - 0.5I(n) + 0.00167I(n)S(n)$$
$$S(n+1) = S(n) - 0.00167S(n)I(n)$$
$$I(0) = 3, S(0) = 997, R(0) = 0$$

The SIR Model can be solved iteratively and viewed graphically. Let's iterate the solution and obtain the graph, Figure 2.17, to observe the behavior to obtain some insights.

The worse of the flu epidemic occurs around week 8, at the maximum of the infected graph. The maximum number is slightly larger than 400, from the table it is 427. After 25 weeks, slightly more than 9 persons never get the flu. You will be asked to check for sensitivity to the coefficient in the exercise set.

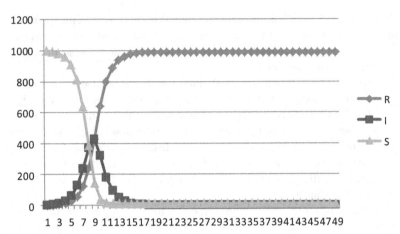

FIGURE 2.17
Plot of SIR model over time.

2.6 Technology Examples for Discrete Dynamical Systems

Using DDS is an interesting and productive approach. The use of computer technology is essential to the methods described in this article. It allows for interactions between instructors and students. It provides a means to use technology in a nonstandard way. It provides another way to educated students concerning squares and cubes using discrete mathematics. In this section, we present technology and examples of solving linear and nonlinear discrete dynamical systems.

2.6.1 Excel for Linear and Nonlinear DDS

Let's consider a discrete dynamical system such as $a(n + 1) = 0.5a(n)$, with $a(0) = 100$.

Steps to iterate and graph in Excel are as follows:

Step 1. Open a new worksheet and name is DDS or some appropriate name.
Step 2. Label the following columns as n and $a(n)$ in cell *a1* and *b1*.
Step 3. In cells *a2* and *b2* input the initial condition by putting in 0 in cell *a2* and 100 in cell *b2*.
Step 4. In cell *a3* type *= 1 + cell a2.*
Step 5. In cell *b3* type *= 0.5*cell b2.*
Step 6. Highlight cells *a3* and *b3* and drag the curser down to fill in cells as far as desired or needed, in this case to about a4:b16.

Step 7. Highlight cells a1:b16, INSERT scatterplot to obtain the graph.

Step 8. Interpret the results.

A screenshot of the model and results is provided in Figure 2.18, where we see that our DDS tends to zero over time.

In Figure 2.19, we show the appropriate formulas used.

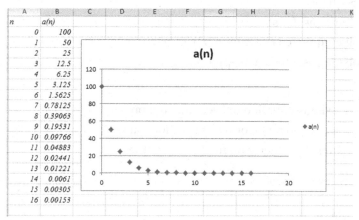

FIGURE 2.18
Screenshot of Excel's solution.

n	a(n)
0	100
=1+A2	=0.5*B2
=1+A3	=0.5*B3
=1+A4	=0.5*B4
=1+A5	=0.5*B5
=1+A6	=0.5*B6
=1+A7	=0.5*B7
=1+A8	=0.5*B8
=1+A9	=0.5*B9
=1+A10	=0.5*B10
=1+A11	=0.5*B11
=1+A12	=0.5*B12
=1+A13	=0.5*B13
=1+A14	=0.5*B14
=1+A15	=0.5*B15
=1+A16	=0.5*B16
=1+A17	=0.5*B17

FIGURE 2.19
Screenshot of Excel formulas used in our example.

2.6.2 Maple for Linear and Nonlinear DDS

In Maple, DDS are referred to as recursion equations. One might obtain closed form solutions, if they exist. One might also iterate and graph the behavior of the recursion equation. We will use both commands and libraries from Maple such as *with(plots)* and we will add some commands to our Maple toolbox, **rsolve** and **seq**.

rsolve - **recurrence equation solver**

Calling Sequence

rsolve(**eqns, fcns**)

rsolve(**eqns, fcns**, 'genfunc'(**z**))

rsolve(**eqns, fcns**, 'makeproc')

Parameters

eqns – single equation or a set of equations,

fcns – function name or set of function names,

z – name, the generating function variable.

seq - **create a sequence**

Calling Sequence

seq(**f,i=m..n**)

seq(**f,i=x**)

Parameters

f   – any expression,

i   – name,

m, n – numerical values,

x   – expression.

Many of the models that we will solve have closed form solutions so that we can use the command **rsolve** t o obtain the closed solution, and then we can use the sequence command (**seq**) to obtain the numerical values in the solution. Many dynamical systems do not have closed form analytical solutions so we cannot use the rsolve and seq commands to obtain solutions. When this occurs, we will write a small program using PROC to obtain the numerical solutions. To plot the solution to the dynamical systems, we will use plot commands to plot the sequential data pairs. We will illustrate all these commands in our example.

Example 2.8 Solve the DDS $a(n + 1) = 0.75\ a(n)$ B 100, $a(0) = 0$. Determine the value of $a(72)$.

Let's illustrate the iterative technique for analyzing a DDS in Maple. Figure 2.20 shows the graphical representation of the solution.

We type the following commands:

```
>restart;
>drug:=rsolve({a(n+1)=.75*a(n)+100,a(0)=0}, a(n));
```

$$drug := 400 - 400 \left(\frac{3}{4}\right)^n$$

```
>L:=limit(drug,n=infinity);
```

$$L := 400$$

```
>with(plots):
>pointplot({seq([i,–400*(3/4)^i+400],i=0..48)});
>drug_table:=seq(–400.0*(0.75)^i+400.,i=0..48);
```

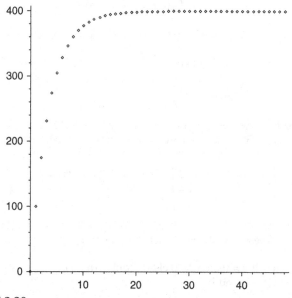

FIGURE 2.20
Behavior of drugs in our systems.

$drug\_table: = 0., 100.000, 175.00000, 231.4375000, 305.0781250,$
    $328.8085938, 346.604453, 359.9548340, 377.4745941, 383.1059456,$
    $387.3294592, 390.4970944, 392.8728208, 394.6546156, 395.9909617,$
    $396.9932213,\ 397.7449160, 398.3086870, 398.7315152, 399.0486364,$
    $399.2864773, 399.4648580,\ 399.5986435, 399.6989826, 399.7742370,$
    $399.8306777, 399.8730083, 399.9047562, 399.9285672, 399.9464254,$
    $399.9598190, 399.9698643, 399.9773982, 399.9830487,$
    $399.9872865, 399.9904649, 399.9928487, 399.9946365,$
    $399.9959774, 399.9969830,\ 399.9977373, 399.9983030, 399.9987272,$
    $399.9990454, 399.9992841, 399.9994630, 399.9995973$

Interpretation of Results:

The DDS shows that the drug reaches a value where change stops and eventually the concentration in the bloodstream levels at 400 mg. If 400 mg is both a safe and effective dosage level, then this dosage schedule is acceptable.

Using Maple for a System of DDS

Again, we use Systems of DDS with rsolve, numerical, and plotting for the problem

$$PS(n) = 0.65\ PS(n-1) + 0.25\ PM(n-1),$$

$$PM(n) = 0.35\ PS(n-1) + 0.75\ PM(n-1),$$

$$PS(0) = 2{,}000,$$

$$PM(0) = 1{,}500.$$

>dds: = rsolve({PS(n) = .65*PS(n − 1) + .25*PM(n − 1), PM(n) = .35*PS(n − 1) + .75*PM(n − 1), PS(0) = 2000, PM(0) = 1500}, {PM, PS});

$$dds: = \left\{ PM(n) = -\frac{1625}{3}\left(\frac{2}{5}\right)^{n} + \frac{6125}{3}, PS(n) = \frac{1625}{3}\left(\frac{2}{5}\right)^{n} + \frac{4375}{3} \right\}$$

> plot({-(1625/3)*(2/5)^n + 6125/3, (1625/3)*(2/5)^n + 4375/3},
n = 0..15,thickness = 3, title = `Student Vouchers`);

> public: = n-> if n = 0 then 2000 else.65*public(n − 1) + .25*magnent(n − 1) end if;

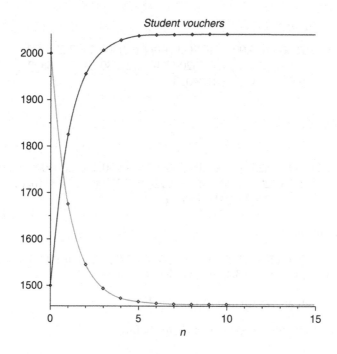

$$public := n \rightarrow \textbf{if } n{=}0 \textbf{ then } 2000 \textbf{ else}$$
$$0.65\,public(n-1) + 0.25\,magnent(n-1)$$
$$\textbf{end if}$$

>magnent: = n-> if n = 0 then 1500 else 0.35*public(n – 1) +
.75*magnent(n – 1) end if;

$$magnent := n \rightarrow \textbf{if } n{=}0 \textbf{ then } 1500 \textbf{ else}$$
$$0.35\,public(n-1) + 0.75\,magnent(n-1)$$
$$\textbf{end if}$$

> seq([public(n),magnent(n)],n = 0..10);

$$[2000,1500],[1675.00,1825.00],[1545.0000,1955.0000],$$
$$[1493.000000,2007.000000],[1472.200000,2027.800000],$$
$$[1463.880000,2036.120000],[1460.552000,2039,448000],$$
$$[1459.220800,2040.779200],[1458.688320,2041.311680],$$
$$[1458.475328,2041.524672],[1458.390131,2041.609869]$$

```
> u: = seq(public(n),n = 0..10);
```

$$u := 2000, 1675.00, 1545.0000.1493, 000000, 1472.200000,$$
$$1463.880000, 1460.552000, 1459.220800, 1458.688320,$$
$$1458.475328, 1458.390131$$

```
> w: = seq(magnent(n),n = 0..10);
```

$$w := 1500, 1825.00, 1955.0000, 2007.000000, 2027.800000,$$
$$2036.120000, 2039.448000, 2040.779200, 2041.311680,$$
$$2041.524672, 2041.609869$$

```
> with(plots):
```

```
> a: = plot({-(1625/3)*(2/5)^n + 6125/3, (1625/3)*(2/5)^n + 4375/
3},n = 0..15,thickness = 3, title = `Student Vouchers`):
```

```
> b: = pointplot({seq([n,public(n)],n = 0..10)}):
```

```
> c: = pointplot({seq([n,magnent(n)],n = 0..10)}):
```

```
> display(a,b,c);
```

```
>pointplot({seq([public(n),magnent(n)],n = 0..20)});
```

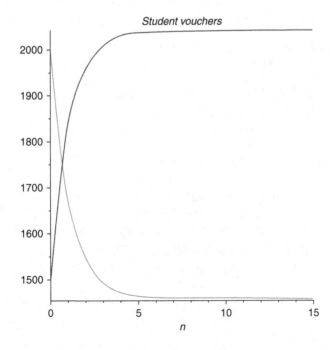

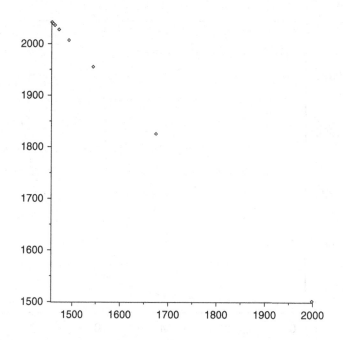

2.6.3 R for Linear and Nonlinear DDS

Example 2.9 Population Dynamics

Given the DDS, $N[t+1] = \lambda\, N[t]$, where λ is $(1 + r)$.

We open R studio and we are going to use the **For Loop** to address this problem. We type the following commands:

> generations <- 10

> N <- numeric(generations)

> lambda <- 2.1

> N [1] <- 3

> for (t in 1: (generations-1)) {N [t + 1] <- lambda* N [t]}

> N

[1] 3.0000 6.3000 13.2300 27.7830 58.3443 122.5230 257.2984

[8] 540.32661134.68582382.8401

> plot(0:(generations-1),N, type = "o", xlab = "Time", ylab = "Pop Size")

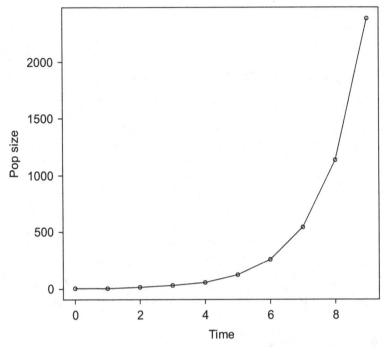

FIGURE 2.21
Screenshot from R.

We see unbounded growth in the plot shown in Figure 2.21.

Example 2.10 Repeat Example 2.8 $a(n + 1) = 0.75\ a(n) + 100$, $a(0) = 0$
using *R.*

We type the following commands:

```
>gener<-20
> D<-numeric(gener)
> lam<-0.75
> D[1]<-0
> for (t in 1:(gener-1)) {D[t + 1]<-lam*D[t] + 100}

> D
[1] 0.0000 100.0000 175.0000 231.2500 273.4375 305.0781 328.8086 346.6064
[9] 359.9548 369.9661 377.4746 383.1059 387.3295 390.4971 392.8728394.6546
[17] 395.9910 396.9932 397.7449 398.3087
> plot(0:(gener-1),D,type = "o", xlab = "Time", ylab = "Drug_in_Sys")
```

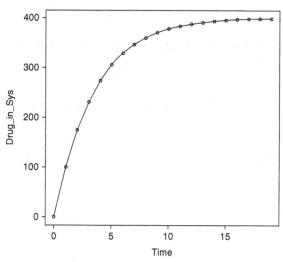

FIGURE 2.22

Screenshot from R solution graph for $a(n + 1) = 0.75\, a(n) + 100$, $a(0) = 0$.

In R, we can see that the drug becomes stable at approximately 400 units as shown in Figure 2.22.

We now present a drug dosage model analytically.

Logistics Growth

Let's modify this model to a nonlinear model using a logistics growth DDS in R.

Given the DDS, $N[t + 1] = N[t] + r\,\lambda\, N[t](1 - N[t]/K)$, we have the following R script using the **function** command.

DDSL <- function(K,r,N0,generations)

+ {N <- c(N0,numeric(generations-1))

+ for (t in 1:(generations-1)) N [t + 1]<- {N [t] + r*N [t]* (1– (N [t]/K))}

+ return(N)}

> Output <-DDSL(K = 1000,r = 1.5,N0 = 10,generations = 30)

> *generations* <-30

> *plot(0:(generations-1),Output, type = 'o', xlab = "time", ylab = "Population")*

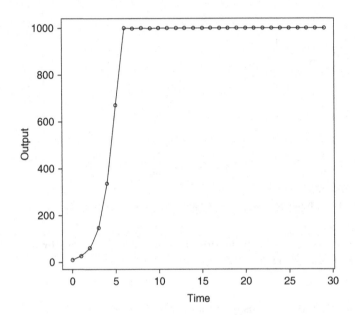

The plot shows the exponential growth at the beginning and then level off to the carrying capacity, $K = 1,000$.

Next, we present a system of DDS using *R*.

Systems of DDS using a multiple **for** loop command.

```
> nn<-11
> a[1]  <-100
> b[1]<-150
> for (t in 1: (nn–1)) {
> for (tt in 1: (nn–1)){
```

```
>plot(a, type = "l", col = "green")
>par(new = TRUE)
> plot(b, type = "l", col = "red", axes = FALSE)
```

```
> for (t in 1:(nn-1)) {
+ for (tt in 1: (nn-1)) {
+ a[t+1] <-.6*a[t]+.3*b[t]
+ b[tt+1] <-.4 *a[tt]+.7*b[tt]
```

```
+}
+}
> a
```
[1] 100.0000 105.0000 106.5000 106.9500 107.0850 107.1255
107.1376107.1413
[9] 107.1424 107.1427 107.1428
```
> b
```
[1] 150.0000 145.0000 143.5000 143.0500 142.9150 142.8745
142.8623 142.8587
[9] 142.8576 142.8573 142.8572

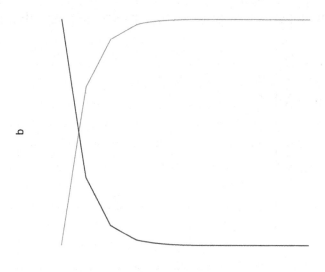

Index

Steady state probabilities

```
> a[1] <- 1
> b[1] <-0
> for (t in 1:(nn-1)) {
+ for (tt in 1: (nn-1)) {

    + a[t + 1]<-.6*a[t] + .3*b[t]
    + b[tt + 1]<-.4*a[tt] + .7*b[tt]
    +}
    +}

> a
```
[1] 1.0000000 0.6000000 0.4800000 0.4440000 0.4332000 0.4299600
0.4289880
[8] 0.4286964 0.4286089 0.4285827 0.4285748

> b
[1] 0.0000000 0.4000000 0.5200000 0.5560000 0.5668000 0.5700400
0.5710120
[8] 0.5713036 0.5713911 0.5714173 0.5714252

0.42857 and 0.57143 respectively.

The matrix multiplied by

times $\begin{bmatrix} 100 \\ 150 \end{bmatrix} = \begin{bmatrix} 107.1428 \\ 142.8572 \end{bmatrix}$ as before.

Getting multiple plots: here are some suggested R commands:

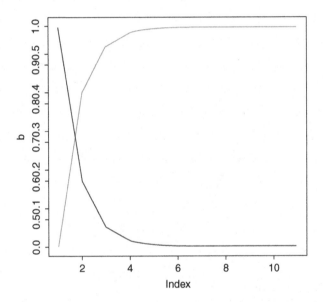

plot.new()

plot.window(xlim = range(x1),ylim = range(y1))

lines(x1,y1)
axis(1); axis(2); box()

plot.window(xlim = range(x2),ylim = range(y2))

lines(x2,y2)

axis(1); axis(2); box()

```
## Using the `deSolve` package
library(deSolve)

## Time
t <- seq(0, 100, 1)

## Initial population
N0 <- 10

## Parameter values
params <- list(r = 0.1, K = 1000)

## The logistic equation
fn<- function(t, N, params) with(params, list(r * N * (1 - N/ K)))

## Solving and plot in the solution numerically
out <- ode(N0, t, fn, params)
plot(out, lwd = 2, main = "Logistic equation\nr = 0.1, K = 1000, N0 = 10")

## Plotting the analytical solution
with(params, lines(t, K * N0 * exp(r * t)/ (K + N0 * (exp(r * t) - 1)), col = 2, lwd = 2))
```

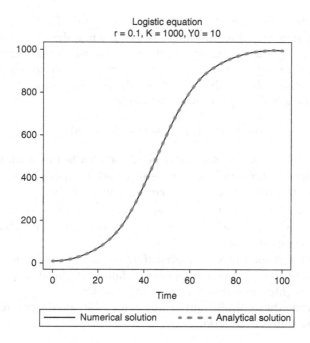

2.7 Exercises

Consider the model $a(n + 1) = r\, a(n)(1 - a(n))$. Let $a(0) = 0.2$. Determine the numerical and graphical solution for the following values of r. Find the pattern in the solution.

1. $r = 2$.

2. $r = 3$.

3. $r = 3.6$.

4. $r = 3.7$.

For problems 5–8 find the equilibrium value by iteration and determine if it is stable or unstable.

5. $a(n + 1) = 1.7\, a(n) - 0.14\, a(n)^2$.

6. $a(n + 1) = 0.8\, a(n) + 0.1\, a(n)^2$.

7. $a(n +1) = 0.2\, a(n) - 0.2\, a(n)^3$.

8. $a(n + 1) = 0.1\, a(n)^2 + 0.9\, a(n) - 0.2$.

9. Consider spreading a rumor through a company of 1,000 employees all working in the same building. We assume that the spread of a rumor is similar to the spread of a contagious disease in that the number of people hearing the rumor each day is proportional to the product of the number hearing the rumor and the number who have not heard the rumor. This is given by the formula:

$$r(n + 1) = r(n) + 1000\, k\, r(n) - k\, r(n)^2,$$

where k is the parameter that depends on how fast the rumor spreads. Assume $k = 0.001$ and further assume that four people initially know the rumor. How soon will everyone know the rumor?

10. Determine the equilibrium values of the bass and trout model presented in Section 2.2. Can these levels ever be achieved and maintained? Explain.

11. Test the fast food models with different starting conditions summing to 14,000 students. What happens? Obtain a graphical output and analyze the graph in terms of long-term behavior.

    (a) Find the equilibrium values for the Predator–Prey Model presented in Section 2.5.

(b) In the Predator–Prey Model, presented in Section 2.5, determine the outcomes with the following sets of parameters.
   (i)   Initial foxes are 200 and initial rabbits are 400.
   (ii)   Initial foxes are 2,000 and initial rabbits are 10,000.
   (iii) Birth rate of rabbits increases to 0.1.
(c) In the SIR model, presented in Section 2.5, determine the outcome with the following parameters changing:
   (i)   The flu lasts 1 week.
   (ii)   Initially 5 are sick and 10 the next week.
   (iii) The flu lasts 4 weeks.
   (iv) There are 4,000 students in the dorm and 5 are initially infected and 30 more the next week.

2.8 Projects

1. Consider the contagious disease as the Ebola virus. Use the internet to find out how deadly this virus actually is. Now consider an animal research laboratory in Restin, VA., a suburb of Washington, DC, with population 856,900 people. A monkey with the Ebola virus has escaped its captivity and infected one employee (unknown at the time) during its escape. This employee reports to University hospital later with Ebola symptoms. The Infectious Disease Center (IDC) in Atlanta gets a call and begins to model the spread of the disease. Build a model for the IDC with the following growth rates to determine the number infected after 2 weeks:

   a. $k = 0.00025$
   b. $k = 0.000025$
   c. $k = 0.00005$
   d. $k = 0.000009$

   List some ways of controlling the spread of the virus.

2. Consider the spread of a rumor concerning termination among 1000 employees of a major company. Assume that the spreading of a rumor is similar to the spread of contagious disease in that the number hearing the rumor each day is proportional to the product of those who have heard the rumor and those who have not heard the rumor. Build a model for the company with the following rumor growth rates to determine the number having heard the rumor after 1 week:

   a. $k = 0.25$
   b. $k = 0.025$
   c. $k = 0.0025$

    d. $k = 0.00025$

List some ways of controlling the spread of the rumor.

3. Lions and spotted hyena: Predict the number of lions and spotted hyena in the same environment at a function of time

Assumptions:

The variables: $L(n)$ = number of lions at the end of period n.

              $H(n)$ = number of hyenas at the end of period n.

Assume the Model:

$$L(n + 1) = 1.2\, L(n) - 0.001\, L(n)H(n).$$

$$H(n + 1) = 1.3\, H(n) - 0.002\, H(n)\, L(n).$$

(1) Find the equilibrium values of the system.
(2) Iterate the system from the following initial conditions and determine what happens to the lions and the spotted hyenas in the long term.

Lions	Spotted hyena
150	200
151	199
149	210
10	10
100	100

4. It is getting close to election day. The influence of the new Independent Party is of concern to the Republicans and Democrats. Assume that in the next election that 75% of those who vote Republican vote Republican again, 5% vote Democratic, and 20% vote Independent. Of those that voted Democratic before, 20% vote Republican, 60% vote Democratic, and 20% vote Independent. Of those that voted Independent, 40% vote Republican, 20% vote Democratic, and 40% vote Independent.

    a. Formulate and write the system of discrete dynamical systems that models this situation.

    b. Assume that there are 399,998 voters initially in the system, how many will vote Republican, Democratic, and Independent in the long run? (Hint: you can break down the 399,998 voters in any manner that you desire as initial conditions.)

    c. (New scenario) In addition to the above, the community is growing. (18-year-olds + new people − deaths − losses to the community, etc.) Republicans predict a gain of 2,000 voters between elections. Democrats estimate a gain of 2,000 voters between elections. The Independents estimate a gain of 1,000 voters between elections. If this rate of growth continues, what will be the long-term distribution of the voters?

2.9 References and Suggested Further Reading

Albright, B. (2010). *Mathematical Modeling with Excel.* Jones and Bartlett. Sudberry, MA.

Alfred, U. (1967). Sums of squares of consecutive odd integers. *Mathematics Magazine,* 40(4): 194–199.

Allman, E. and J. Rhodes (2004). *Mathematical Models in Biology: An Introduction.* Cambridge University Press, Cambridge, UK.

Arney, D., F. Giordano, and J. Robertson (2002). *Mathematical Modeling with Discrete Dynamical Systems.* McGraw Hill, Boston, MA.

Fox, W. P. (2010). Discrete combat models: Investigating the solutions to discrete forms of Lanchester'scombat models, *International Journal of Operations Research and Information Systems* (IJORIS), 1(1): 16–34.

Fox, W. P. (2012a). Discrete combat models: Investigating the solutions to discrete forms of Lanchester's Combat Models. In *Innovations in Information Systems for Business Functionality and Operations Management,* IGI Global & SAGE Publishers, pp. 106–122.

Fox, W. P. (2012b). Mathematical modeling of the analytical hierarchy process using discrete dynamical systems in decision analysis, *Computers in Education* Journal, 3(3): 27–34.

Fox, W. P. (2012c). *Mathematical Modeling with Maple.* Cengage Publishing, Boston, MA.

Fox, W. P. and P. J. Driscoll (2011). Modeling with dynamical systems for decision making and analysis. *Computers in Education Journal* (COED), 2(1): 19–25.

Giordano. F. R., W. P. Fox, and S. Horton (2014). *A First Course in Mathematical Modeling,* 5th ed. Cengage Publishing, Boston, MA.

Leyendekker, J. and A. Shannon (2015). The odd-number sequence: Squares and sums. *International Journal of Mathematical Education in Science and Technology,* 46(8): 1222–1228.

Sandefur, J. (1990). *Discrete Dynamical System: Theory and Applications.* Oxford University Press,New York, NY.

Sandefur, J. (2002). *Elementary Mathematical Model: A Dynamic Approach,* 1st Edition. Brooks-Cole Publishers, Belmont, CA.

Sandefur, J. (2003). *Elementary Mathematical Modeling: A Dynamic Approach.* Thomson Publishing, Belmont, CA.

3

Modeling with Differential Equations

Objectives

1. Understand the typical models of ordinary differential equation.
2. Understand the analytical solution techniques for separable and linear ODEs.
3. Know the numerical methods.
4. Understand the importance of slope fields.

Consider a sports sky diving parachutist. We want to build a model on how far the parachutist free falls before they open their parachute. We might initially assume the person is falling from rest and we choose to neglect all resistive forces. We might choose to neglect a variable because we want to investigate the problem without the influence of that variable. For instance, we might want to study the falling body by neglecting buoyancy but considering the effects of drag within the earth's atmosphere. We can simplify the submodel for the resistive force by making a reasonable assumption that the drag force equals some constant times the speed of the falling body or we might decide to try a constant time the speed squared of the falling body. Next, we want to construct submodels for propulsion, drag, and buoyancy forces. The propulsion force acting on the falling body from rest is due to gravity. This gravitational attraction in turn depends on the mass of the falling body and its distance above the surface of the earth. Thus

Propulsion force = F_p
$\qquad\qquad\quad$ = gravitational attraction
$\qquad\qquad\quad$ = f(*mass, distance*).

Next, consider the resistive force, which is the sum of the drag and buoyancy forces:

Drag = F_d
   = f(*speed, air density, cross-sectional area of the body, shape of the body*)

Buoyancy = F_b
   = f(*air density, density of the body*).

A free-body diagram displays all the forces acting on the body with a coordinate system indicating directions.

From the free-body diagram (Figure 3.1), it can be seen that the total force F acting on the body is given by

$$F = F_p + F_d + F_b.$$

We assume that Newton's second law applies, $F = ma$ (itself a mathematical model). If we assume to neglect all resisting forces, we find

$$F = ma = -mg + 0 + 0$$

or

$my'' = -mg$, y'' is a second derivative and subject to initial conditions $y(0) = h, y'(0) = 0$.

Here m is the mass of the body, g is the acceleration due to gravity, and y is the distance of the body measured from the surface of the earth. The acceleration a is the second derivative of y with respect to time, and it is this idea of change that led to a differential equation. Since the propulsion force is acting down, then it is negative.

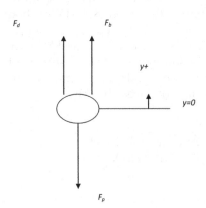

FIGURE 3.1
Free body diagram.

If we were to include the force due to air resistance as a constant times speed, then $F_r = kv$. We then would have the model,

$$F = F_p + F_d + F_b$$

$$F = ma = m\, y'' = -mg + kv$$

with the same initial conditions of $y(0) = h$, $y'(0) = 0$.

In this chapter, we will consider a variety of models, concepts, and techniques that give the reader some of the basic tools needed in solving and analyzing first-order differential equations. Since differential equations are used often in mathematical modeling, we devote a more thorough approach in this chapter.

- **Section 3.1** provides some models that we will solve and analyze. These models come from a variety of disciplines in science and engineering, including chemistry, physics, biology, fluids mechanics, Newtonian Mechanics, environmental engineering, and financial mathematics. Modeling techniques are discussed that help determine the necessary coefficients in the models.

- **Section 3.2** introduces slope fields. Slope fields provided qualitative viewing of solutions of first-order differential equations. Recall that we said this book discusses the modeling of change. Slope fields provide information about rates of change since they are derived from derivative information.

- **Section 3.3** covers the analytical method of solution from Excel. Existence and uniqueness are addresses for conceptual and theoretical basis. The focus is on the initial-value problem and the question as to whether or not the initial value problem has a unique solution and whether the new functions can be defined by the initial-value problem.

- **Section 3.4** is devoted to various numerical techniques. Euler, Improved Euler, and Runge-Kutta techniques are introduced as techniques to approximate solutions. Since problems may arise in numerical methods, some errors are discussed as well as methods to compare long-term behavior of the solutions to determine qualitative accuracy of solutions.

3.1 Applied First-Order Models

In this section, we introduce many mathematical models from a variety of disciplines. Our emphasis in this section is building the mathematical model and its associated differential equation that will be solved later in the chapter. Recall we have previously discussed the modeling process. In this section, we will confine ourselves to the first three steps: (1) identifying the problem, (2) assumptions and variables, and (3) building the model.

Example 3.1 Radioactive Decay

A British physicist Lord Ernest Rutherford (1871–1937) established that an atom consists of a nucleus surrounded by electrons. The nucleus itself consists of protons and neutrons and two nuclei with identical numbers of protons and identical numbers of neutrons belonging to the same nuclear species or nuclide.

The accidental discovery of radioactivity by Henri Becquerel in 1896 followed quite closely by that of x-rays. Becquerel had left a substance containing uranium on a photographic plate wrapped in black paper. After the plate was developed, it showed images of crystals of uranium compounds. The uranium had emitted penetrating rays, and Becquerel coined the word radioactive to describe this property. Later research by Becquerel, Madam Currie, and Rutherford led to the discovery that there are three different kinds of radioactive emissions: alpha, beta, and gamma radiation. Ordinarily it is of interest or concern to know how long it will take for a certain radioactive substance to lose some fraction of its total mass through the emission process. Let's construct a model to provide these answers.

Assumptions: Assume we have an object that if exposed to a certain substance will emit radiation. Assume that the emissions are all the same kind and that they all decay by the same process (alpha, beta, and gamma). The decaying process is such that once a particular nucleus decays, it cannot repeat the process again. Moreover, the probability of its decaying in any time interval is the same. Suppose that at time t there are N undecayed nuclei present. Then the number of nuclei that decay during the time interval from t to t +Δt must be proportional to the product of N and Δt. That is,

$$\Delta N = -kN\Delta t,$$

where k > 0 is the proportionality coefficient, called the decay constant. Let us assume that the number of undecayed nuclei is a continuous function of time. Then dividing both sides of the last equation by Δt and passing to the limit as $\Delta t \rightarrow 0$ results in the following differential equation:

$$\frac{dN}{dt} = -kN$$

For a given radioactive element, the number k is determined from experimental data. We desire information concerning the radioactive decay of an element.

Problem Identification:
Determine a relationship between time and amount of radiation emitting from a specific substance after exposure.

Assumptions:
We assume the rate in which an item decays is proportional to the amount of that item at any time, t. The model is accurate over a short period of time. Half-life is the measure of stability of a radioactive substance. The half-life is simply the time, t_i, takes one-half of the atoms to disintegrate. The longer the half-life, the more stable the item is.

Model Construction:
Ordinarily, it is of interest to know how long it takes an object emitting radiation to lose some fraction of its total mass or how much of its mass is remaining after a specified amount of time. Many radioactive materials disintegrate at a rate proportional to the amount present. For example, if x is the radioactive material and $Q(t)$ is the amount of radioactive material present at time, t, then the rate of change of $Q(t)$ with respect to t is given by $\dfrac{dQ}{dt} = -kQ$, where k represents the proportionality constant and $k > 0$. Let us refer to $Q(0) = Q_0$ as the initial quantity of radioactive material at any arbitrary time, $t = 0$.

Finding the proportionality constant k from the data in Table 3.1.

TABLE 3.1

Example 3.1 Radioactive decay data

Time, t	Number of Nuclei, N
0	100.0
1	58.5
2	34.2
3	20.0
4	11.7
5	7.0
6	4.0
7	2.3
8	1.4

Let's estimate using the law of radioactive decay. The law of radioactive decay leads to solution:

$$Q(t) = Q_0\, e^{-kt}.$$

From the data we know initially that $Q(0) = 100$ grams. After 8 hours, there is 1.4 grams remaining. So

$$Q(8) = 1.4 = 100\, e^{-k(8)}$$

$$.014 = e^{(-8k)}$$

$$ln(014)/-8 = k$$

$$k = 0.5335.$$

If we use this method then we might use the model, $Q(t) = Q_0\, e^{-0.5335\,t}$.

If we use all the data, then we must create the change in the number of nuclei and then plot versus N.

The slope (using linear regression without an intercept) will be our value for k.

We plot the difference versus N and find the slope as -0.4152. If we use this approach our model is

$$Q(t) = Q_0 e^{(-0.4152\,t)}.$$

From a modeling perspective, if you have all the data, or a substantial amount of data, build the proportionality model (see Giordano et al. for a good discussion of proportionality) to find the slope. If you only have a few data points, then use the model form to estimate the value of k. We do note that the estimates are moderately close.

TABLE 3.2

Difference in the number of nuclei

Time	Number of nuclei, N	Difference
0	100.0	-41.5
1	58.5	-24.3
2	34.2	-14.2
3	20.0	-8.3
4	11.7	-4.7
5	7.0	-3.0
6	4.0	-1.7
7	2.3	-0.9
8	1.4	

Example 3.2 Newton's Law of Cooling

You are working as an assistant for the Crime Scene Investigators at a potential homicide. A body was found in a trash dumpster and you were called to the site immediately. The ambient temperature is approximately 68°F. The body temperature is 77 degrees. After one hour, the body temperature is 74°F.

Problem Identification:
Build a mathematical modeling relating a dead body's cooling temperature to time in order to find the time of death.

Assumptions:
A simplified list could include normal body temperature, average seasonal temperature outdoors, constant indoor temperature, and no other outside sources of heat or cooling.

Modeling Construction:
Newton's Law of Cooling states that the surface temperature of an object changes at a rate proportional to its relative temperature. That is, the difference between its temperature and the surrounding environment temperature. Let $T(t)$ = the temperature of the object at time t.

According to Newton's Law: $\dfrac{dT}{dt} = k(T - T_s)$ where k is the proportionality constant.

Example 3.3 Mixtures

Consider an initial amount of a substance, say barley measured in pounds, is dissolved in a large vat in order to make beer. The solution is pumped in at one rate, well-mixed, and pumped out at the same rate. We want to know the concentration of the substance in the vat after time.

Problem Identification:
Build a mathematical modeling relating the concentration of barley in the vat to time, t.

Assumptions:
We know the size of the vat, 300 gallons. The initial amount dissolved is 50 pounds. The pumping rate is constant both into and out of the vat. No other substances are interacting with the vat and the pumps.

Model Construction:

The rate of change of concentration is measured as the difference between the rate pumped in and the mixture being pumped out of the vat. Let $C(t)$ be the concentration or amount of barley in the vat after time, t, $\dfrac{C(t)}{dt} = R_{in} - R_{out}$.

Example 3.4 Population Models

Consider an experiment measuring the growth of a yeast culture. An analysis yields the growth rate as approximately 0.6. We might try $\dfrac{dP}{dt} = 0.6P$ as our ODE. Note that this model will predict a population that increases forever. We might think that is not possible.

3.1.1 Model Refinement: Modeling Births, Deaths, and Resources

In both births and deaths during a period are proportional to the population, then the change in population itself should be proportional to the population, as was illustrated above. However, certain resources (food, for instance) can support only a maximum population level rather than one that increases indefinitely. As these maximum levels are approached, growth should slow moving toward the carrying capacity.

Suppose we estimate the carrying capacity to be 665. Nevertheless, as p_n approaches 665, the change does slow considerably. Since $665 - p_n$ does get smaller as p_n approaches 665, consider the following model: $\dfrac{dP}{dt} = k(665 - P)P$

If we accept the proportionality argument, we can estimate the slope of the line approximating the data to be about $k \approx 0.00082$, which gives the model:

$$p_{n+1} - p_n = 0.00082(665 - p_n)p_n.$$ Converting to an ODE yields

$$\frac{dP}{dt} = 0.008271 \cdot P \cdot (665 - P)$$

Example 3.5 The Spread of a Contagious Disease

Suppose there are 400 students in a college dormitory and that one or more of the students has a severe case of the flu. Let i_n represent the number of infected students after n time periods. Assume some interaction between those infected and those not infected is required to pass the disease. If all are susceptible to the disease, $(400 - i_n)$ represents those susceptible but not yet infected. If those infected remain contagious, we may model the change in infected as a proportionality of the product of those infected by those susceptible but not yet infected or,

$$\Delta i_n = i_{n+1} - i_n = k i_n (400 - i_n)$$

In this model the product $i_n(400 - i_n)$ represents possible interactions between those infected and those not infected. A fraction k would now become infected. There are many refinements to the above model. For example, we can consider that a segment of the populations is not susceptible to the disease, that the infection period is limited, that infected students are removed from the dorm to prevent interaction with uninfected students, and so forth.

$$\frac{di}{dt} = ki(400 - i), k > 0$$

3.1.2 Some Classical First-Order Models in Differential Equations

Decay:

$$\frac{dQ}{dt} = -kQ, k > 0$$

Growth:

$$\frac{dQ}{dt} = kQ, k > 0$$

Newton's Law of Cooling:

$$\frac{dT}{dt} = k(T - T_0), k > 0$$

Chemical Mixtures, Drug Dissemination, Dialysis:

$$\frac{dA}{dt} = R_{in} - R_{out}$$

L-R Series Circuit:

$$L\frac{dI}{dt} + Ri = E(t)$$

Falling Bodies, Newton's Law of Σ Forces = 0

$$m\frac{dv}{dt} = F_{weight} + F_{resistance} = mg - kv, k > 0$$

Chemical Reactions:

$$\frac{dX}{dt} = k(\alpha - X)(\beta - X)...$$

Spread of a Disease or Rumor:

$$\frac{dN}{dt} = kN(L - N), k > 0, L > 0$$

Filling a Tank:
   Tank with cross-sectional area A, filled from the top at rate K, while draining from the bottom through an orifice of area α, with height of the tank h:

$$\frac{dh}{dt} = \frac{\left(k - \alpha a\sqrt{2gh}\right)}{A}$$

3.2 Slope Fields and Qualitative Assessments of Autonomous First-Order ODE

We have used the modeling process to create a number of models that are important to differential equations. However, sometimes we can predict outcomes of models graphically without actually solving them. This is true when we are looking for behavior rather than specific values. For example, will a specific population survive is a question that can be answered graphically. In calculus, we used derivative information to analyze a function or just to gain more information. Knowing where a function is increasing or decreasing, locations of extrema, concavity, and so forth can be important information.

   Interpreting the derivative as the slope of the line tangent to the graph of the function is useful in gaining information about the solution to the differential equation. From this information, it is possible to sketch qualitatively the solution curves to the equations from any real initial condition. These graphs provide considerable information regarding the solution and these graphs provide piratical benefits as well. Since most real differential equations cannot be solved analytically, other techniques, such as graphical analysis are needed to analyze the behavior of solutions.

Slope Field

Suppose the first-order differential equation $\frac{dy}{dt} = f(t, y)$ is given. Each time an initial value $y(t_0) = y_0$ is specified, the solution curve is required to pass through the coordinate point (t_0, y_0) and we also know the value of slope at that point, $\frac{dy}{dt} = f(t_0, y_0)$. Therefore, it is possible to draw a short line segment with the correct slope through each point (t, y) in the $t - y$ plane. The result is known as the slope field of the first-order differential equation. We can say that the slope field of a differential equation is the set of all line segments with the correct slope or mini-tangent lines.

Let's sketch the possible solution curves of the differential equation, $\frac{dy}{dt} = \frac{(t-y)}{2}$ without solving. Consider the information in Table 3.3 provides specific points in the plane and the slope at each point.

Slope fields are much easier to visualize with the use of technology and we will demonstrate this with an Excel macro for slope fields. The base version of Excel does not have the capability to do slope fields but we did find an Excel macro program to generate Slope fields, as seen in Figure 3.2 and which we will make available by request from anyone. Once the macro is started the first step in the process is to click the Modify the Slope Field button to make changes to the differential equation to be viewed.

We modified the slope field to $dy/dx = (x - y)/2$.

A complete slope field over the region $-2 \leq x \leq 2$ is seen in the Figure 3.3.

TABLE 3.3

Points in the plane and corresponding slope

Point	Slope
(0,1)	-0.50
(0,2)	-1.00
(1,0)	0.50
(2,0)	1.00
(-1,0)	-0.50
(-2,0)	-1.00
(1,1)	0.00
(-1,1)	-1.00
(2,2)	0.00
(1,2)	-0.50

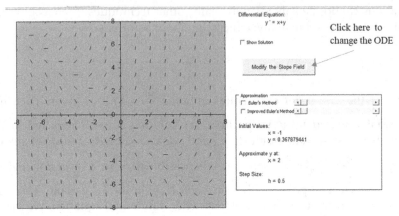

FIGURE 3.2
Screenshot of macro program for slope field in Excel.

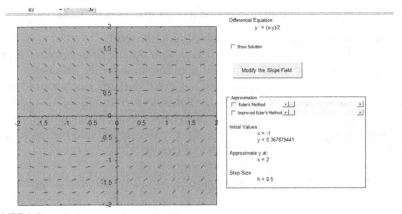

FIGURE 3.3
Screenshot of slope field for $dy/dx = (x - y)/2$ from $-2 \leq x \leq 2$.

Example 3.6 $y' = y - t^2$

We obtain the slope field as shown in Figure 3.4

Example 3.7 $\dfrac{dy(t)}{dt} = y(t)\big(2 - y(t)\big)$

Slope fields provide us with possible solution curves for a differential equation as well as information about stable rest points, if they exist.

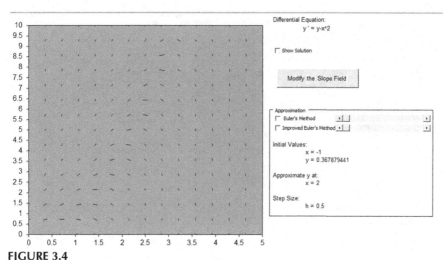

FIGURE 3.4

Screenshot of slope field for $y' = y \cdot (2 - y)$.

Autonomous Qualitative Assessment

In many modeling applications, the instantaneous rate of change $\dfrac{dy}{dt}$ is assumed to be proportional to some function of the dependent variable y alone. These are called autonomous differential equations, $\dfrac{dy}{dt} = f(y)$. In these cases, it is possible to analyze the nature of the solution curve by analyzing $\dfrac{dy}{dt}$ versus y. Let's explore this qualitative assessment.

Rest Points

Qualitative information concerning the solutions can often be obtained from the differential equation. For example, consider the differential equation describing the Malthusian model of population growth discussed earlier: $\dfrac{dP}{dt} = kP, \ k > 0$. Thus, the Malthusian model assumes a simple proportionality between growth rate and population (see Figure 3.6). Since $P > 0$ and $k > 0$, you can see that $\dfrac{dP}{dt} > 0$, which means that the population curve $P(t)$ is everywhere increasing. Moreover, the smaller the value of k, the less rapid is the growth in the population over time. For a fixed value of k, as P increases so does its rate of change $\dfrac{dP}{dt}$. Thus, the solution curves must appear

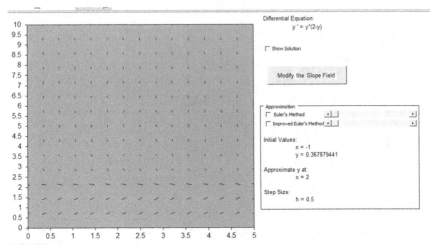

FIGURE 3.5

Screenshot of Excel's slope field and a slope field from Excel for $dy/dx = y*(2 - y)$.

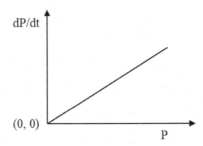

FIGURE 3.6

Graph of dP/dt versus P.

qualitatively as depicted in Figure 3.7. Assuming k is constant, at each population level P the derivative $\frac{dP}{dt}$ is constant. Therefore, all the solution curves are horizontal translations of one another. The initial population $P(0) = P_0$ distinguishes these various curves.

Next, recall the refined population model: $\frac{dP}{dt} = r(M - P)P$. The graph of $\frac{dP}{dt}$ versus P is the parabola shown in Figure 3.8. Since the second derivative $P'' = rP'(M - 2P)$ is zero when $P = \frac{M}{2}$, positive for $P < \frac{M}{2}$, and negative for $P > \frac{M}{2}$, we conclude that $\frac{dP}{dt}$ is at a maximum at the population level. Note that for $0 < P < M$, the derivative $\frac{dP}{dt}$ is positive and the population $P(t)$

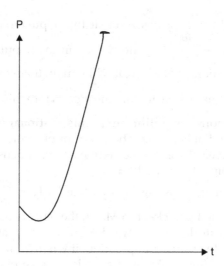

FIGURE 3.7
Plot of *P* versus *t*.

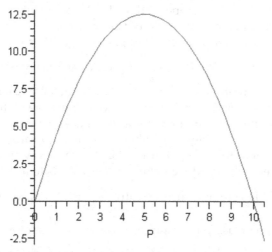

FIGURE 3.8
Plot of $\dfrac{dP}{dt}$ versus *P* assuming that $M = 10$.

is increasing; for $M < P$, $\dfrac{dP}{dt}$ is negative and $P(t)$ is decreasing. Moreover, if $P > \dfrac{M}{2}$, the second derivative P'' is positive and the population curve is concave upward for $P > \dfrac{M}{2}$, P'' is negative and the population curve is concave downward.

The parabola in Figure 3.8 has zeros at the population levels $P = 0$ and $P = M$. At those levels, $\dfrac{dP}{dt} = 0$, so no change in the population P can occur. That is, if the population level is at 0, it will remain there for all time; if it is at the level M, it will remain there for all time. Points for which the derivative $\dfrac{dP}{dt}$ at 0 are called **rest points, equilibrium points, stationary points,** or **critical points** of the differential equation. The behavior of solutions near rest points are of significant interest to us. Let's examine what happens to the population P is near the rest points $P = M$ and $P = 0$.

Suppose the population P is slightly less than M. Then $\dfrac{dP}{dt}$ is positive, so the population increases and gets closer to M. On the other hand, if $P > M$, $\dfrac{dP}{dt} < 0$, so the population will decrease toward M. Thus, no matter what positive value is assigned to the starting population, it will tend to the limiting value M as time tends to infinity. We say that M is an **asymptotically stable rest point**, because whenever the population level is perturbed away from that level it tends to return there again. However, if the starting population is not at the level M the population $P(t)$ cannot reach M in a finite amount of time. This fact follows from the property that $P = M$ is a solution to the limited growth equation and the two solutions cannot cross.

Next, consider the rest point $P = 0$ in Figure 3.8. If P is perturbed slightly away from 0 so that $M > P > 0$, then $\dfrac{dP}{dt}$ is positive and the population increases toward M. In this situation we say that $P = 0$ is an **unstable rest point**, because any population not starting at that level tends to move away from it.

In general, equilibrium solutions to differential equations are classified as stable or unstable according to whether, graphically, nearby solutions stay close to or converge to the equilibrium, or diverge away from the equilibrium, respectively, as the independent variable t tends to infinity.

At $P = \dfrac{M}{2}$, P'' is zero and the derivative $\dfrac{dP}{dt}$ is at a maximum. Thus, the population P is increasing most rapidly when $P = \dfrac{M}{2}$ and a point of inflection occurs in the graph there. These features give each solution curve its characteristic "S" or sigmoid shape. From the information we have obtained, the family of solutions to the limited growth model must appear (approximately) as shown in Figure 3.8. Notice again that at each population level P, the derivative $\dfrac{dP}{dt}$ is constant, so that the solution curves are horizontal translations of one another. The curves are distinguished by the initial population level $P(0) = P_0$. In particular, note the solution curve when $\dfrac{M}{2} < P_0 < M$ in Figure 3.9.

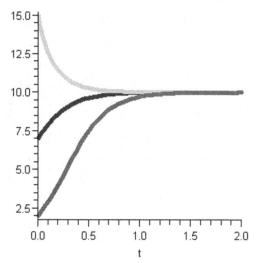

FIGURE 3.9
Possible plots of P versus t with $M = 10$.

3.2.1 Maple for Modeling Slope Fields

Maple has a toolbox called "with(DETools)." By using the command >
with(DETools), we enable the ability to use all the DEtools available in Maple.
This allows the use of DEplots and phase portraits at a minimum.

The first command that we illustrate is DEplot, which produces a
slope filed.

Let's use Maple to obtain the slope filed for $dy/dt = y(2 - y)$ with various
initial conditions such as $y(0) = 0.5$, $y(0) = 1$, and $y(0) = 4$. This is shown in
Figure 3.10. We also produce a few sample solution curves within the slope
field plot.

In Maple the commands are:

```
>with(DETools):
>ode: = diff(y(t),t = y(t)*(2 – y(t));
Inc: = {y(0),0,5,y(0) = 1,y(0) = 4};
>DEplot(ode,y(t), t = 0..10, ic, color = black, thickness = 3, title = "DE Plot");
```

3.2.2 R for Modeling Slope Fields

R will also produce slope fields. We need to obtain from the CRAN library
deSolve and open its library. Here is the necessary R code to produce slope
fields.

> *DEplot(ode, y(t), t* = 0 ..10*, inc, color* = *black, thickness* = 3*, title* = `*DE Plot* `);

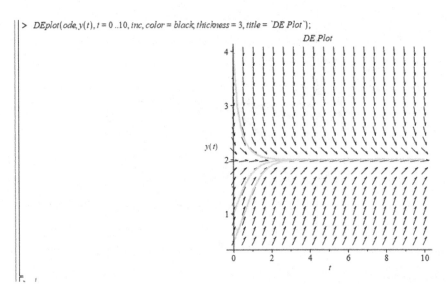

FIGURE 3.10
Maple ODE slope field.

Our differential equation

diff<-function(x,y)

{

return(x/y)

}

Line function

The Line
    <-function(x1,y1,slp,d)

    {

    z = slope*(d – x1) + y1

    return(z)

    }

```
# Domains

  x = seq(-20,20,0.5)

  y = seq(-20,20,0.5)

# Points to draw our graph

  f = c(-5,5)

  h = c(-5,5)

  plot(f,h,main = "Slope field")

# Let's generate the slope field

  for(j in x)

  {

  for(k in y)

  {

  slope = diff(j,k)

  domain = seq(j-0.07,j + 0.07,0.14)

  z = TheLine(j,k,slope,domain)

  arrows(domain[1],z[1],domain[2],z[2],length = 0.08)

  }
  }

  library(phaseR)
      apma1 <- function(t, y, parameters){
      a <- parameters[1]
      dy<- a*((y/3) – 360)
      list(dy)

}
apma1.flowField <- flowField(apma1, x.lim = c(0, 10),
y.lim = c(1070, 1090), parameters = c(1),
```

points = 9, system = "one.dim",

add = FALSE, xlab = "time", ylab = "P",

main = "Mice Population")

grid()

Assume our model is $dP/dt = P/3 - 360$. Figure 3.11 illustrates R studio and the slope field.

Our equilibrium is unstable as seen in Figure 3.11.

3.3 Analytical Solution to First-Order ODEs

3.3.1 Separable Ordinary Differential Equations

The differential equation of the form $\dfrac{dy}{dt} = f(t, y)$ is called separable, if $f(t,y) = h(t) \cdot g(y)$. That is, $\dfrac{dy}{dt} = h(t) \cdot g(y)$. You need to able to write the function $f(t,y)$ as $h(t) \cdot g(y)$.

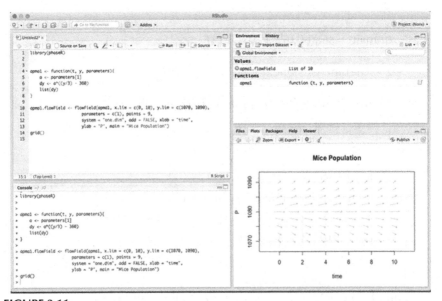

FIGURE 3.11
Screenshot from R studio.

Example 3.8 $\dfrac{dy}{dt} = t \cdot y$

This is separable because the function $f(t,y) = h(t) \cdot g(y) = t \cdot y$.

Example 3.9 $\dfrac{dy}{dt} = t + 9y$

This is not separable because $f(t,y) = t + 9y$ cannot be written as the product of two functions, $h(t) \cdot g(y)$.

Sometimes this is more complicated to see as in the following example.

Example 3.10 $\dfrac{dy}{dt} = 8ty - 12y - 2t + 3$

This can be written as $\dfrac{dy}{dt} = (2t - 3) \cdot (4y - 1)$ and is separable in this form.

In order to solve separable ODEs, we perform the following steps:

(1)   Separate the differential equation into a clear form of

$$\frac{dy}{dt} = h(t) \cdot g(y)$$

(2)   Rewrite the differential equation as

$$\frac{dy}{g(y)} = h(t)dt$$

(3)   Integrate both sides of the expression and include the constant of integration, C,

$$\int \frac{dy}{g(y)} = \int h(t) \cdot dt$$

to obtain the general solution:

$$G(y) = H(t) + C.$$

(4)   If you are given an IVP, use the initial conditions to find the particular solution.

Example 3.11 Solve $\dfrac{dy}{dt} = t \cdot y$

$$\frac{dy}{dt} = tdt$$

$$\int \frac{dy}{y} = \int tdt$$

$$ln|y| + C = \frac{t^2}{2} + C$$

$$ln|y| = \frac{t^2}{2} + C$$

$$e^{ln|y|} = e^{\frac{t^2}{2}+C}$$

$$y = ce^{\frac{t^2}{2}}$$

If the ODE $\dfrac{dy}{dt} = t \cdot y$ passes through the point (1, 1), then we use (1, 1) as the initial condition.

$$1 = ce^{\frac{1}{2}}$$

$$c = \frac{1}{e^{\frac{1}{2}}} = e^{-5} = 0.60653$$

$$y = 0.60653e^{\frac{t^2}{2}}$$

We plot the solution below shown in Figure 3.12:

Plot of $y = 0.60653e^{\frac{t^2}{2}}$.

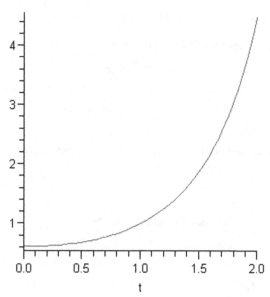

FIGURE 3.12
Solution plot of $y = 0.60653e^{\frac{t^2}{2}}$.

Example 3.12 Radioactive Decay

Consider the following radioactive decay model, $\dfrac{dQ}{dt} = -kQ,\ k > 0$ with a radioactive isotope with a half-life of 16 days. You wish to have 30 grams at the end of 25 days.

$$\frac{dQ}{dt} = -kdt$$

$$ln|Q| = -kt + C$$

$$Q = ce^{-kt}.$$

Since the half-life is 16 days, then

$$Q = ce^{-kt}$$

$$\frac{Q}{2} = ce^{-16k}$$

$$k = \frac{ln|2|}{16} = .043322$$

$$Q = ce^{-.043322t}.$$

We want 30 grams after 25 days, so

$$30 = ce^{(-.043322 \cdot 25)}$$

$$c = \frac{30}{e^{-.043322.25}} = 88.61g$$

Initially, we should start with 88.61 grams of the radioactive substance.

The model is $Q = 88.61 \cdot e^{-.043322t}$.

Example 3.13 Malthusian Population Model

The notional population in the US is provided in Table 3.3.

The model that we will initially use is our simple growth model, $\frac{dP}{dt} = kP, \ k > 0$.

Solution:

$$\frac{dP}{p} = kt$$

$$\int \frac{dP}{p} = \int kt$$

$$ln|P| = kt + C$$

TABLE 3.3

Notional population of the United States

Year	Population in millions
1990	249.4
1991	252.2
1992	254.9
1993	257.7
1994	260.2
1995	262.7
1996	265.1
1997	267.7
1998	270.2
1999	272.6
2000	280.5

$$P = ce^{kt}.$$

Our initial population in 1990 was 249.4 so $P(0) = 249.4$.

$$P = 249.4e^{kt}.$$

We need to find a good estimate for the growth parameter k.
We find the estimate for k is 0.00493.

$$P = 249.4e^{0.00493t}.$$

Predicting the population in 2020 yields

$$P = 249.4e^{0.00493 \cdot 20} = 275.24 \text{ million people}$$

There are two problems with this model. First, it is not realistic to assume that populations tend to infinity. Second, the predictions of this model always underestimate the real populations. We could use a refinement to this model which will be suggested later that does not yield the slow exponential growth toward infinity.

Example 3.14 Newton's Law of Cooling

Suppose a cold can of soda is taken from a refrigerator and placed in a warm classroom and the temperature is measured periodically. The temperature of soda is initially 40° Fahrenheit and the room temperature is a cozy 72° Fahrenheit.

Newton's Law of Cooling gives

$$\frac{dT}{dt} = -k(T - T), T(0) = 40$$

$$\frac{dT}{dt} = -k(72 - T), T(0) = 40$$

$$\frac{dT}{(72 - T)} = -kdt$$

$$\int \frac{dT}{(72 - T)} = \int -kdt$$

$$ln|72 - T| = -kt + C$$

Assume $k = .05$.

$$72 - T = Ce^{-5t}$$

When $T(0) = 40, C = 32$

$$T = 72 - 32e^{-.05t}.$$

After 10 minutes the soda's temperature is

$$T(10) = 72 - 32 \cdot e^{-.05 \cdot 10} = 52.591.$$

After 30 minutes the soda's temperature is

$$T(10) = 72 - 32 \cdot e^{-.05 \cdot 30} = 64.860.$$

Note the temperature of the soda warms and eventually will reach the temperature of the room, a cozy 72° Fahrenheit (but not cozy for drinking a soda).

Example 3.15 Chemical Mixtures

Initially 50 pounds of salt is dissolved in a 300-gallon tank of water. A brine solution is pumped into the tank at a rate of 3 gallons per minute, and a well-stirred mixed solution is then pumped out at a rate of 3 gallons per minute. We want to determine the amount of salt present at any time t if the concentration of the entering solution is 2 pounds per gallon.

$$\frac{dA}{dt} = R_{in} - R_{out}$$

$$R_{in} = \frac{3 \ gallons}{1 \ minute} \cdot \frac{2 \ pounds}{1 \ gallon} = 6 \ pounds \ per \ minute$$

$$R_{out} = \frac{3 \ gallons}{1 \ minute} \cdot \frac{\dfrac{A}{300 \ pounds}}{1 \ gallon} = \frac{A}{100} \ pounds \ per \ minute$$

$$\frac{dA}{dt} = 6 - \frac{A}{100}, A(0) = 50$$

We will rewrite the problem as:

$$\frac{dA}{dt} = .01(600 - A), A(0) = 50$$

This is separable.

$$\frac{dA}{600 - A} = .01dt$$

$$\int \frac{dA}{600 - A} = \int .01dt$$

$$-\ln(600 - A) = .01t + C$$

$$\ln(600 - A) = -.01t + C$$

$$A = 600 + ce^{-.01t}.$$

Using the initial condition, $A(0) = 50$

$$C = -550$$

$$A = 600 - 550e^{-.01t}.$$

Example 3.16 Refined Populations Model

Consider the logistic growth model defined for blue crabs in Venezuela,
$\frac{dB}{dt} = .25B(10 - B), B(0) = 4$. ($B$ is in millions of bushels.)

$$\frac{dB}{B(10 - B)} = .25dt$$

$$\int \frac{dB}{B(10 - B)} = \int .25dt$$

$$\frac{A}{B} + \frac{D}{10 - B} = 1$$

$$A \cdot (10 - B) + B \cdot D = 1 + 0B$$

$$10A = 1, A = \frac{1}{10}$$

$$-A + D = 0, A = D, D = \frac{1}{10}$$

So, $\int \frac{dB}{B(10 - B)} = \int \frac{1}{10B} dB + \frac{1}{10} \int \frac{1}{(10 - B)} dB = \frac{1}{10}[\ln(B) - \ln(10 - B)]$

$$\frac{1}{10}[\ln(B) - \ln(10 - B)] = .25t + C$$

Multiply both sides by 10,

$$[ln(B) - ln(10 - B)] = 2.5t + C$$

Solve for C, using $B(0) = 4$

$$ln(4) = ln(6) = C$$

$$C = -.405465.$$

Using the laws of logarithms and exponents, we obtain

$$\frac{B}{10 - B} = .666666 \cdot e^{2.5t}$$

We simplify to

$$B = \frac{6.66666 \cdot e^{2.5t}}{1 + .66666 \cdot e^{2.5t}} \text{ or } B = \frac{6.6666666}{.666666 + e^{-2.5t}}$$

The result is known as a logistics curve. It approaches the maximum sustainable amount of 6.666 million bushels.

3.3.2 Linear Differential Equations

Let's consider differential equations of the form, $a_1(t)\frac{dy}{dt} + a_0(t, y) = g(t)$, which can be simplified by dividing every term by $a_1(t)$ to obtain $\frac{dy}{dt} + P(t)y = f(t)$. We seek a solution to this differential equation. Using the exact method, $dy + P(t)y - f(t))dt = 0$, it is probably not true that $\frac{\partial M}{\partial y} = \frac{N}{\partial t}$. If there exists a function $u(t)$ such that $\frac{\partial}{ty}u(t) = \frac{\partial}{\partial y}[u(t) \cdot (P(t)y - f(t))]$, then we will have an exact equation. $\frac{\partial}{ty}u(t) = \frac{\partial}{\partial y}[u(t) \cdot (P(t)y - f(t))]$ reduces to $\frac{du}{dt} = u \cdot P(t)$. This is now separable so

$$\frac{du}{u} = P(t)dt$$

$$ln|u| = \int P(t)dt$$

$$u = e^{\int P(t)dt}$$

Now, this function $u = e^{\int P(t)dt}$ is called the **integrating factor** for the linear equation. We will use this concept of the integrating factor in the linear solution method, also called **variation of parameter**.

Step 1: Put the differential equation in the form: $\dfrac{dy}{dt} + P(t)y = f(t)$.

Step 2: Find the integrating factor: $e^{\int P(t)dt}$.

Step 3: Multiply through the differential equation by the integrating factor

$$e^{\int P(t)dt} \frac{dy}{dt} + e^{\int P(t)dt} \cdot P(t)y = e^{\int P(t)dt} \cdot f(t).$$

Step 4: This simplifies to $\dfrac{d}{dt}\left[y \cdot e^{\int P(t)dt} \right] = e^{\int P(t)dt} \cdot f(t)dt.$

Step 5: Integrate both sides, include constant of integration:

$$\int \frac{d}{dt}\left[y \cdot e^{\int P(t)dt} \right] = \int e^{\int P(t)dt} \cdot f(t)dt$$

$$y \cdot e^{\int P(t)dt} = \int e^{\int P(t)dt} \cdot f(t)dt + C$$

Step 6: Solve for y.

Example 3.17

Let's take a look at our example from slope fields, $\dfrac{dy}{dt} = \dfrac{t-y}{2}$.

Step 1: $\dfrac{dy}{dt} + \dfrac{y}{2} = \dfrac{t}{2}, P(t) = \dfrac{1}{2}$.

Step 2: The integrating factor is $e^{\int \frac{dt}{2}} \cdot u = e^{\frac{t}{2}}$.

Steps 3 & 4: $e^{\frac{t}{2}} \cdot \dfrac{dy}{dt} + e^{\frac{t}{2}} \dfrac{y}{2} = e^{\frac{t}{2}} \dfrac{t}{2}$ simplifies to $\dfrac{d}{dt}\left[y \cdot e^{\frac{t}{2}} \right] = \dfrac{t}{2} \cdot e^{\frac{t}{2}}.$

Step 5: $\int \dfrac{d}{dt}\left[y \cdot e^{\frac{t}{2}} \right] = \int \dfrac{t}{2} \cdot e^{\frac{t}{2}} dt$ (use integration by parts).

$$y \cdot e^{\frac{t}{2}} = t \cdot e^{\frac{t}{2}} - 2 \cdot e^{\frac{t}{2}} + C$$

Step 6: $y = t - 2 + C \cdot e^{\frac{-t}{2}}.$

Example: Newton's Law of Cooling

$\dfrac{dT}{dt} = -k\left(T_m - T_s\right)$, where $k > 0$, T_s = constant temperature of the surroundings,

   T_m = temperature of the mass being cooled, and
   $T(0) = \alpha$ (the initial temperature of the mass).
   Let's assume that $k = 0.190$, $T_s = 70$, $T(0) = 300$. Now,

$\dfrac{dT}{dt} = -.190\left(T_m - 70\right), T(0) = 300$. We notice that this equation is both

separable and linear. We will use the linear method.

Step 1: Standard form is $\dfrac{dT}{dt} + .190(T_m) = .190 \cdot (70) = 13.3, \ \ T(0) = 300$

Step 2: $P(t) = .190$

$$u = e^{\int P(t)dt} = e^{.19t}$$

Steps 3 & 4: $\dfrac{d}{dt}\left[T_m \cdot e^{.19t}\right] = 13.3 \cdot e^{.19t}$.

Step 5: $\dfrac{d}{dt}\left[T_m \cdot e^{.19t}\right] = 13.3 \cdot e^{.19t}$

$$T_m \cdot e^{.19t} = 70e^{.19t} + C$$

Step 6: $T_m = 70 + Ce^{-.19t}$
   Use $T(0) = 300$
   $300 + 70 + C$

   $C = 230$, so $T_m = 70 + 230 + e^{-.19t}$

We note that the $\lim\limits_{t \to \infty} 70 + 230 \cdot e^{-.19t} = 70$ (the room temperature) as expected.

   Of our technology choices, only Maple will provide closed form analytical solutions. We provide the commands at the end of this chapter, however, not all ODES are in fact solvable in closed form. We might try a numerical method as we will discuss in the next section.

3.4 Numerical Methods for Solutions to First-Order ODEs with Technology

As mentioned in the last section, not all first order differential equations are solvable in closed form. In this section we briefly introduce three numerical techniques: Euler's Method, Improved Euler's Method, and Runge-Kutta. We will discuss the technique and its algorithm and we will illustrate the Excel commands with our examples.

3.4.1 Euler's Method

From the point of view of a mathematician, the *ideal* form of the solution to an initial value problem would be a **formula** for the solution function. After all, if this formula is known, it is usually relatively easy to produce any other form of the solution you may desire, such as a **graphical solution**, or a **numerical solution** in the form of a table of values. You might say that a formulaic solution contains the recipes for these other types of solution within it. Unfortunately, as we have seen in our studies already, obtaining a formulaic solution is not always easy, and in many cases is absolutely impossible.

So we often have to "make do" with a numerical solution, i.e. a table of values consisting of points which lie along the solution's curve. This can be a perfectly usable form of the answer in many applied problems, but before we go too much further, let's make sure that we are aware of the shortcomings of this form of solution.

By its very nature, a numerical solution to an initial value problem consists of a **table of values** which is **finite** in length. On the other hand, the true solution of the initial value problem is most likely a whole continuum of values, i.e. it consists of an *infinite* number of points. Obviously, the numerical solution is actually leaving out an infinite number of points. The question might arise, "With this many holes in it, is the solution good for anything at all?" To make this comparison a little clearer, let's look at a very simple specific example starting with Euler's Method.

In mathematics and computational science, the **Euler method**, named after Leonhard Euler, is a numerical procedure for solving ordinary (ODEs) with a given initial value. It is the most basic kind of explicit method for numerical integration for ordinary differential equations.

Given the differential equation:

$$\frac{dy}{dt} = g(t, y), y(t_0) = y_0, t_0 \leq t \leq b.$$

Euler's Method

Step 1. Pick step size, h so that the interval, $[b-t_0]/n = h$, is divided evenly.
Step 2. Start at $y(t_0) = y_0$.
Step 3. Compute $y_{n+1} = y_n + h^* g(t_n, y_n)$.
Step 4. Let $t_{n+1} = t_n + h, n = 0, 1, 2,...$
Step 5. Continue until $t_n = b$.
STOP

Example 3.18

$$\frac{dy}{dt} = .25 \cdot y \cdot t, y(0) = 2.$$

We want to estimate the solution to y(3) using a numerical approach. The interval is [0,3].
A few steps by hand:
Step size of 1 and the point (t,y).

When t = 0, y = 2 This is the initial condition: (0, 2)
When t = 0 + h = 1, y = 2 + h*0.25*t*y = 2 + (1)*(0.25)(0)(2) = 2   (1, 2)
When t = 1 + h = 2, y = 2 + h*0.25*t*y = 2 + (1)*(0.25)(1)(2.5) = 2.5   (2, 2.5)
When t = 3, y = 2.5 + h*0.25*t*y = 2.5 + (1)*(0.25)(2)(2.5) = 3.75   (3, 3.75).

Geometrically, this is depicted in Figure 3.13.

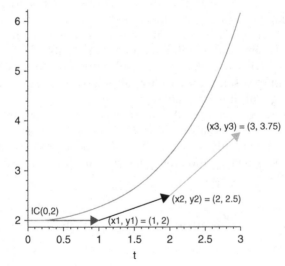

FIGURE 3.13
Geometric depiction of Euler's method.

TABLE 3.4

Performance comparison

t	Y (approximate, Ya)	Y (exact, Ye)	Error = $|Ye\text{-}Ya|$	Percent error = $\dfrac{100 \cdot |Ye - Ya|}{Ye}$
0	2.00	2.000	0.000	0
1	2.00	2.266	0.266	11.73%
2	2.50	3.297	0.797	24.17%
3	3.75	6.160	2.410	39.12%

Euler's method				
Iteration	y	Change equation	Y actual	% Error
)	2	Initial value	=2*EXP(A8^2/8)	=100*(ABS(D8-B8))/D8
=A8+G$6	=B8+G$6*(0.25*B8*A8)		=2*EXP(A9^2/8)	=100*(ABS(D9-B9))/D9
=A9+G$6	=B9+G$6*(0.25*B9*A9)		=2*EXP(A10^2/8)	=100*(ABS(D10-B10))/D10
=A10+G$6	=B10+G$6*(0.25*B10*A10)		=2*EXP(A11^2/8)	=100*(ABS(D11-B11))/D11
=A11+G$6	=B11+G$6*(0.25*B11*A11)		=2*EXP(A12^2/8)	=100*(ABS(D12-B12))/D12
=A12+G$6	=B12+G$6*(0.25*B12*A12)		=2*EXP(A13^2/8)	=100*(ABS(D13-B13))/D13
=A13+G$6	=B13+G$6*(0.25*B13*A13)		=2*EXP(A14^2/8)	=100*(ABS(D14-B14))/D14

Euler's method					St.
Iteration	y	Change equation	Y actual	% Error	
0	2	Initial value	2	0	
0.5	2		2.063487	3.076676552	
1	2.125		2.266297	6.2347041	
1.5	2.390625		2.64957	9.773078825	
2	2.838867188		3.297443	13.9070006	
2.5	3.548583984		4.368402	18.76699325	
3	4.657516479		6.160434	24.39628916	

FIGURE 3.14
Screenshot of geometric depiction of Euler's method in Excel.

Note that we have under-estimated every value between t = 0 and t = 3.

The exact solution is $y(t) = 2 \cdot e^{\frac{t^2}{8}}$

We can track how well we did in a table (Table 3.4):

Notice that as we move farther away from the Initial Condition y(0) = 2, the worse our estimate becomes.

Can we do better?

Maybe we could improve by changing the step-size. In this example, let us make the step size 0.5 instead of 1.

We will use Excel to obtain our output. We will apply the Euler algorithm with a step size of $h = 0.5$ (Figure 3.14).

We can once again generate a table (Table 3.5) to demonstrate our performance:

TABLE 3.5

Update of performance comparison

t	Y (approximate, Ya)	Y (exact, Ye)	Error = | Ye-Ya |	Percent error = $\dfrac{100 \cdot \lvert Ye - Ya \rvert}{Ye}$
0.0	2.000	2.000	0.000	0
0.5	2.000	2.063	0.063	3.05%
1.0	2.125	2.263	0.138	6.11%
1.5	2.391	2.650	0.259	9.77%
2.0	2.839	3.297	0.459	13.91%
2.5	3.549	4.368	0.820	18.76%
3.0	4.658	6.160	1.503	24.84%

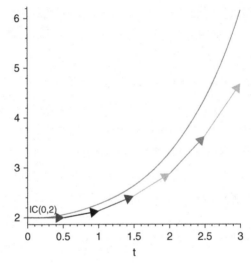

FIGURE 3.15
Graphical depiction of Euler's method.

Out plot is shown in Figure 3.15.

We are still underestimating at all points. We can do better by making this step-size small. (Note: it is not always true the making the step size smaller improves the estimates.) So we might try a step size of $h = 0.1$ (Figure 3.16).

Graphical Comparisons

First, we plot the actual solution (if we have one) then the numerical solution. Second, we will overlay them on one plot as in Figure 3.17. Otherwise, compare numerical solution to either qualitative plots or slope field plots.

Euler's method					St
Iteration	y	Change equation	Y actual	% Error	
0		2	Initial value	2	0
0.1	2		2.002502	0.124921908	
0.2	2.005		2.010025	0.249998961	
0.3	2.015025		2.022627	0.375849739	
0.4	2.030137688		2.040403	0.503086604	
0.5	2.050439064		2.063487	0.632315677	
0.6	2.076069553		2.092056	0.764136776	
0.7	2.107210596		2.126329	0.899143322	
0.8	2.144086781		2.166574	1.037922201	
0.9	2.186968517		2.213106	1.181053604	
1	2.236175309		2.266297	1.329110824	
1.1	2.292079691		2.326575	1.482660025	
1.2	2.355111883		2.394435	1.642259984	
1.3	2.425765239		2.470442	1.808461794	
1.4	2.50460261		2.555243	1.981808546	
1.5	2.592263701		2.64957	2.162834985	
1.6	2.68947359		2.754256	2.352067127	
1.7	2.797052533		2.870244	2.550021872	
1.8	2.915927266		2.998605	2.75720657	
1.9	3.047143993		3.140548	2.974118585	
2	3.191883333		3.297443	3.201244825	
2.1	3.351477499		3.470842	3.439061257	
2.2	3.527430068		3.662504	3.688032404	
2.3	3.721438722		3.874425	3.948610829	
2.4	3.935421448		4.108866	4.221236595	
2.5	4.171546735		4.368402	4.506336723	
2.6	4.432268406		4.655956	4.804324629	
2.7	4.720365853		4.97486	5.115599563	
2.8	5.038990548		5.328912	5.440546025	
2.9	5.391719886		5.722451	5.779533189	
3	5.782619578		6.160434	6.132914317	

FIGURE 3.16

Screenshot of geometric depiction of Euler's method ($h = 0.1$) in Excel.

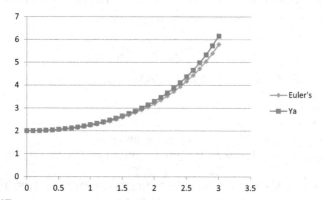

FIGURE 3.17

Overlay of Euler's method ($h = 0.1$) and actual solution to $dy/dx = 0.25\, x\, y$.

The second method that we will discuss is Improved Euler's method, also known as Heun's method.

3.4.2 Improved Euler's Method (Heun's Method)

The accuracy of Euler's method is improved by using an average of two slopes in the tangent line approximation. Use the slope obtained at the beginning of the step and the slope obtained at the end of the step in order to improve our accuracy.

Given $\frac{dy}{dt} = g(t,y), y(t_0) = y_0, t_0 \le t \le b$

Improved Euler's Method Algorithm

Step 1.   Pick step size, h so that the interval, $[b - t_0]/n = h$, is divided evenly.
Step 2.   Start at $y(t_0) = y_0$.
Step 3.   Let $t_{n+1} = t_n + h$.
Step 4.   Compute $y_{n+1} = y_n + (h/2)* [g(t_n, y_n) + g(t_{n+1}, y_{n+1})]$.
Step 5.   Continue until $t_{n+1} = b$.
STOP

Step 4 uses the formula:

$$y_{n+1} = y_n + \frac{h}{2}(k_1 + k_2)$$
$$k_1 = f(t_n, y_n)$$
$$k_2 = f(t_{n+1}, y_n + h*k_1)$$

Example 3.19 $y' = .25\ t\ y, y(0) = 2$

We want to estimate the solution to y(3) using a numerical approach.

$$k_1 = .25 t_n y_n$$

$$k_2 = .25\ t_{n+1}(y_n + h*k_1).$$

A few steps of Improved Euler's Method by hand:

Step size of h = 1.
Point (t,y).
When $t_0 = 0$, $y_0 = 2$. This is the initial condition: (0, 2).
When $t_1 = t_0 + h = 1$, $y_1 = 2 + (1/2)*[(.25*(0)*(2) + 0.25*(1)*(2 + 0.25*(0)*2)$
              = 2.25   (1, 2.25).
When $t_2 = t_1 + h = 2$, $y_2 = 2.25 + (1/2)*(.25(1)(2.25) +.25*(2)(2.25 + (1).25*(1)(2.25))$
              = 3.23437   (2, 3.234375).

TABLE 3.6

Performance comparison with the Heun method

t	Y (approximate, Ya)	Y (exact, Ye)	Error = $\lvert$Ye-Ya$\rvert$	Percent error = $\dfrac{100 \cdot \lvert Ye - Ya \rvert}{Ye}$
0	2.000	2.000	0.000	0
1	2.250	2.266	0.016	0.71%
2	3.234	3.297	0.063	1.90%
3	5.862	6.160	0.298	4.83%

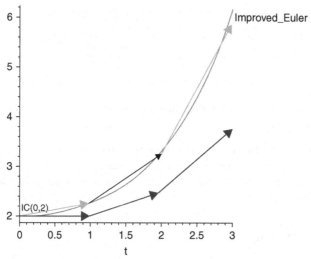

FIGURE 3.18
Graphical representation of the improved Euler method.

When $t_3 = 2 + h = 3$, $y_3 = 3.234375 + (1/2)*((0.25)*(2)*(3.234375) + (0.25)*(3)*$
$(3.234375 + 0.25*(2)*(3.234375)) = 5.862305$ $(3, 5.862305)$.

The exact solution is $y(t) = 2 \cdot e^{\frac{t^2}{8}}$.

We can once again generate a table (Table 3.6) to demonstrate the performance of the Heun Method:

Notice that as we move farther away from the Initial Condition $y(0) = 2$, the worse our estimate becomes. Also note that these estimates are "better" estimates than Euler's Method. Think about why.

The Plot

We apply the algorithm of Improved Euler's method in Excel and demonstrated in Figure 3.18.

Improved Euler's method								
			Enter step size					0.5
				k1		k2		
Iteration	Y		h	g(x,y)	u(n+1)	g(x*,y*)	Yactual	% Error
0	2	Initial Value	=$I$3	=0.25*A7*B7	=B7+$I$3*E7	=0.25*(A7+D7)*(F7)	=2*EXP(A7^2/8)	=100*(ABS(B7-H7)/H
=A7+$I$3	=B7+($I$3/2)*(E7+G7		=$I$3	=0.25*A8*B8	=B8+$I$3*E8	=0.25*(A8+D8)*(F8)	=2*EXP(A8^2/8)	=100*(ABS(B8-H8)/H
=A8+$I$3	=B8+($I$3/2)*(E8+G8		=$I$3	=0.25*A9*B9	=B9+$I$3*E9	=0.25*(A9+D9)*(F9)	=2*EXP(A9^2/8)	=100*(ABS(B9-H9)/H
=A9+$I$3	=B9+($I$3/2)*(E9+G9		=$I$3	=0.25*A10*B10	=B10+$I$3*E10	=0.25*(A10+D10)*(F1	=2*EXP(A10^2/8)	=100*(ABS(B10-H10)
=A10+$I$3	=B10+($I$3/2)*(E10+		=$I$3	=0.25*A11*B11	=B11+$I$3*E11	=0.25*(A11+D11)*(F1	=2*EXP(A11^2/8)	=100*(ABS(B11-H11)
=A11+$I$3	=B11+($I$3/2)*(E11+		=$I$3	=0.25*A12*B12	=B12+$I$3*E12	=0.25*(A12+D12)*(F1	=2*EXP(A12^2/8)	=100*(ABS(B12-H12)
=A12+$I$3	=B12+($I$3/2)*(E12+		=$I$3	=0.25*A13*B13	=B13+$I$3*E13	=0.25*(A13+D13)*(F1	=2*EXP(A13^2/8)	=100*(ABS(B13-H13)

Improved Euler's method								
			Enter step size					0.5
				k1		k2		
Iteration	Y		h	g(x,y)	u(n+1)	g(x*,y*)	Yactual	% Error
0	2	Initial Value	0.5	0	2	0.25	2	0
0.5	2.0625		0.5	0.257813	2.191406	0.547852	2.063487	0.047823
1	2.263916		0.5	0.565979	2.546906	0.95509	2.266297	0.105056
1.5	2.644183		0.5	0.991569	3.139968	1.569984	2.64957	0.203292
2	3.284571		0.5	1.642286	4.105714	2.566071	3.297443	0.390341
2.5	4.336661		0.5	2.710413	5.691867	4.2689	4.368402	0.726607
3	6.081489		0.5	4.561117	8.362047	7.316791	6.160434	1.281484

FIGURE 3.19
Screenshot of geometric depiction of Heun method in Excel.

First, we plot the actual solution (if we have one) then the numerical solution. Second, we will overlay them on one plot shown in Figure 3.19. Otherwise, compare numerical solution to either qualitative plots or slope field plots.

Runge-Kutta Methods

When we use the average of the estimates of the derivatives at the end points we can improve the approximation to the solution. A class of approximation techniques that estimate derivatives at various points within an interval and then computes a *weighted average* is the Runge-Kutta Method, named for two German mathematicians.

The Runge-Kutta Methods are classified by order, where the order depends upon the number of slope estimates used at each step. A very popular method is the fourth-order Runge-Kutta Method.

3.4.3 Fourth-Order Runge-Kutta Method

For solving $dy/dt = g(t,y)$, $y(t_0) = y_0$ over an interval.

Step 1. First divide the interval $x_0 \leq x \leq b$ into p sub-intervals using equally spaced points. This yields the step-size, $h = (b - x_0)/p$.

Step 2. For n = 1,2,3..p obtain the following sequence of approximations

$$y_{n+1} = y_n + \frac{K_1 + 2K_2 + 2K_3 + K_4}{6}, where$$

$$K_1 = g(t_n, y_n)h$$
$$K_2 = g(t_n + h/2, y_n + K_1/2)h$$
$$K_3 = g(t_n + h/2, y_n + K_2/2)h$$
$$K_4 = g(t_n + h, y_n + K_3)h$$

Example by hand:

Example 3.20 $y' = 0.25\ t\ y$, $y(0) = 2$

We want to estimate the solution to y(3) using a numerical approach. A few steps of Runge-Kutta Method by hand:

Step size of $h = 1$.

When $t = 0$, $y = 2$. This is the initial condition: (0, 2)

When t = 0 + h = 1,

$K1 = h*g(t(n), y(n) = 1*0.25*0*2 = 0$
$K2 = h*g(t(n + h/2), y(n) + K1/2) = 1*0.25*(0 + 1/2)*(2 + 0) = 0.25$
$K3 = h*g(t(n + h/2), y(n) + K2/2) = 1*.25*(0 + 1/2)*(2 + .125) = 0.265625$
$K4 = h*g(t(n + h), y(n) + K3) = 1*0.25*(0 + 1)*(2 + 0.265625) = 0.56640625$
$y(1) = 2 + (1/6)*(0 + 2*0.25 + 2*0.265625 + 0.56640625) = 2.266276.$

When $t = 1 + h = 2$,

$K1 = 0.566569$ $K2 = 0.956085$ $K3 = 1.029119$ $K4 = 1.647698$

$y(2) = 2.266276 + (1/6)*(0.566569 + 2*0.956085 + 2*1.029119$
$+ 1.647698) = 3.297055.$

When $t = 2 + h = 3$

$K1 = 1.648528$, $K2 = 2.575825$, $K3 = 2.865605$, $K4 = 4.621995$

$y(3) = 3.297055 + (1/6)* (1.648528 + 2*2.575825 + 2*2.86560 + 4.621995)$
$= 6.155952.$

The exact solution is $y(t) = 2 \cdot e^{\frac{t^2}{8}}$.

TABLE 3.7

Performance comparison with the Runge-Kunta method

t	Y (approximate, Ya)	Y (exact, Ye)	Error = $\lvert$Ye-Ya$\rvert$	Percent error = $\dfrac{100 \cdot \lvert Ye - Ya \rvert}{Ye}$
0	2.000	2.000	0.000	0
1	2.266	2.266	0.000	0.00%
2	3.297	3.297	0.000	0.04%
3	6.156	6.160	0.004	0.45%

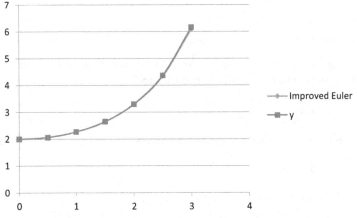

FIGURE 3.20

Overlay of actual solution and solution via improved Euler's method.

We can once again generate a table (Table 3.7) to demonstrate the perform-ance of the Runge-Kunta Method:

Notice that as we move farther away from the Initial Condition y(0) = 2, the worse our estimate becomes. Also note that these estimates are much better estimates than either Euler's Method or Improved Euler's. Think about why?

The results are depicted in Figure 3.20 and the RK4 Method in Excel is depicted in Figure 3.21.

Graphical Comparisons

First, we plot the actual solution (if we have one) then the numerical solution. Second, we will overlay them on one plot. The plot looks good. Otherwise, compare numerical solution to either qualitative plots or slope field plots.

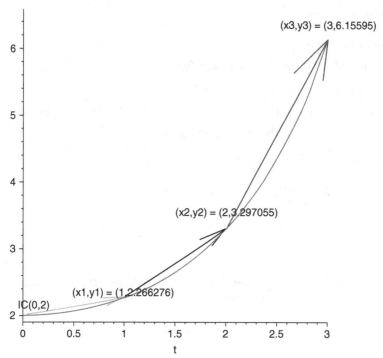

FIGURE 3.21
The RK4 plot.

3.5 Technology Examples for Ordinary Differential Equations

We have previously shown and illustrated methods with Excel. We will provide the commands and examples in both Maple and R in this section. It is important to repeat and stress that only Maple can provide closed form analytical solutions, if they exist.

3.5.1 Maple for Ordinary Differential Equations

Numerical Methods and Plots

In Maple we will use the with(DEtools): command to access all the tools for Chapter 3. We will illustrate slope fields, analytical solutions to ODEs (if they exist), and numerical methods.

In Maple we must first call the library with(DEtools). Then we enter the differential equation and enter the commands to get a specific plot.

Command Sequence:

with(DEtools):

deq: = diff(y(t),t) = (t − y(t))/2,y(t));

DEplot(deq,y(t), t = −2..2, y = 0..2);

A complete slope field over the region $-2 \leq t \leq 2$ is seen in Figure 3.22.

>example_1: = diff(y(t),t) = (t − y(t))/2;

>DEplot(example_1,y(t),t = −2..2,y = 0..2);

$$example\ 1 := \frac{d}{dt}y(t) = \frac{t}{2} - \frac{1}{2}y(t)$$

>

>

We use Maple to produce the slope field plot. Although knowledge of how to graph slope fields is important, you will not want to graph complete slope

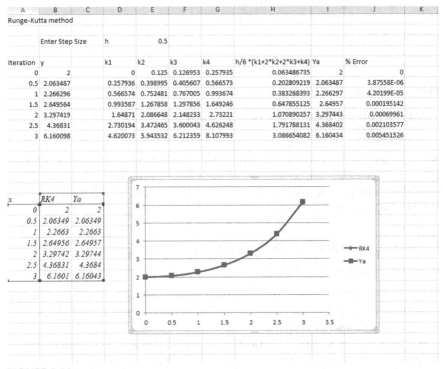

FIGURE 3.22
Screenshot for RK4 method in Excel.

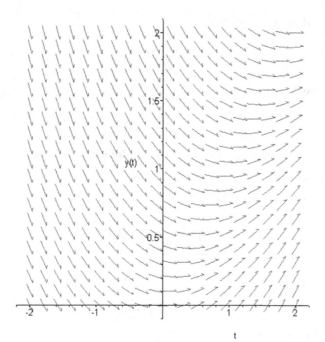

FIGURE 3.23
Slope field for $\dfrac{dy}{dt} = \dfrac{(t-y)}{2}$.

fields without technology. A solution curve is then tangent to the slope line at each point through which the curve passes. Thus, the slope field gives a visual representation of what a family of possible solution curves of the differential equation looks like. Several solution curves are now depicted in Figure 3.23 for $\dfrac{dP}{dt} = 0.008271 \cdot P \cdot (665 - P)$ with starting points (0, 2) and (0.900). The possible solution curves are seen in Figure 3.23.

We examine the slope field and we notice that the solution curves do not cross tangent lines.

Analytical Solutions in Maple

Example 3.21 Solve $\dfrac{dy}{dt} = t \cdot y$

First, we will solve using Maple without initial conditions.
>restart;
>with(DETools):

>eqn: = diff(y(t),t) = y(t)*t;

$$eqn := \frac{d}{dt} y(t) = y(t)t$$

>dsolve(eqn,y(t));

$$y(t) = \_CI e^{\left(\frac{t^2}{2}\right)}$$

Now, we solve again after we introduce the initial condition *y(1) = 1*.

>with(DETools):

>eqn: = diff(y(t),t) = y(t)*t;

$$eqn := \frac{d}{dt} y(t) = y(t)t$$

>

>inits: = y(1) = 1;

$$inits := y(1) = 1$$

>dsolve({eqn,inits},y(t));

$$y(t) = \frac{e^{\left(\frac{t^2}{2}\right)}}{e^{(1/2)}}$$

We plot the solution below from [−1,1].

>plot(exp(t^2/2)/exp(1/2),t = −1..1).

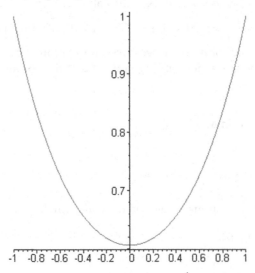

Example 3.22 Refined Population Model in Maple

Consider the logistic growth model defined for blue crabs in Venezuela,

$$\frac{dB}{dt} = .25B(10 - B), B(0) = 4. \text{ (B is in millions of bushels.)}$$

>eqn: = diff(B(t),t) = 0.25*B(t)*(10-B(t));

$$eqn := \frac{d}{dt}B(t) = 0.20\,B(t)(10 - (t))$$

>inits: = B(0) = 4;

$$inits := B(0) = 4$$

>dsolve({eqn,inits},B(t));

$$B(t) = \frac{20}{2 + 3\,e^{\left(\frac{-5t}{2}\right)}}$$

>plot(20/(2 + 3*exp(–5/2*t)), t = 0..5).

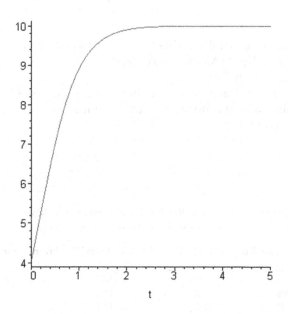

Numerical Methods

We will use Maple to obtain our numerical output. In the methods, we might pick foreuler for Euler's methods, heunform for huen's methods, or RK4 for Runge-Kutta 4.

>ans2: = dsolve({eqn, y(0) = 2}, numeric,

    method = classical[foreuler],

    output = array([0,0.5,1,1.5,2,2.5,3]),

    stepsize = 0.5);

$$
ans2 = \begin{bmatrix} & [t, y(t)] \\ \begin{bmatrix} 0 & 2. \\ 0.5 & 2. \\ 1 & 2.1250 \\ 1.5 & 2.3906250 \\ 2 & 2.83886718750 \\ 2.5 & 3.54858398437500 \\ 3 & 4.6575516479492187500 \end{bmatrix} \end{bmatrix}
$$

We can once again generate a table (Table 3.8) to demonstrate the performance of the Runge-Kunta Method and Maple:

Out plot is shown in Figure 3.24.

We are still underestimating at all points. We can do better by making this step-size small. (Note: it is not always true that making the step size smaller improves the estimates.)

TABLE 3.8

Performance comparison with the Runge-Kunta method and Maple

t	Y (approximate, Ya)	Y (exact, Ye)	Error = $\|Ye\text{-}Ya\|$	Percent error = $\dfrac{100 \cdot \|Ye - Ya\|}{Ye}$
0.0	2.000	2.000	0.000	0
0.5	2.000	2.063	0.063	3.05%
1.0	2.125	2.263	0.138	6.11%
1.5	2.391	2.650	0.259	9.77%
2.0	2.839	3.297	0.459	13.91%
2.5	3.549	4.368	0.820	18.76%
3.0	4.658	6.160	1.503	24.84%

Example 3.23 with Maple

1. Set UP Maple

   >restart;
   >with(DEtools):with(plots):with(linalg):

2. Enter your ODE

   >eqn: = diff(y(t),t) = .25*t*y(t);

   $$eqn := \frac{d}{dt}y(t) = 0.25\,t\,y(t)$$

3. If possible, solve the ODE Analytically & Plot

   >dsolve({eqn,y(0) = 2},y(t));

   $$y(t) = 2e^{\left(\frac{t^2}{8}\right)}$$

   >plot(2*exp(1/8*t^2),t = 0..3, thickness = 3);

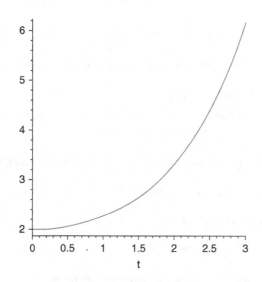

Note you have to copy and paste the output equation from **dsolve** into the pot command.

4.   Numerical Method– Internal and Classic-Output only

You will need to specify both the step-size you want and the output array you want to see.

>**Digits: = 20:**

ans2: = dsolve({eqn, y(0) = 2}, numeric,

        method = classical[foreuler],

        output = array([0,0.5,1,1.5,2,2.5,3]),

        stepsize = 0.5);

$$ans\,2:\begin{bmatrix} & [t,y(t)] \\ \begin{bmatrix} 0 \\ 0.5 \\ 1 \\ 1.5 \\ 2 \\ 2.5 \\ 3 \end{bmatrix} & \begin{matrix} 2. \\ 2. \\ 2.1250 \\ 2.3906250 \\ 2.8388718750 \\ 3.54858398437500 \\ 4.657516479492187500 \end{matrix} \end{bmatrix}$$

5.   Numerical Output for Analysis: graphical and percent error

First get the numerical output as a procedure as follows

>dsol2 = dsolve({eqn, y(0) = 2},

        numeric,method = classical[foreuler], stepsize = 1,

        output = listprocedure);

$$dsol2 := \left[t = proc(t) \right]...end\,proc), y(t) = (proc(t)...end\,proc)]$$

Prepare for numerical output

    >fy2: = eval(y(t),dsol2);

$$fy2 := proc(t)...end\,proc$$

View output to compare to above output (numerical)

    >seq(fy2(i),i = 0..3);

    2., 2., 2.50, 3.7500

Prepare exact solution for output format:

>actual_y2: = evalf(subs(t = i,2*exp(1/8*t^2)));

$$actual\_y2:=2.e^{(0.1250000000000000000\, i^2)}$$

View actual output

>seq(actual_y2(i),i = 0..3);

2., 2.2662969061336526336, 3.2974425414002562936.6.1604336978360624900

Put into an array that lists **[t, numerical y(t), actual y(t), Percent Error]**

>array([seq([i,fy2(i),evalf(subs(t = i, 2*exp(1/
8*t^2))),evalf(100*abs(fy2(i)-actual_y2(i))/actual_y2(i))],i = 0..3)]);

$$
\begin{bmatrix}
0 & 2. & 2. & 0. \\
1 & 2. & 2.2662969061336526336 & 11.750309741540459711 \\
2 & 2.50 & 3.2974425414002562936 & 24.183667535920822047 \\
3 & 3.7500 & 6.1604336978360624900 & 39.127662370309425663ass
\end{bmatrix}
$$

6.   Graphical Comparisons

First, we plot the actual solution (if we have one) then the numerical
solution. Second, we will overlay them on one plot. Otherwise, compare numerical solution to either qualitative plots or slope field plots.

>plot(2*exp(1/8*t^2),t = 0..3);

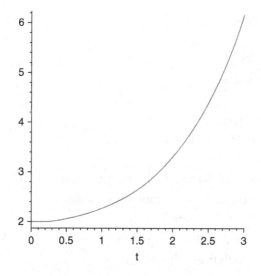

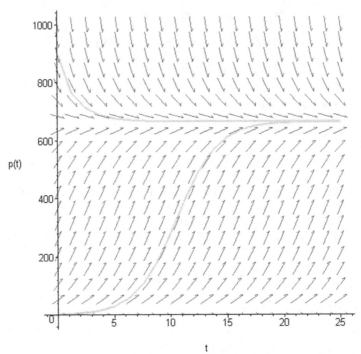

FIGURE 3.24
Slope field for $\dfrac{dP}{dt} = 0.008271 \cdot P \cdot (665 - P)$ with starting points (0.2) and (0.900).

3.5.2 R to Solve Ordinary Differential Equations

```
## Using the `deSolve` package
   library(deSolve)
## Time
   t <- seq(0, 100, 1)
## Initial population
   N0 < -4
## Parameter values
   params <- list(r = .25, K = 10)
## The logistic equation
   fn<- function(t, N, params) with(params, list(r * N * (1 - N/ K)))
## Solving and ploting the solution numerically
   out <- ode(N0, t, fn, params)
   plot(out, lwd = 2, main = "Logistic equation\nr = 0.25, K = 10, N0 = 4")
## Plotting the analytical solution
```

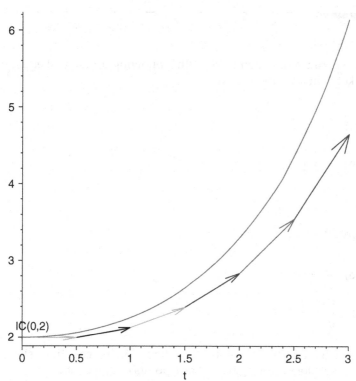

FIGURE 3.25
Graphical overlay with hand solution.

with(params, lines(t, K * N0 * exp(r * t)/ (K + N0 * (exp(r * t) − 1)), col = 2,
lwd = 2))

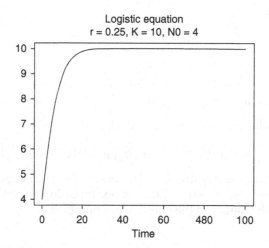

3.6 Exercises

1. Construct a direction field for the following differential equations and sketch in a solution curve.

   (a) $\dfrac{dy}{dx} = y$.

   (b) $\dfrac{dy}{dx} = x$.

   (c) $\dfrac{dy}{dx} = x + y$.

   (d) $\dfrac{dy}{dx} = x - y$.

   (e) $\dfrac{dy}{dx} = xy$.

   (f) $\dfrac{dy}{dx} = \dfrac{1}{y}$.

2. Sketch a number of solutions to the following equations showing the correct slope, concavity, and any points of inflection.

   (a) $\dfrac{dy}{dx} = (y + 2)(y - 3)$.

   (b) $\dfrac{dy}{dx} = y^2 - 4$.

   (c) $\dfrac{dy}{dx} = y^3 - 4$.

   (d) $\dfrac{dy}{dx} = x - 2y$.

3. Analyze graphically the equation $\dfrac{dy}{dt} = ry$, when $r < 0$. What happens to any solution curve as t becomes large?

4. Develop graphically the following models. First graph $\dfrac{dP}{dt}$ versus P, then obtain various graphs of P versus t by selecting different initial values $P(0)$ (as in our population example in the text). Identify and discuss the nature of the equilibrium points in each model.

(a) $\dfrac{dP}{dt} = a - bP, \quad a,b > 0.$

(b) $\dfrac{dP}{dt} = P(a - bP), \quad a,b > 0.$

(c) $\dfrac{dP}{dt} = k(M - P)(P - m), \quad k, M, m > 0.$

5. Get Euler estimates for y(3) using the ODE $y' = y + t$ for step sizes of 0.1, 0.05, and 0.01. Compute the percent error for each.

6. Use Euler's method by hand $y' = y + 1$, $y(0) = 1$ with step size of 0.25 to estimate y(.5). Then do Euler's Method using Excel and compare your results to make sure you have the correct solution. Compute the % error since you can find this exact solution. Then change the step sizes to 0.1 and then 0.05 and use Excel to estimate y(.5).

7. Consider $v' = 32 - 1.6$ v, v(0) = 0. Using Euler's method in Excel and a step size of 0.05 estimate v(2), what is the terminal velocity? Keep stepping out until you approximate the terminal velocity to 3 decimal places of accuracy.

8. Try **by hand** $y' = y + 1$, y(0) = 1 with step size of 0.25 to estimate y(.5). Then do Improved Euler's Method using Excel and compare your results to make sure you have the correct solution. Compute the % error since you can find this exact solution.

9. Consider $v' = 32 - 1.6$ v, v(0) = 0. Using heunform in Excel and a step size of 0.05 estimate v(2). What is the terminal velocity? Keep stepping out until you approximate the terminal velocity to 3 decimal places of accuracy.

10. Try **by hand** $y' = y + 1$, y(0) = 1 with step size of 0.25 to estimate y(0.5). Then do RK4 Method using Excel and compare your results to make sure you have the correct solution. Compute the % error since you can find this exact solution.

11. Consider $v' = 32 - 1.6$ v, v(0) = 0. Using RK4 in Excel and a step size of 0.05 estimate v(2), what is the terminal velocity? Keep stepping out until you approximate the terminal velocity to 3 decimal places of accuracy.

3.7 Projects

1. **The Spread of a Contagious Disease**

    Given the following ODE model for the spread of a communicable disease as:

    $$\frac{dN}{dt} = .25N(10 - N), \ N(0) = 2$$

    where N is in 100s

    (a) Since this is an autonomous ODE perform a complete graphical analysis:

        (i)   Plot dN/dt versus N. Find and label all rest points (equilibrium points).
        (ii)   Find the value where the rate of change of the disease is the fastest. Why is it this value that you provided?
        (iii) Plot N versus t from the following initial conditions:
        (iv) $N(0) = 2$, $N(0) = 7$, and $N(0) = 14$.
        (v)   Describe the *stability* of each rest point (equilibrium point).

    (b) Obtain a slope field plot of this ODE from Excel.
    (c) Solve this ODE using separable variables. (Hint: You will need partial fraction decomposition as well.) Insure you find the value of arbitrary constant C using the initial condition $N(0) = 2$.
    (d) Compute the time, t, when N is changing the fastest using the initial condition $N(0) = 2$. Recall, you have found the value of N already.
    (e) Use Euler's Method with step sizes of h = *0.1* and then $h = 1$ to approximate the solutions to the ODE for $N(.5)\&N(5)$. Find the relative error or absolute differences.

    Plot Euler's approximations for $h = 0.1$ and $h = 1.0$ and compare these graphs to your graphical analysis in part (a) and your other plots. Briefly discuss these plots – you may compare and contrast these plots. What happened with Euler's Method? (Any opinions?)

2. **Chemical Reactions**

    Purpose: To model the changing amounts of salt dissolved in brine as an initial value problem, and to use the computer to obtain graphical and numerical solutions.

Background: A brine solution is a solution of salt in water. If the brine is in a tank equipped with fill and drainpipes, then the total amount of dissolved salt in the tank varies as the concentration in the inflow stream changes and the inflow and outflow rates are adjusted. The amount of salt in the tank can be modeled by appeal to the

Balance law: *New rate of change = rate in − rate out.*

The term "rate in" refers to the rate at which salt is added to the brine by means of the inflow stream and the term "rate out" is the rate at which salt leaves the tank through the outflow pipe. It is assumed throughout that the inflow stream is instantaneously mixed with the brine in the tank so that at any time the concentration of salt in the tank is uniform. Each of these terms (rate in and rate out) is a product of the appropriate brine flow rate with corresponding salt concentrations.

Consider a tank with capacity of 4,000 liters holds 2,000 liters of brine that contains 50 kg of dissolved salt. Brine with a salt concentration of 0.2 kg/liter is piped into the tank at a rate of 40 liters/min. Well-mixed brine is drawn off at a rate of a liters/min.

$$\frac{dx}{dt} = (.2)(40) + \frac{x}{2000 + (40 - a)t}a$$

Requirement 1. Show that the above is the general ODE; find the appropriate initial condition for this Brine problem.

Requirement 2. Solve the ODE leaving a as a parameter.

Requirement 3. Solve the ODE and obtain plots when the "rate out" $a = 30$, 40, and 50.

Requirement 4. **(BULK SALT)** Suppose that 200 kg of bulk salt is placed in a tanking holding 2,000 liters of brine already containing x_0 kg of dissolved salt. Suppose that the saturation level of brine is 300 kg of salt. The model representing this is:

$$\frac{dx}{dt} = .01(200 - x)(300 - x), x(0) = x_0$$

Solve this ODE, and obtain a plot for various choice of x_0, ($x_0 = 25$, 50, 100, 200).

Requirement 5. Bimolecular Chemical Reactions.

Background: In a bimolecular chemical reaction of two species interact and create one or more products. Consider the reaction, *A+B->C+D*.

The Law of Conservation: $\dfrac{dC}{dt} = \dfrac{dD}{dt} = -\dfrac{dA}{dt} = -\dfrac{dB}{dt}$.

The Law of Mass Action: The rate of an elementary reaction is proportional to the product of the concentrations of the reactants.

The ODE,

$$\frac{dx}{dt} = k_1(a+c-x)(b+c-x), x(0) = c$$

Consider the following reversible chemical reaction,

$$\frac{dx}{dt} = (0.7-x)(0.4-x) - 0.1x^2, x(0) = c$$

Solve qualitatively (autonomous). Solve this ODE (at least numerically) and show the curve and discuss the equilibrium values. Let $c = 0.2$. This reaction is only valid for $0 \le c \le 0.4$.

3. **Harvesting a Species**

Consider harvesting the blue crab in South Carolina. There has been many newspaper reports about the declining populations of blue crabs and the difficulty in harvesting these crabs. Let's model the situation and analysis some "what ifs."

The basic Balance law for Harvesting is:

$$P'(t) = Pr(1-\frac{P(t)}{k})P(t) - H(t),$$

where r is the intrinsic rate coefficient (growth rate if > 0) and k represents the carrying capacity (or saturation level).

Requirement 1. In the absence of harvesting ($H(t) = 0$), the ODE is autonomous. Let $r = 0.3$, $k = 12$, and $P(0) = 5$.

(a) Perform a qualitative assessment of this situation since it is autonomous. List and classify each equilibrium value.
(b) Solve this ODE.
(c) Plot the solution to this ODE and compare to your qualitative solution. Briefly discuss similarities and/or differences.
(d) Solve numerically using Euler's Method and obtain a plot of the numerical solution. Compare to the analytical solution.

Requirement 2. Now consider light harvesting where $H(t)$ = 0.4 (assume r and k are the same values as before).

(a) Obtain a slope field plot and discuss any equilibrium values observed.
(b) Find all equilibrium values of this ODE.
(c) Solve this ODE. Assume initial conditions of $P(0)$ = 19, $P(0)$ = 8, and $P(0)$ = .8.
(d) Solve numerically using Euler's Method with $P(0)$ = 5.

Requirement 3. Now consider a heavier harvesting where $H(t)$ = 1.5.

(a) Obtain a slope field plot and discuss any equilibrium values observed.
(b) Find all equilibrium values of this ODE.
(c) Solve this ODE. Assume initial conditions of $P(0)$ = 19, $P(0)$ = 8, and $P(0)$ = .8.
(d) Solve numerically using Euler's Method with $P(0)$ = 5.

Requirement 4. Determine which of the models is more likely to represent the current situation for blue crabs in the Chesapeake Bay.

3.8 References and Suggested Further Reading

Abell, M. L. and J. P. Braselton (2000). *Differential Equation with Maple V.* 2nd Edition. Academic Press,San Diego, CA.

Barrow, D., A. Belmonte, A. Boggess, J. Bryant, T. Kiffe, J. Morgan, M. Rahe, K. Smith, and M. Stecher (1998). *Solving Differential Equations with Maple V Release 4.* Brooks-Cole Publisher, Pacific Grove, CA.

Fox, W. (2013). *Mathematical Modeling with Maple.* Cengage Publishers, Boston, MA.

Fox, W. and W. Bauldry (2019). *Problem Solving with Maple.* Taylor & Francis, Boca Raton, FL.

Giordano, F. R. and M. D. Wier (1991). *Differential Equations: A Modeling Approach.* Addison-Wesley Publishers, Reading, MA.

Giordano, F. R., W. Fox, and S. Horton (2012). *A First Course in Mathematical Modeling,* 5th ed., Brooks-Cole, Boston, MA.

Zill, D. (2015). *A First Course in Differential Equations,* 10th ed., Cengage Publishers, Boston, MA.

4

Modeling System of Ordinary Differential Equation

Objectives

1. Be able to set and solve a system of ODEs.
2. Apply appropriate technology to solve a system of ODEs.
3. Understand direction fields and phase portraits.
4. Apply qualitative assessment to autonomous systems.

4.1 Introduction

Interactive situations occur in the study of economics, ecology, electrical engineering, mechanical systems, control systems, systems engineering, and so forth. For example, the dynamics of population growth of various species is an important ecological application of applied mathematics.

Consider a friend who recently retired and bought rural land in South Carolina. His desire is to have a fishing pond and his favorite fish to catch are bass and trout. He finds he has a fair size freshwater pond on his land but it contains no fish. He takes a water sample to the local fish and game authority and they analyze his water and they conclude that the water can sustain a fish population. He visits the local fish hatchers where they provide him the growth rates of bass and trout in isolation, call these values r and s, respectively. The experts tell him that bass and trout have the same food sources in the water and will compete for the oxygen in the water as well as the food for survival. The experts estimate the interactions rates between the bass and trout for survival and call these rates m and n respectively. We desire to build a mathematical model to help our friend determine if the pond can sustain both specifies of fish. This leads to a competitive hunter system of differential equations. We will revisit a similar scenario later in the chapter.

In this chapter, we will consider a variety of models, concepts, and techniques that give the reader some of the basic tools needed in solving and analyzing systems of differential equations.

- **Section 4.2** provides some models that we will solve and analyze. These models come from a variety of disciplines in science and engineering, including chemistry, physics, biology, fluids mechanics, Newtonian Mechanics, environmental engineering, and financial mathematics. Modeling techniques are discussed that help determine the necessary coefficients in the models.
- **Section 4.2** introduces phase portraits. Phase portraits provided qualitative viewing of solutions of systems of differential equations.
- **Section 4.3** covers some analytical methods in Maple for solving systems of differential equations with constant coefficients in both the homogeneous and non-homogeneous cases.
- **Section 4.4** is devoted to numerical techniques for obtaining numerical tables and plots of solutions to systems of differential equations that do not have closed form analytical solutions.

4.2 Applied Systems of Differential Equations

In this section, we introduce many mathematical models from a variety of disciplines. Our emphasis in this section is building the mathematical model, or expression, that will be solved later in the chapter. Recall previously that we discussed the modeling process. In this section, we will confine ourselves to the first three steps of the modeling process: (1) identifying the problem, (2) assumptions and variables, and (3) building the model.

Example 4.1 Economics: Basic Supply and Demand Models

Suppose we are interested in the variation of the price of a specific product. It is observed that a high price for the product attracts more suppliers. However, if we flood the market with the product the price is driven down. Over time there is an interaction between price and supply. Recall the "tickle me Elmo" from Christmas a few years ago.

Problem Identification: Build a model for price and supply for a specific product.

Assumptions and Variables:

Assume the price is proportional to the quantity supplied. Also assume the change in the quantity supplied is proportional to the price. We define the following variables.

$P(t)$ = the price of the product at time, t

$Q(t)$ = the quantity supplied at time, t.

We define two proportionality constants as a and b. The constant a is negative and represents a decrease in price as quantity increases.

With our limited assumptions, the model could be

$$\frac{dP}{dt} = -aQ$$

$$\frac{dQ}{dt} = bP$$

Example 4.2 An Electrical Network

Electrical networks with more than one loop give rise to systems of differential equations. Consider the electrical network displayed in Figure 4.1 where there are two resisters and two inductors. We apply Kirchhoff's law (the sum of the voltage drops in a closed circuit is equal to the impressed voltage) to each loop. We assume that no other factors interact with the flow of electricity in this circuit.

Loop ABEF:

$$E(t) = i_1 R_1 + L_1 \frac{di_2}{dt}$$

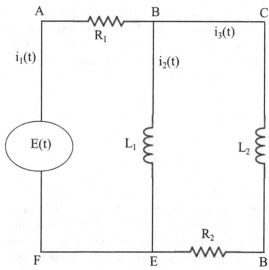

FIGURE 4.1

Electrical circuit.

Loop ABCDEF:

$$E(t) = i_1 R_1 + L_2 \frac{di_3}{dt} + i_3 R_2$$

We know that $i_1(t) = i_2(t) + i_3(t)$. We substitute this expression for i_1 into the laws to obtain the model:

$$\frac{di_2}{dt} = -\frac{R_1}{L_1} i_2 - \frac{R_1}{L_1} i_3 + \frac{E(t)}{L_1},$$

$$\frac{di_3}{dt} = -\frac{R_1}{L_2} i_2 - \frac{R_1 + R_2}{L_2} i_3 + \frac{E(t)}{L_2}$$

Example 4.3 Competition between Species

Imagine a small fish pond supporting both trout and bass. Let $T(t)$ denote the population of trout at time t and $B(t)$ denote the population of bass at time t. We want to know if both can coexist in the pond. Although population growth depends on many factors, we will limit ourselves to basic isolated growth and the interaction with the other competing species for the scarce life-support resources.

We assume that the species grow in isolation. The level of the population of the trout or the bass, $B(t)$ and $T(t)$, depend on many variables such as their initial numbers, the amount of competition, the existence of predators, their individual species birth and death rates, and so forth. In isolation we assume the following proportionality models (following the same arguments as the basic populations models that we have discussed before) to be true where the environment can support an unlimited number of trout and/or bass. Later, we might refine this model to incorporate the limited growth assumptions of the logistics model:

$$\frac{dB}{dt} = mB$$

$$\frac{dT}{dt} = aT$$

Next, we modify the proceeding differential equations to take into account the competition of the trout and bass for living space, oxygen, and food supply. The effect is that the interaction decreases the growth of the species. The interaction terms for competition led to decay rate that we call n for bass and b for trout. This leads to following simplified model:

$$\frac{dB}{dt} = mB - nBT$$

$$\frac{dT}{dt} = aT - bBT$$

If we have the initial stocking level, B_0 and T_0, we determine how the species coexist over time.

If the model is not reasonable, we might try logistic growth instead of isolated growth. Logistic growth in isolation were discussed in first-order ODEs models as a refinement.

Example 4.4 Predator–Prey Relationships

We now consider a model of population growth for two species in which one animal is hunted by another animal. An example of this might be wolves and rabbits where the rabbits are the primary food source for the wolves.

Let $R(t)$ = the population of the rabbits at time t and $W(t)$ = the population of the wolves at time t.

We assume that rabbits grow in isolation but are killed by the interaction with the wolves. We further assume that the constants are proportionality constants.

$$\frac{dR}{dt} = a \cdot R - b \cdot R \cdot W$$

We assume that the wolves will die out without food and grow through their interaction with the rabbits. We further assume that these constants are also proportionality constants.

$$\frac{dW}{dt} = -m \cdot W + n \cdot R \cdot W$$

Example 4.5 Diffusion Models

Diffusion through a membrane leads to a first-order system of ordinary linear differential equations. For example, consider the situation in which two solutions of substance are separated by a membrane of permeability P. Assume the amount of substance that passes through the membrane at any particular time is proportional to the difference between the concentrations of the two solutions. Therefore, if we let x_1 and x_2 represent the two concentrations, and V_1 and V_2 represent their corresponding volumes, then the system of differential equations is given by:

$$\frac{dx_1}{dt} = \frac{P}{V_1}(x_2 - x_1)$$

$$\frac{dx_2}{dt} = \frac{P}{V_2}(x_1 - x_2).$$

where the initial amounts of x_1 and x_2 are given.

If this model does not yield satisfactory results in terms of realism we might try a refinement of the Diffusion Model as follows:

Diffusion through a double-walled membrane, where the inner wall has permeability P_1 and the outer wall has permeability P_2 with $0 < P_1 < P_2$. Suppose the volume of the solution within the inner wall is V_1 and between the two walls is V_2. We let x represent the concentration of the solution within the inner wall and y, the concentration between the two walls. This leads to the following system:

$$\frac{dx}{dt} = \frac{P_1}{V_1}(y - x)$$

$$\frac{dy}{dt} = \frac{1}{V_2}(P_2(C - y) + P_1(x - y))$$

$$x(0) = 2, y(0) = 1, C = 10.$$

Example 4.6 Insurgencies Models

As we look around the world, we see many conflicts involving insurgencies. We have the political faction (usually the status quo or the new regime) battling the insurgents or the rebels that are resisting the change or the political status. This also can be seen from history if we look at our own Revolutionary War.

In insurgency operations, we find the following assumptions: They are messy, grass roots fights that are confused and brutally contested. We find the definition of the enemy is loosely defined. We find that positive control of the forces is usually weak. There are few rules of engagement (they are often permissive). There are political divisions that are deep seated that leave little room for compromise.

Further as we consider building a mobilization model, we assume that growth is subject to the same laws as any other natural or man-made population (basic growth or logistical growth as discussed before). Additionally, there are three considerations: pool of potential recruits, number of recruiters, and the transformation rate.

We assume logistical growth and our systems could look like:

$X(t)$ = insurgency
$Y(t)$ = regime

$$\frac{dX}{dt} = a \cdot (k_1 - X) \cdot X$$

$$\frac{dY}{dt} = b \cdot (k_2 - Y) \cdot Y$$

Where:
a measures insurgency growth rate,
b measures regime growth rate,
k_1 and k_2 are the respective carrying capacities.

Example 4.7 S-I-R Models in a Pandemic

Consider a disease that is spreading throughout the Unites States such as the new flu or COVID-19. The CDC is interesting in know and experimenting with a model for this new disease prior to it actually becoming a "real" epidemic or part of the pandemic. Let us consider the population being divided into three categories: susceptible, infected, and removed. We make the following assumptions for our model:

- No one enters or leaves the community and there is no contact outside the community.
- Each person is either susceptible, S (able to catch this new flu); infected, I (currently has the flu and can spread the flu); or removed, R (already has the flu and will not get it again that includes death).
- Initially every person is either S or I.
- Once someone gets the flu this year they cannot get it again.
- The average length of the disease is 2 weeks, after which the person is deemed infected and can spread the disease.
- Our time period for the model will be per week.

The model we will consider is the SIR model (Allman and Rhodes, 2003). Let's assume the following definition for our variables.

$S(n)$ = number in the population susceptible after period n.
$I(n)$ = number infected after period n.
$R(n)$ = number removed after period n.

Let's start our modeling process with $R(n)$. Our assumption for the length of time someone has the flu is 2 weeks. Thus, half the infected people will be removed each week,

$$\frac{dR}{dt} = 0.5 * I(t)$$

The value, 0.5, is called the removal rate per week. It represents the proportion of the infected persons who are removed from infection each week. If real data is available, then we could do "data analysis" in order to obtain the removal rate.

$I(t)$ will have terms that both increase and decrease its amount over time. It is decreased by the number that are removed each week, $0.5*I(tn)$. It is increased by the numbers of susceptible that come into contact with an infected person and catch the disease, $aS(t)I(t)$. We define the rate, a, as the rate in which the disease is spread or the transmission coefficient. We realize this is a probabilistic coefficient. We will assume, initially, that this rate is a constant value that can be found from initial conditions.

Let's illustrate as follows. Assume we have a population of 1,000 students in the dorms. Our nurse found on 3 students reporting to the infirmary initially. The next week, 5 students came into the infirmary with flu like symptoms. $I(0) = 3$, $S(0) = 997$. In week 1, the number of newly infected is 30.

$$5 = a\, I(n)S(n) = a\,(3)*(995)$$

$$a = 0.00167.$$

Let's consider $S(t)$. This number is decreased only by the number that becomes infected. We may use the same rate, a, as before to obtain the model:

$$\frac{dS}{dt} = -0.00167 \cdot S(t) \cdot I(t)$$

Our coupled SIR model is shown in the systems of differential equations below:

$$\frac{dR}{dt} = 0.5I(t)$$
$$\frac{dI}{dt} = -0.5I(t) + 0.00167I(t)S(t)$$
$$\frac{dS}{dt} = -0.00167S(t)I(t)$$
$$I(0) = 3, S(0) = 997, R(0) = 0$$

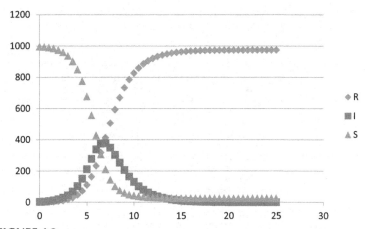

FIGURE 4.2

SIR model of the spread of a disease.

The SIR Model above can be solved iteratively and viewed graphically. Let's iterate the solution and obtain the graph (Figures 4.2–4.3) to observe the behavior to obtain some insights.

In this example, we see that the maximum number of inflected persons occurs at about day 7.

Everyone survives and everyone will get the flu, which might not be the case in COVID-19, where we actually have a S-I-R-D model. In a S-I-R-D model a percent of the infected die. Under the current COVID-19 scenarios about 1–2% die from the disease.

4.3 Qualitative Assessment of Autonomous Systems of First-Order Differential Equations

Consider the system of differential equations:

$$\frac{dx}{dt} = f(x, y)$$
$$\frac{dy}{dt} = g(x, y)$$

These are autonomous systems because they do not include t.

The solution is a pair of parametric equations, $x = x(t), y = y(t)$. The solution is also a curve that varies over time. We call the solution curve a *trajectory*, *path*, or *orbit*. The x-y plane is called the phase-plane. We can also obtain plots

	A	B	C	D	E	F	G
1	SIR Model		Step Size :	0.5			
2							
3	t	R	I	S	R'	I'	S'
4	0	0	3	997	1.5	3.49497	-4.99497
5	0.5	0.75	4.747485	994.5025	2.373743	5.510972	-7.88471
6	1	1.936871	7.502971	990.5602	3.751485	8.660195	-12.4117
7	1.5	3.812614	11.83307	984.3543	5.916534	13.53551	-19.452
8	2	6.770881	18.60082	974.6283	9.300412	20.97483	-30.2752
9	2.5	11.42109	29.08824	959.4907	14.54412	32.06541	-46.6095
10	3	18.69315	45.12094	936.1859	22.56047	47.98299	-70.5435
11	3.5	29.97338	69.11244	900.9142	34.55622	69.42529	-103.982
12	4	47.25149	103.8251	848.9234	51.91254	95.2805	-147.193
13	4.5	73.20776	151.4653	775.3269	75.73267	120.384	-196.117
14	5	111.0741	211.6573	677.2686	105.8287	133.5639	-239.393
15	5.5	163.9884	278.4393	557.5723	139.2197	120.0479	-259.268
16	6	233.5983	338.4633	427.9385	169.2316	72.6536	-241.885
17	6.5	318.2141	374.7901	306.9959	187.395	4.753497	-192.149
18	7	411.9116	377.1668	210.9216	188.5834	-55.7305	-132.853
19	7.5	506.2033	349.3016	144.4952	174.6508	-90.3619	-84.2889
20	8	593.5287	304.1206	102.3507	152.0603	-100.078	-51.982
21	8.5	669.5588	254.0815	76.3597	127.0407	-94.6401	-32.4006
22	9	733.0792	206.7614	60.15938	103.3807	-82.6082	-20.7725
23	9.5	784.7696	165.4573	49.77312	82.72867	-68.9757	-13.753
24	10	826.1339	130.9695	42.89662	65.48475	-56.1024	-9.38231
25	10.5	858.8763	102.9183	38.20546	51.45914	-44.8926	-6.56651
26	11	884.6058	80.47196	34.92221	40.23598	-35.5428	-4.69313
27	11.5	904.7238	62.70054	32.57564	31.35027	-27.9393	-3.41099
28	12	920.399	48.7309	30.87015	24.36545	-21.8532	-2.51223
29	12.5	932.5817	37.80429	29.61403	18.90214	-17.0325	-1.86963
30	13	942.0327	29.28803	28.67922	14.64402	-13.2413	-1.40273
31	13.5	949.3548	22.66739	27.97785	11.33369	-10.2746	-1.05909
32	14	955.0216	17.53009	27.44831	8.765043	-7.96149	-0.80356
33	14.5	959.4041	13.54934	27.04653	6.774671	-6.16268	-0.61199
34	15	962.7915	10.468	26.74054	5.234001	-4.76654	-0.46747

FIGURE 4.3

Screenshot of SIR model of the spread of a disease in Excel.

of x versus t and y versus t. *Rest points* or *equilibrium points* are points that satisfy both $f(x, y) = 0$ and $g(x, y) = 0$, simultaneously. Once we have the equilibrium values, we desire information about their stability.

Rules of Stability: We classify equilibrium values as stable, asymptotically stable, or unstable. We define these as follows:

- **Stable:** if a trajectory starts close to a rest point it remains close for all future time.
- **Asymptotically Stable**: if a trajectory starts close then it tends toward the rest point as t→∞.
- **Unstable:** Does not follow either stable or asymptotically stable rules.

The following results are useful in investigating solutions to autonomous systems. We offer these without proof:

1. There is at most one trajectory through any point in the phase-plane.
2. A trajectory that starts at a point other than a rest point cannot reach a rest point in a finite amount of time.
3. No trajectory can cross itself unless it is a closed curve. If it is a closed curve, then it is a periodic solution.
4. Implications and properties of motion from a starting point (not a rest point):
   a. will move along the same path regardless of starting time;
   b. cannot return to a starting point unless motion is periodic;
   c. can NEVER cross another trajectory;
   d. can only approach and never can reach a rest point.

Consider the following autonomous competitive hunter system of differential equations:

$$\frac{dy}{dt} = .24 \cdot x(t) - .08 \cdot y(t) \cdot x(t)$$
$$\frac{dx}{dt} = y(t) \cdot (4.5 - .9 \cdot x(t))$$

Qualitative Graphical Assessment

First, we plot (Figure 4.4) dx/dt and dy/dt respectively. We want to see where $dx/dt = 0$ and $dy/dt = 0$ simultaneously. In the equation $dx/dt = 0$ we find that either $y(t) = 0$ (which is the x-axis) or $x = 5$ (the horizontal line). In the equation $y(t) = 0$, we find either $x(t) = 0$ (the y axis) or the line $y(t) = 3$. There are two equilibrium points. They are the points (0,0) where the x axis

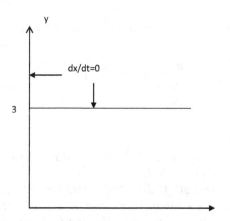

FIGURE 4.4
Plot of dx/dt and dy/dt.

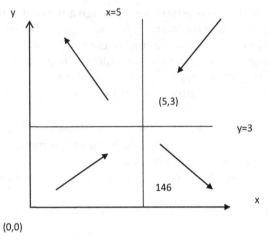

FIGURE 4.5
Plot of dx/dt and dy/dt and direction of movement.

and y axis intersect and the point (5,3) where the vertical line $x = 5$ intersect the horizontal line $y = 3$.

Analysis shows that both (0,0) and (5,3) are unstable equilibrium values (Figure 4.5).

4.4 Solving Homogeneous and Non-homogeneous Systems

We can solve systems of differential equations of the form:

$$\frac{dx}{dt} = ax + by + g(t)$$

$$\frac{dy}{dt} = mx + ny + h(t)$$

where (1) a, b, m , n are constants and (2) the functions $g(t)$ and $h(t)$ can either be 0 or functions of t with real coefficients.

When $g(t)$ and $h(t)$ are both 0 then the system of differential equations is called a homogeneous system, otherwise it is non-homogeneous. We will begin with homogeneous systems. The method we will use involves eigenvalues and eigenvectors.

Example 4.8 Consider the following homogeneous system: with initial conditions:

$$x' = 2x - y + 0$$

$$y' = 3x - 2y + 0$$

$$x(0) = 1, y(0) = 2.$$

Basically, if we rewrite the system of differential equation in matrix form:

$$X' = Ax,$$
where

$$A = \begin{bmatrix} 2 & 1 \\ 3 & -2 \end{bmatrix}$$

$$X' = \begin{bmatrix} \dfrac{dx}{dt} \\ \dfrac{dy}{dt} \end{bmatrix}, x = \begin{bmatrix} x \\ y \end{bmatrix}$$

then we can solve $X' = Ax$. This form is highly suggestive of the first-order separable equation that we saw in the previous chapter. We can assume the solution to have a similar form: $X = Ke^{\lambda t}$, where λ is a constant and X and K are vectors. The values of λ are called **eigenvalues** and the components of K are the corresponding **eigenvectors**. We note that a full discussion of the theory and applications of eigenvalues and eigenvectors can be found in linear algebra textbooks as well as many differential equations textbooks.

Since we have a 2 × 2 system there are two linearly independent solutions that we call X_1 and X_2. The *complementary solution* or *general solution* $X = c_1 X_1 + c_2 X_2$ where c_1 and c_2 are arbitrary constants. We use the initial conditions to find specific values for c_1 and c_2.

The following steps can be used when we have real distinct eigenvalues:

Step 1. Set up the system as a matrix, $X' = AX, X(0) = X_0$

Step 2. Find the eigenvalues, K_1 and K_2.

Step 3. Find the corresponding eigenvectors, K_1 and K_2.

Step 4. Set up the complementary solution $X_c = c_1 X_1 + c_2 X_2$ where

$$X_1 = K_1 e^{\lambda_1 t}$$

$$X_2 = K_2 e^{\lambda_2 t}$$

Step 5. Solve for c_1 and c_2. and rewrite the solution for X_c.

Step 2. Finding the eigenvalues. We set up the characteristic polynomial
by finding the determinant of $A - \lambda I = 0$.

$$\det\begin{bmatrix} 2-\lambda & -1 \\ 3 & -2-\lambda \end{bmatrix} = 0$$

$$(2-\lambda)(-2-\lambda) + 3 = 0$$

$$\lambda^2 - 1 = 0$$

$$\lambda = 1, -1.$$

Step 3. Finding the eigenvectors.

We substitute each solution for l back in $A*k = 0$ and solve the system of
equations for k, the eigenvectors.
   Let $\lambda = 1$.
   Let k_1 and k_2 be the components of eigenvector $\mathbf{K_1}$.

$$k_1 - k_2 = 0$$

$$3k_1 - 3k_2 = 0.$$

We arbitrarily make $k_1 = 1$ thus $k_2 = 1$.

$$\mathbf{K_1} = [1,1]$$

Let $\lambda = -1$.

Let k_1 and k_2 be the components of eigenvector $\mathbf{K_1}$.

$$3k_1 - k_2 = 0$$

$$3k_1 - k_2 = 0.$$

We arbitrarily make $k_2 = 3$ thus $k_1 = 1$.

$$\mathbf{K_2} = [1,3].$$

Step 4. We set up the complementary solution.

$\mathbf{X_c} = c_1\,\mathbf{X_1} + c_2\,\mathbf{X_2}$, where

$$X_1 = K_1 e^{\lambda_1 t}$$
$$X_2 = K_2 e^{\lambda_2 t}$$

$$X_C = c_1 \begin{bmatrix} 1 \\ 1 \end{bmatrix} e^t + c_2 \begin{bmatrix} 1 \\ 3 \end{bmatrix} e^{-t}$$

We find the complimentary solution by setting $\mathbf{X_c}$ = initial condition:
   Since we only had a homogeneous system, we will solve for c_1 and c_2 now using the initial conditions, $x(0) = 1$, $y(0) = 2$.
   We solve the system:

$$c_1 + c_2 = 1$$

$$c_1 + 3c_2 = 2.$$

We find c_1 and c_2 both equal 0.5.
The particular solution is

$$X_C = 0.5\begin{bmatrix}1\\1\end{bmatrix}e^t + 0.5\begin{bmatrix}1\\3\end{bmatrix}e^{-t}$$

We might plot the solutions to the components X_1 and X_2, each a function of t. We note that both solutions grow without bound as $t\to\infty$, Figure 4.6.

Example 4.9 Complex Eigenvalues (eigenvalues of the form $\lambda = a \pm bi$)

We note here that we do not use the form $e^{a \pm bi}$ and that complex eigenvalues always appear in conjugate pairs. The key to finding two real linearly independent solutions from complex solutions is Euler's identity:

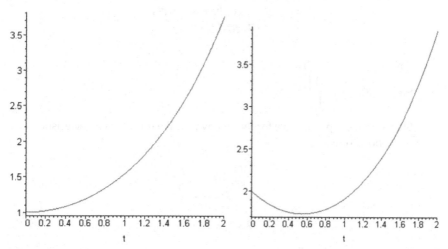

FIGURE 4.6
Plots of $x(t)$ and $y(t)$.

$$e^{i\theta} = \cos\theta + i\sin\theta$$

We can rewrite the solutions for X_1 and X_2 using Euler's identity.

$$Ke^{\lambda t} = Ke^{(a+bi)} = Ke^{at}(\cos bt + i\sin bt)$$
$$Ke^{\lambda t} = K * e^{(a-bi)} = K * e^{at}(\cos bt - i\sin bt)$$

Consider the following steps as a summary when we get complex eigenvalues.

Step 1. Find the complex eigenvalues, $\lambda = a \pm bi$

Step 2. Find the complex eigenvector, K.

$$K = \begin{bmatrix} u_1 + iv_1 \\ u_2 + iv_2 \end{bmatrix}$$

Step 3. Form the real vectors

$$B_1 = \begin{bmatrix} u_1 \\ u_2 \end{bmatrix}$$
$$B_2 = -\begin{bmatrix} v_1 \\ v_2 \end{bmatrix}$$

Step 4. Form the linearly independent set of real solutions:

$$X_1 = e^{at}(B_1 \cos bt + B_2 \sin bt)$$
$$X_2 = e^{at}(B_2 \cos bt - B_1 \sin bt)$$

Step 1. $X' = \begin{bmatrix} 6 & -1 \\ 5 & 4 \end{bmatrix} X$.

Step 2. We set up and solve for the eigenvalue. We solve the characteristic polynomial

$$(6-\lambda)(4-\lambda) + 5 = 0$$

$$29 - 10\lambda + \lambda^2 = 0$$

$$\lambda = 5 \pm 2I.$$

We find the eigenvalues are $5 + 2I$ and $5 - 2I$.

Step 3. We find the eigenvectors we substituting λ as we did before. We then create the two vectors B1 and B2.

Let $\lambda = -5 + 2I$.
Let k_1 and k_2 be the components of eigenvector K_1.

$$(1 - 2I)k_1 - k_2 = 0$$

$$5k_1 + (1 - 2I)k_2 = 0$$

We arbitrarily make $k_2 = (1 - 2I)$ thus $k_1 = 1$.

$$K1 = [1, 1 - 2I]$$

$$B1 = \text{real}(K1) = [1,1], B2 = \text{Imaginary}(K1) = [0-2].$$

$$B1 = [1,1]$$

$$B2 = [0, -2]$$

By substitution, we find the complementary solution:

$$X_c = c_1 e^{5t}\left(\begin{bmatrix} 1 \\ 1 \end{bmatrix}\cos(2t) - \begin{bmatrix} 0 \\ -2 \end{bmatrix}\sin(2t)\right) + c_2 e^{5t}\left(\begin{bmatrix} 0 \\ -2 \end{bmatrix}\cos(2t) + \begin{bmatrix} 1 \\ 1 \end{bmatrix}\sin(2t)\right)$$

Since we only had a homogeneous system, we will solve for c_1 and c_2 now using the initial conditions, $x(0) = 1$, $y(0) = 2$.
We get two equations

$$c_1 = 1 \text{ and}$$

$$c_1 - 2c_2 = 2,$$

whose solutions are $c_1 = 1$, $c_2 = -0.5$.

$$X_p = e^{5t}\left(\begin{bmatrix} 1 \\ 1 \end{bmatrix}\cos(2t) - \begin{bmatrix} 0 \\ -2 \end{bmatrix}\sin(2t)\right) - 0.5e^{5t}\left(\begin{bmatrix} 0 \\ -2 \end{bmatrix}\cos(2t) + \begin{bmatrix} 1 \\ 1 \end{bmatrix}\sin(2t)\right)$$

Again, we obtain plots of X_1 and X_2 as functions of t, Figure 4.7.

Example 4.10 Repeated Eigenvalues Solution

When eigenvalues are repeated, we must find a method to obtain independent solutions. The following is a summary for repeated real eigenvalues.

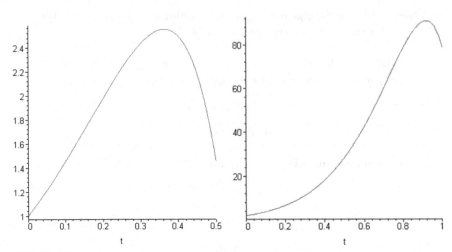

FIGURE 4.7
Plots of X_1 and X_2 as functions of t.

Step 1. Find *the repeated eigenvalues,* $\lambda_1 = \lambda_2 = \lambda$.

Step 2. One solution is

$$X_1 = Ke^{\lambda t}$$

and the second linearly independent solution is given by

$$X_2 = Kte^{\lambda t} + Pe^{\lambda t}$$

where the components of P must satisfy the system

$$(a - \lambda)p_1 + bp_2 = k_1$$
$$cp_1 + (d - \lambda)p_2 = k_2$$

and

$$\begin{bmatrix} a & b \\ c & d \end{bmatrix} = A$$

Step 1. $X' = \begin{bmatrix} 3 & -18 \\ 2 & -9 \end{bmatrix} X$.

Step 2. Solve the characteristic equation $(3 - \lambda)(-9 - \lambda) + 36 = 0$.

We find we have repeated roots and $\lambda = -3, -3$.

Step 3. We find K easily as the vector. Then we solve for the vector P.

$$\begin{bmatrix} 6 & -18 \\ 2 & -6 \end{bmatrix} \begin{bmatrix} p_1 \\ p_2 \end{bmatrix} = \begin{bmatrix} 3 \\ 1 \end{bmatrix}.$$

We find p_1 and p_2 must solve $p_1 - 3p_2 = 1/2$ or $2p_1 - 6p_2 = 1$. We select $p_1 = 1$ then $p_2 = 1/6$.

Our complementary solution is

$$X_c = c_1 \begin{bmatrix} 3 \\ 1 \end{bmatrix} e^{-3t} + c_2 \left(\begin{bmatrix} 3 \\ 1 \end{bmatrix} te^{-3t} + \begin{bmatrix} 1 \\ 1 \\ \frac{1}{6} \end{bmatrix} e^{-3t} \right)$$

Since we only had a homogeneous system, we will solve for c_1 and c_2 now using the initial conditions, $x(0) = 1$, $y(0) = 2$.

We obtain two equations:

$$3c_1 + c_2 = 1$$

$$c_1 + (1/6)c_2 = 2$$

$$c_1 = 5/3 \; c_2 = -4.$$

$$X_c = \frac{5}{3} \begin{bmatrix} 3 \\ 1 \end{bmatrix} e^{-3t} - 4 \left(\begin{bmatrix} 3 \\ 1 \end{bmatrix} te^{-3t} + \begin{bmatrix} 1 \\ 1 \\ \frac{1}{6} \end{bmatrix} e^{-3t} \right)$$

We plot the solutions shown in Figure 4.8.

4.5 Technology Examples for Systems of Ordinary Differential Equations

4.5.1 Excel for System of Ordinary Differential Equations

We can use the techniques described in Chapter 3 to input a predator–prey system and initial conditions in Excel and then iterate to determine the numerical estimates using Euler's method for a predator–prey system of ODEs (Figure 4.9).

We can plot this data (Figure 4.10) to gain a visual qualification of the results:

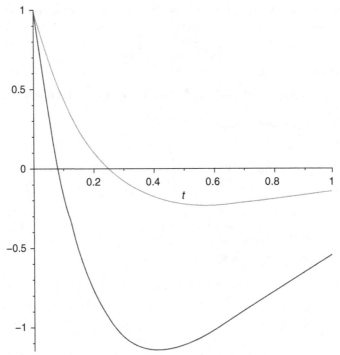

FIGURE 4.8

Plot of $X_c = \dfrac{5}{3}\begin{bmatrix}3\\1\end{bmatrix}e^{-3t} - 4\left(\begin{bmatrix}3\\1\end{bmatrix}te^{-3t} + \begin{bmatrix}1\\1\\ \frac{1}{6}\end{bmatrix}e^{-3t}\right)$.

The power of Euler's method is two-fold. First it is easy to use and second as a numerical method it can be used to estimate a solution to a system of differential equations that does not have a closed form solution.

Assume we have the following predator–prey system that does not have a closed form analytical solution:

$$\frac{dx}{dt} = 3x - xy$$

$$\frac{dy}{dt} = xy - 2y$$

$$x(0) = 1, y(0) = 2$$

$$t_0 = 1, \Delta t = .1$$

We will obtain an estimated the solution using Euler's method (Figure 4.11).

We experiment and find that when we plot $x(t)$ versus $y(t)$ we have an approximately closed loop (Figure 4.12).

	A	B	C	D	E	F	G	H
1	t	x	y	x'	y'		step size	0.1
2	0	3	6	-3	-9			
3	0.1	2.7	5.1	-2.1	-6.9			
4	0.2	2.49	4.41	-1.35	-5.19			
5	0.3	2.355	3.891	-0.717	-3.789			
6	0.4	2.2833	3.5121	-0.1743	-2.6319			
7	0.5	2.26587	3.24891	0.29979	-1.66629			
8	0.6	2.295849	3.082281	0.722985	-0.84988			
9	0.7	2.368148	2.997293	1.109856	-0.14843			
10	0.8	2.479133	2.98245	1.4725	0.465867			
11	0.9	2.626383	3.029036	1.821077	1.01577			
12	1	2.808491	3.130613	2.164246	1.520001			
13	1.1	3.024915	3.282613	2.509519	1.994123			
14	1.2	3.275867	3.482026	2.86355	2.451234			
15	1.3	3.562222	3.727149	3.232369	2.902515			
16	1.4	3.885459	4.017401	3.621576	3.357694			
17	1.5	4.247617	4.35317	4.036511	3.825404			
18	1.6	4.651268	4.73571	4.482383	4.313498			
19	1.7	5.099506	5.16706	4.964398	4.82929			
20	1.8	5.595946	5.649989	5.48786	5.379773			
21	1.9	6.144732	6.187967	6.058263	5.971794			
22	2	6.750558	6.785146	6.681383	6.612208			
23	2.1	7.418697	7.446367	7.363356	7.308016			
24	2.2	8.155032	8.177168	8.11076	8.066488			
25	2.3	8.966108	8.983817	8.930691	8.895273			
26	2.4	9.859177	9.873345	9.830843	9.802509			
27	2.5	10.84226	10.8536	10.81959	10.79693			
28	2.6	11.92422	11.93329	11.90609	11.88795			
29	2.7	13.11483	13.12208	13.10032	13.08582			
30	2.8	14.42486	14.43067	14.41326	14.40165			
31	2.9	15.86619	15.87083	15.8569	15.84762			
32	3	17.45188	17.45559	17.44445	17.43702			

FIGURE 4.9

Screenshot of Euler's method for a system of equations in Excel.

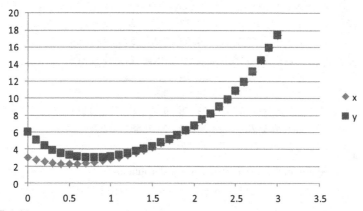

FIGURE 4.10

Visualization of Euler's method for a system of equations.

	A	B	C	D	E	F
1	Predator–Prey				step size	0.1
2						
3						
4	t	x	y	x'	y'	
5	0	1	2	1	-2	
6	0.1	1.1	1.8	1.32	-1.62	
7	0.2	1.232	1.638	1.677984	-1.25798	
8	0.3	1.399798	1.512202	2.082618	-0.90763	
9	0.4	1.60806	1.421439	2.538421	-0.55712	
10	0.5	1.861902	1.365727	3.042856	-0.1886	
11	0.6	2.166188	1.346867	3.580997	0.223833	
12	0.7	2.524288	1.36925	4.116482	0.717881	
13	0.8	2.935936	1.441038	4.577012	1.348719	
14	0.9	3.393637	1.57591	4.832844	2.196247	
15	1	3.876921	1.795535	4.669617	3.370078	
16	1.1	4.343883	2.132543	3.768134	4.998431	
17	1.2	4.720697	2.632386	1.735396	7.161922	
18	1.3	4.894236	3.348578	-1.70602	9.691575	
19	1.4	4.723634	4.317735	-6.2245	11.75993	
20	1.5	4.101184	5.493728	-10.2272	11.54333	
21	1.6	3.07846	6.648062	-11.2304	7.16967	
22	1.7	1.955419	7.365029	-8.53546	-0.32834	
23	1.8	1.101873	7.332195	-4.77353	-6.58524	
24	1.9	0.62452	6.67367	-2.29428	-9.1795	
25	2	0.395092	5.75572	-1.08876	-9.2374	
26	2.1	0.286216	4.83198	-0.52434	-8.28097	
27	2.2	0.233782	4.003883	-0.23469	-7.07173	
28	2.3	0.210313	3.29671	-0.0624	-5.90008	
29	2.4	0.204072	2.706702	0.059854	-4.86104	
30	2.5	0.210058	2.220598	0.16372	-3.97474	
31	2.6	0.22643	1.823124	0.26648	-3.23344	
32	2.7	0.253078	1.49978	0.379672	-2.62	
33	2.8	0.291045	1.23778	0.512885	-2.11531	
34	2.9	0.342334	1.026249	0.675681	-1.70118	

FIGURE 4.11
Screenshot of predator–prey system without a closed solution.

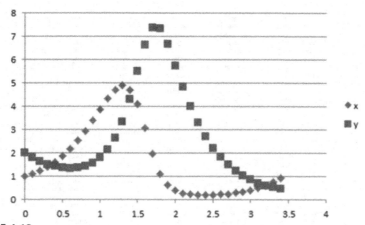

FIGURE 4.12
Status of predator and prey over time.

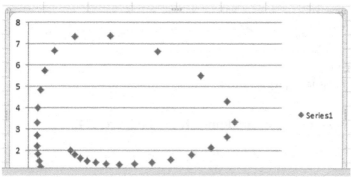

FIGURE 4.13
Visualization of predator–prey system without a closed solution.

We can use the improved Euler method and Runge-Kutta 4 methods to iterate solutions to systems of differential equations as well (Figure 4.13). The vector version of the iterative formula for Runge-Kutta 4 is

$$X_{n+1} = X_n + \frac{h}{6}(K_1 + 2K_2 + 2K_3 + K_4)$$

where

$$K_1 = f(t_n, X_n)$$
$$K_2 = f(t_n + \frac{h}{2}, X_n + \frac{h}{2}K_1)$$
$$K_3 = f(t_n + \frac{h}{2}, X_n + \frac{h}{2}K_2)$$
$$K_4 = f(t_n + h, X_n + hK_3)$$

F	G	H	I	J	K	L	M	N	O	P	Q	R
0	1	2	1	-2	1.40625	-1.53125	1.40083313	-1.49068	1.853389	-1.05743	1.352815	1.620
0.25	1.352815	1.62078	1.865829	-1.04894	2.395461874	-0.61666	2.406171251	-0.53682	2.957776	-0.06785	1.953934	1.4781
0.5	1.953934	1.478124	2.973647	-0.06809	3.559131579	0.478565	3.507217468	0.613372	3.873961	1.355323	2.828114	1.6227
0.75	2.828114	1.622753	3.895012	1.343824	4.008715145	2.354794	3.605186993	2.548218	2.760482	3.908135	3.739918	2.2501
1	3.739918	2.250169	2.804307	3.915109	1.065328238	5.726925	0.131550815	5.555632	-2.41111	6.451376	3.856041	3.6223
1.25	3.856041	3.622319	-2.39969	6.723172	-5.201533183	6.944342	-4.777875042	5.414699	-5.25925	3.291979	2.705301	5.0695
1.5	2.705301	5.069537	-5.59872	3.57555	-5.046704346	0.030126	-4.30099017	0.377774	-3.5274	-1.9104	1.546071	5.172
1.75	1.546071	5.17291	-3.35947	-2.34813	-2.116455	-4.26392	-2.101581357	-3.33372	-1.36718	-4.24976	0.997625	4.2648
2	0.997625	4.264862	-1.26186	-4.27499	-0.613531086	-4.32777	-0.666655149	-4.01833	-0.21628	-3.8114	0.829353	3.2324
2.25	0.829353	3.232421	-0.19276	-3.78402	0.193730508	-3.29679	0.15336757	-3.2333	0.49971	-2.74481	0.871068	2.4162
2.5	0.871068	2.416212	0.508519	-2.72774	0.864306935	-2.2109	0.842178659	-2.18456	1.222144	-1.71745	1.085386	1.8647
2.75	1.085386	1.864707	1.232231	-1.70549	1.671324227	-1.25612	1.672636091	-1.20512	2.15995	-0.77617	1.50539	1.5562
3	1.50539	1.556202	2.17348	-0.76971	2.736717798	-0.32547	2.742550268	-0.23115	3.290015	0.286239	2.189641	1.4896
3.25	2.189641	1.489672	3.307076	0.282503	3.839501735	0.919605	3.725069425	1.074422	3.875302	1.970869	3.119288	1.7497
3.5	3.119288	1.749732	3.899947	1.958454	3.626480237	3.204787	3.035529146	3.381605	1.570144	4.874101	3.902376	2.5832
3.75	3.902376	2.583287	1.626169	4.914384	-0.81121603	6.732986	-1.615074597	6.168175	-3.93709	6.182251	3.603897	4.1207
4	3.603897	4.120744	-4.03905	6.609247	-6.033473627	5.436721	-5.130434543	4.078903	-4.96865	1.651571	2.29825	5.2575
4.25	2.29825	5.257913	-5.18925	1.568175	-4.047997084	-1.91109	-3.618602708	-1.0427	-2.78335	-3.03033	1.327175	4.9508
4.5	1.327175	4.950841	-2.58911	-3.33105	-1.539887252	-4.51842	-1.572723071	-3.79529	-0.93588	-4.26618	0.920917	3.9414
4.75	0.920917	3.941481	-0.86703	-4.25319	-0.333004912	-4.04905	-0.382799778	-3.85003	0.017351	-3.49965	0.825863	2.96
5	0.825863	2.96019	0.032877	-3.47567	0.39362982	-2.95517	0.358083144	-2.91447	0.703406	-2.4204	0.919184	2.2253
5.25	0.919184	2.225384	0.712015	-2.40523	1.084072392	-1.90897	1.068655423	-1.8781	1.475985	-1.42866	1.189745	1.7500
5.5	1.189745	1.750049	1.487123	-1.41799	1.963305758	-0.982	1.970043344	-0.91917	2.489306	-0.48305	1.683209	1.5124
5.75	1.683209	1.512408	2.503927	-0.47912	3.089081537	-0.00552	3.079766476	0.104828	3.584996	0.697224	2.450984	1.5297
6	2.450984	1.529772	3.603507	0.689903	4.015541142	1.456708	3.803782904	1.631278	3.614241	2.716367	3.403334	1.929
6.25	3.403334	1.929032	3.644863	2.707076	2.826998237	4.214995	2.044002269	4.314312	-0.02979	5.757572	3.959879	2.9925
6.5	3.959879	2.992501	0.029695	5.86494	-2.876054802	7.31559	-3.265356831	6.252574	-4.89023	5.209561	3.245572	4.5846

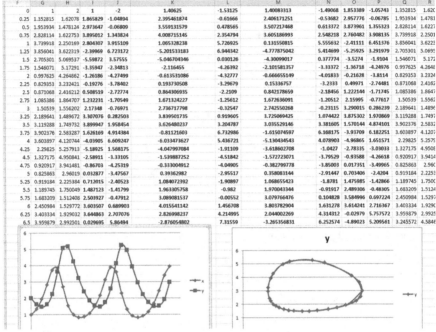

FIGURE 4.14
Screenshot of predator–prey system in Excel.

We repeat our example with Runge-Kutta 4 (Figure 4.14).

$$\frac{dx}{dt} = 3x - xy$$

$$\frac{dy}{dt} = xy - 2y$$

$$x(0) = 1, y(0) = 2$$

$$t_0 = 0, \Delta t = 0.25$$

4.5.2 Maple for System of Ordinary Differential Equations

Again, we start by accessing the *with(DEtools):* command.

Maple allows us to find the homogeneous solution in matrix form. The following commands illustrate this:

```
> with(DEtools):with(plots):with(linalg):
> M: = array([[2,-1],[3,-2]]);
> lambda: = eigenvects(M);
> homsol: = matrixDE(M,t);
```

$$M := \begin{bmatrix} 2 & -1 \\ 3 & -2 \end{bmatrix}$$
$$\lambda = [1,1,\{[1,1]\}],[-1,1,\{[1,3]\}]$$
$$homsol = \left[\begin{bmatrix} e^t & e^{(-t)} \\ e^t & 3e^{(-t)} \end{bmatrix}, [0,0] \right]$$

> phi: = homsol[1];

$$\phi := \begin{bmatrix} e^t & e^{(-t)} \\ e^t & 3e^{(-t)} \end{bmatrix}$$

We find the complimentary solution:
The matrix(2, 1, [c1,c2]) means that we want a 2 row 1 columns matrix of c_1, c_2.

>Xc: = multiply(phi, matrix(2,1,[c1,c2]));

$$Xc = \begin{bmatrix} e^t \; cl + e^{(-t)} \; c2 \\ e^t \; cl + 3e^{(-t)} \; c3 \end{bmatrix}$$

Since we only had a homogeneous system, we will solve for c_1 and c_2 now using the initial conditions, $x(0) = 1, y(0) = 2$.

> eq1: = evalf(subs(t = 0, exp(t)*c1+exp(-t)*c2));

$$eq1 := 1.cl + 1.c2$$

> eq2: = evalf(subs(t = 0,exp(t)*c1+3*exp(-t)*c2));

$$eq2 := 1.cl + 3.c2$$

> solve({1 = eq1,2 = eq2},{c1,c2});

$$\{c2 = 0.5000000000, cl = 0.5000000000\,\}$$

>Xg: = multiply(phi, matrix(2,1,[.5,.5]));

$$Xg := \begin{bmatrix} 0.5e^t + 0.5e^{(-t)} \\ 0.5e^t + 1.5e^{(-t)} \end{bmatrix}$$

We might plot the solutions to the components X_1 and X_2, each a function of t (Figure 4.15 and Figure 4.16). We note that both solutions grow without bound as $t \to \infty$.

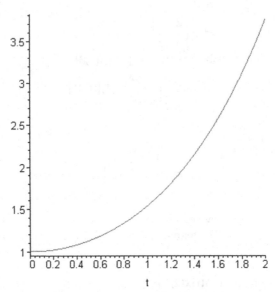

FIGURE 4.15
Plot of (.5*exp(*t*)+.5*exp(-*t*),*t* = 0..2) in Maple.

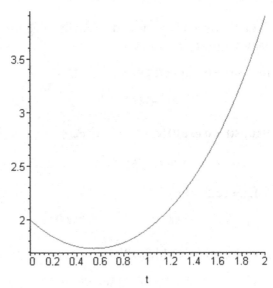

FIGURE 4.16
Plot of (.5*exp(*t*)+1.5*exp(−*t*),t = 0..2) in Maple.

> plot(.5*exp(t)+.5*exp(-t),t = 0..2);
> plot(.5*exp(t)+1.5*exp(-t),t = 0..2);

Example 4.11 Complex Eigenvalues (eigenvalues of the form $\lambda = a \pm bi$)

We note here that we do not use the form $e^{a \pm bi}$ and that complex eigenvalues always appear in conjugate pairs. The key to finding two real linearly independent solutions from complex solutions is Euler's identity:

$$e^{i\theta} = \cos\theta + i\sin\theta$$

We can rewrite the solutions for X_1 and X_2 using Euler's identity.

$$Ke^{\lambda t} = Ke^{(a+bi)} = Ke^{at}(\cos bt + i\sin bt)$$
$$Ke^{\lambda t} = K * e^{(a-bi)} = K * e^{at}(\cos bt - i\sin bt)$$

Consider the following steps as a summary when we get complex eigenvalues.

Step 1. Find the complex eigenvalues, $\lambda = a \pm bi$.

Step 2. Find the complex eigenvector, K.

$$K = \begin{bmatrix} u_1 + iv_1 \\ u_2 + iv_2 \end{bmatrix}$$

Step 3. Form the real vectors:

$$B_1 = \begin{bmatrix} u_1 \\ u_2 \end{bmatrix}$$
$$B_2 = -\begin{bmatrix} v_1 \\ v_2 \end{bmatrix}$$

Step 4. Form the linearly independent set of real solutions:

$$X_1 = e^{at}(B_1 \cos bt + B_2 \sin bt)$$
$$X_2 = e^{at}(B_2 \cos bt - B_1 \sin bt)$$

We find Maple will allow for an easily manipulation of these steps for us:

> with(DEtools):with(plots):with(linalg):
> M: = array([[6,-1],[5,4]]);
> lambda: = eigenvects(M);

> homsol: = matrixDE(M,t);

$$M := \begin{bmatrix} 6 & -1 \\ 5 & 4 \end{bmatrix}$$

$$\lambda := \left[5 + 2I, 1, \{[1, 1 - 2I]\}\right], \left[5 - 2I, 1, \{[1, 1 + 2I]\}\right]$$

$$homsol := \left[\left[e^{(5t)} \begin{array}{cc} e(5t) & \cos(2t) & e^{(5t)}\sin(2t) \\ \cos(2t) + 2e & 2e^{(5t)} & \sin(2t)e^{(5t)}\sin(2t) - 2e^{(5t)}\cos(2t) \end{array}\right]\right]$$

> phi: = homsol[1];
>

$$\phi = \left[\left[e^{(5t)} \begin{array}{cc} e(5t) & \cos(2t) & e^{(5t)}\sin(2t) \\ \cos(2t) + 2e & 2e^{(5t)} & \sin(2t)e^{(5t)}\sin(2t) - 2e^{(5t)}\cos(2t) \end{array}\right]\right]$$

We find the complimentary solution:
    The matrix(2,1,[c1,c2]) means that we want a 2 row 1 columns matrix of c_1, c_2.

>Xc: = multiply(phi, matrix(2,1,[c1,c2]));

$$Xc = \left[(e^{(5t)}\cos(2t) + \begin{array}{c} e^{(5t)}\cos(2t)l + e^{(5t)}\sin(2t)c2 \\ 2e^{(5t)}\sin(2t)cl + \end{array} \quad (e^{(5t)}\sin(2t) - 2e^{(5t)}\cos(2t))c2 \right]$$

Since we only had a homogeneous system, we will solve for c_1 and c_2 now using the initial conditions, x(0) = 1, y(0) = 2.

> eq1: = evalf(subs(t = 0, Xc[1,1]));

$$eq1 := 1.cl$$

> eq2: = evalf(subs(t = 0,Xc[2,1]));

$$eq2 := 1.cl - 2.c2$$

> solve({1 = eq1,2 = eq2},{c1,c2});

$$\{cl = 1., c2 = -0.5000000000 \}$$

>Xg: = multiply(phi, matrix(2,1,[1,-.5]));

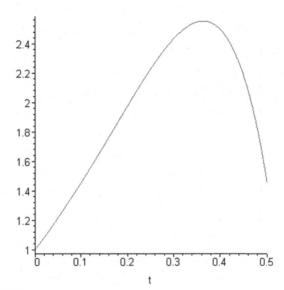

FIGURE 4.17
Plot of (exp(5*t)*cos(2*t)–.5*exp(5*t)*sin(2*t),t = 0..0.5) in Maple.

$$Xg:=\begin{bmatrix} e^{(5t)}\cos(2t)-0.5e^{(5\,t)}\sin(2t) \\ 2.0e^{(5t)}\cos(2t)+1.5e^{(5t)}\sin(2t) \end{bmatrix}$$

Again, we obtain plots of X_1 and X_2 as functions of t (Figure 4.17 and Figure 4.18a).

> plot(exp(5*t)*cos(2*t)-.5*exp(5*t)*sin(2*t),t = 0..0.5);
> plot(2.0*exp(5*t)*cos(2*t)+1.5*exp(5*t)*sin(2*t),t = 0..1);

Example 4.12 Repeated Eigenvalues Solution

When eigenvalues are repeated, we must find a method to obtain independent solutions. The following is a summary for repeated real eigenvalues.

Step 1. Find *the repeated eigenvalues,* $\lambda_1 = \lambda_2 = \lambda$.

Step 2. One solution is

$$X_1 = Ke^{\lambda t}$$

and the second linearly independent solution is given by

$$X_2 = Kte^{\lambda t} + Pe^{\lambda t}$$

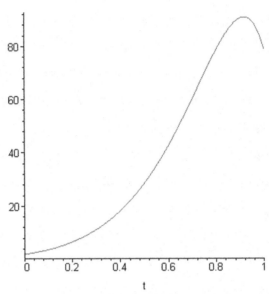

FIGURE 4.18a
Plot of (2.0*exp(5*t)*cos(2*t) + 1.5*exp(5*t)*sin(2*t),t = 0..1) in Maple.

where the components of P must satisfy the system

$$(a - \lambda)p_1 + bp_2 = k_1$$
$$cp_1 + (d - \lambda)p_2 = k_2$$
and
$$\begin{bmatrix} a & b \\ c & d \end{bmatrix} = A$$

We now present an example of this method using Maple. Maple recognizes repeated eigenvalues and places the solution in the correct form.

```
> with(DEtools):with(plots):with(linalg):
> M: = array([[3,-18],[2,-9]]);
> lambda: = eigenvects(M);
> homsol: = matrixDE(M,t);
```

$$M := \begin{bmatrix} 2 & -1 \\ 3 & -2 \end{bmatrix}$$

$$\lambda := [1, 1, \{[1,1]\}], [-1, 1, \{[1,3]\}]$$

$$homsol := \begin{bmatrix} \begin{bmatrix} e^t & e^{(-t)} \\ e^t & 3e^{(-t)} \end{bmatrix}, [0,0] \end{bmatrix}$$

> **phi: = homsol[1];**

$$\phi := \begin{bmatrix} e^{(-3t)} & e^{(-3t)}t \\ \dfrac{1}{3}e^{(-3t)} & \dfrac{1}{3}e^{(-3t)}t - \dfrac{1}{18}e^{(-3t)} \end{bmatrix}$$

We find the complimentary solution:
   The matrix(2,1,[c1,c2]) means that we want a 2 row 1 columns matrix of c_1, c_2.

>**Xc: = multiply(phi, matrix(2,1,[c1,c2]));**

$$Xc := \begin{bmatrix} e^{(-3t)}c1 + e^{(-3t)}tc2 \\ \dfrac{1}{3}e^{(-3t)}c1 + \left(\dfrac{1}{3}e^{(-3t)}t - \dfrac{1}{18}e^{(-3t)} \right)c2 \end{bmatrix}$$

Since we only had a homogeneous system, we will solve for c_1 and c_2 now using the initial conditions, $x(0) = 1$, $y(0) = 2$.

> **eq1: = evalf(subs(t = 0, Xc[1,1]));**

$$eq1 := 1.c1$$

> **eq2: = evalf(subs(t = 0,Xc[2,1]));**

$$eq2 := 0.3333333333 \; c1 - 0.05555555556 \; c2$$

> **solve({1 = eq1,2 = eq2},{c1,c2});**

$$\{c1 = 1, c2 = 30.00000000\}$$

>**Xg: = multiply(phi, matrix(2,1,[1,-30]));**

$$Xg := \begin{bmatrix} e^{(-3t)} - 30e^{(-3t)}t \\ 2e^{(-3t)} - 10e^{(-3t)}t \end{bmatrix}$$

> **plot(exp(-3*t)-30*exp(-3*t)*t,t = 0..0.5); (Figure 4.18b).**

> **plot(2*exp(-3*t)-10*exp(-3*t)*t,t = 0..1); (Figure 4.19).**

   Maple's Use in Nonhomogeneous Systems of Differential Equations for Closed Form Analytical Solutions

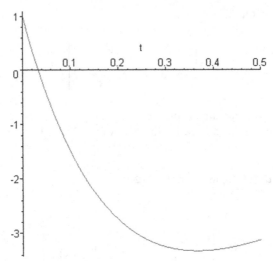

FIGURE 4.18b

Plot of (exp(-3*t)–30*exp(–3*t)*t,t = 0..0.5) in Maple.

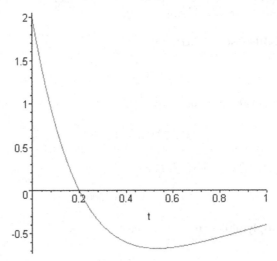

FIGURE 4.19

Plot of (2*exp(-3*t)-10*exp(–3*t)*t,t = 0..1) in Maple.

In the nonhomogeneous form, we need both a complementary solution X_c and a particular solution X_p. The complementary solution is found by solving the homogeneous part of the system. The particular solution is found using the variation of parameters. The following summary of the procedure is provided. Students should consult a differential equations textbook for a more detailed explanation and proof of the procedure.

Step 1. Find the complementary solution by solving for the homogeneous solution of

$$X' = AX$$

and form the linear independent set of solutions, put them as columns into matrix called Φ.

Step 2. Vary the parameters and write the form of the particular solution

$$X_p = u_1(t)X_1(t) + u_2(t)X_2(t)$$

Step 3. Invert the matrix, Φ

$$\Phi^{-1}$$

Step 4. Determine the parameters u_1 and u_2.

$$\begin{bmatrix} u_1 \\ u_2 \end{bmatrix} = \int \Phi^{-1} F(t)dt$$

Step 5. Calculate the particular solution, X_p

$$X_p = \Phi \cdot U = \Phi \cdot \int \Phi^{-1} F(t)dt$$

Step 6. Form the general solution

$$X = X_c + X_p.$$

We provide an example using Maple. Given the following system of non-homogeneous differential equations:

$$x' = 2x - y + 0$$
$$y' = 3x - 2y + 4t$$

$$x(0) = 1, y(0) = 2.$$

Note that $g(t) = 0$ and $h(t) = 4t$.

Part 1: Homogeneous

```
> with(DEtools):with(plots):with(linalg):

> M: = array([[2,-1],[3,-2]]);

> lambda: = eigenvects(M);

> homsol: = matrixDE(M,t);
```

$$M := \begin{bmatrix} 2 & -1 \\ 3 & -2 \end{bmatrix}$$

$$\lambda := [1, 1, \{[1,1]\}], [-1, 1, \{[1,3]\}]$$

$$homsol := \left[\begin{bmatrix} e^t & e^{(-t)} \\ e^t & 3e^{(-t)} \end{bmatrix}, [0,0] \right]$$

> phi: = homsol[1];

$$\phi := \begin{bmatrix} e^t & e^{(-t)} \\ e^t & 3e^{(-t)} \end{bmatrix}$$

The Complimentary Solution:

The matrix(2,1,[c1,c2]) means that we want a 2 row 1 column matrix of c_1, c_2.

>Xc: = multiply(phi, matrix(2,1,[c1,c2]));

$$Xc := \begin{bmatrix} e^t c1 + e^{(-t)} c2 \\ e^t c1 + 3e^{(-t)} c2 \end{bmatrix}$$

If we only had a homogeneous systems, we would solve for c_1 and c_2 now, but since this is a nonhomogeneous system we wait until we get **X = Xc + Xp**.

Finding *Xp*

phi*int(phi^-1*F(t))dt

Nonhomogeneous Systems of Differential Equations

> A: = homsol[1];

$$A: \begin{bmatrix} e^t & e^{(-t)} \\ e^t & 3e^{(-t)} \end{bmatrix}$$

> A1: = inverse(A);

$$A1 := \begin{bmatrix} \dfrac{3}{2} \dfrac{1}{e^t} & -\dfrac{1}{2} \dfrac{1}{e^t} \\ \dfrac{1}{2} \dfrac{1}{e^{(-t)}} & \dfrac{1}{2} - \dfrac{1}{e^{(-t)}} \end{bmatrix}$$

> B: = matrix(2,1,[0,4*t]);

$$B := \begin{bmatrix} 0 \\ 4t \end{bmatrix}$$

> B1: = multiply(A1,B);

$$B1 := \begin{bmatrix} -\dfrac{2t}{e^t} \\ \dfrac{2t}{e^{(-t)}} \end{bmatrix}$$

> B2: = map(int,B1,t);

$$B2 := \begin{bmatrix} \dfrac{2(1+t)}{e^t} \\ \dfrac{2(t-1)}{e^{(-t)}} \end{bmatrix}$$

> B3: = multiply(A,B2);

$$B3 := \begin{bmatrix} 4t \\ 8t-4 \end{bmatrix}$$

> nB3: = simplify(%);

$$nB3 := \begin{bmatrix} 4t \\ 8t-4 \end{bmatrix}$$

> x: = evalm(multiply(A,matrix(2,1,[c1,c2])) + nB3);

$$x := \begin{bmatrix} e^t c1 + e^{(-t)} c2 + 4t \\ e^t c1 + 3e^{(-t)} c2 + 8t - 4 \end{bmatrix}$$

Now, use the initial conditions x1(0) = 1, x2(0) = 2 to find c_1 and c_2.

> solve({c1+c2 = 1,c1+3*c2-4 = 2},{c1,c2});

$$\left\{ c2 = \frac{5}{2}, c1 = \frac{-3}{2} \right\}$$

> x1: = subs({c2 = 2.5, c1 = -1.5},x[1,1]);

$$x1 := -1.5e^t + 2.5e^{(-t)} + 4t$$

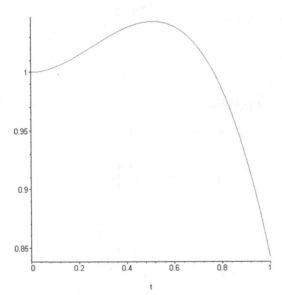

FIGURE 4.20
Plot of $(x1, t = 0..1)$ in Maple.

> x2: = subs({c2 = 2.5, c1 = -1.5},x[2,1]);

$$x2 := -1.5e^{t} + 7.5e^{(-t)} + 8t - 4$$

We can plot each versus t (Figure 4.20 and Figure 4.21).

> plot(x1,t = 0..1);

> plot(x2,t = 0..1);

Phase Portraits
We now examine the phase portrait that provide the same information to us but in a slightly different format using Maple (Figure 4.22).

restart; with(DEtools):with(linalg):
Phase Portrait in Maple

>diffeq1: = diff(y(t),t) = 4.5*y(t)-.9*y(t)*x(t);

$$diffeq1 := \frac{d}{dt}y(t) = 4.5y(t) - 0.9y(t)x(t)$$

>diffeq2: = diff(x(t),t) = .24*x(t)-.08*y(t)*x(t);

$$diffeq2 := \frac{d}{dt}x(t) = 0.24x(t) - 0.08y(t)x(t)$$

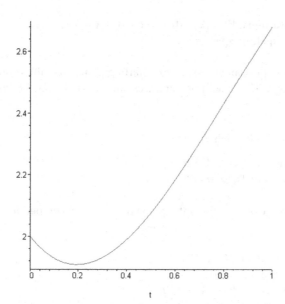

FIGURE 4.21
Plot of $(x2, t = 0..1)$ in Maple.

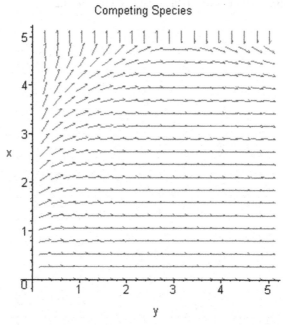

FIGURE 4.22
Plot of phase portrait in Maple.

> **DEplot({diffeq1,diffeq2},[y(t),x(t)],t = -3..3, y = 0..5, x = 0..5,**
title = 'Competing Species');

The phase portrait shows, from a starting point (the initial condition), who survives. The phase portrait traces out a possible solution curve.

Example 4.13 The Fish Pond

$B(t)$ = number of Bass fish after time t
$T(t)$ = number of Trout after t time.

Rate of change of growth = rate in isolation + rate in competition for resources.

$$dB/dt = 0.7B - 0.02 \ B*T$$
$$dT/dt = 0.5 - 0.01 \ B*T$$
Solve $dB/dt = 0$ and $dT/dt = 0$:
$dB/d = 0 = B(0.7 - 0.02T) = 0$ so either $B = 0$ or $T = 35$.
$dT/dt = 0 = T(0.5 - 0.01B) = 0$ so either $T = 0$ or $B = 5.0$.

The equilibrium values are (0,0) and (50,35) (Figure 4.23).

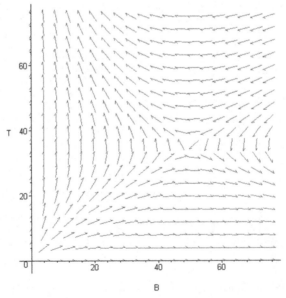

FIGURE 4.23
Plot of Example 4.13 phase portrait in Maple.

>**with(plots):with(DEtools):**

>**eqn1: = diff(B(t),t) = .7*B(t)–.02*B(t)*T(t);**

$$eqn\,1:=\frac{d}{dt}B(t)=0.7B(t)-0.02B(t)T(t)$$

>**eqn2: = diff(T(t),t) = .5*T(t)-.01*B(t)*T(t);**

$$eqn\,2=\frac{d}{dt}T(t)=0.5T(t)-0.01B(t)T(t)$$

>**DEplot([eqn1,eqn2], [B(t),T(t)], t = 0..20, B = 0..75,T = 0..75);**

Again both equilibrium values are not stable. Depending on the starting value one species will dominate over time. Thus, the phase portrait is useful to give us a sense of the possible solutions.

In Maple, we enter the system and initial conditions and then use the dsolve with classical numerical methods. Here is the command sequence to obtain the Euler estimates to our example.

$> ode1 := diff(x(t),t) = 3\cdot x(t) - 2\cdot y(t);$

$$ode1:=\frac{d}{dt}x(t)=3x(t)-2y(t)$$

$> ode2 := diff(y(t),t) = 5\cdot x(t) - 4\cdot y(t);$

$$ode2:=\frac{d}{dt}y(t)=5x(t)-4y(t)$$

$> inits:= x(0) = 3, y(0) = 6;*95$

$> wulersol:= dsolve(\{ode1, ode2, inits\}, numeric,$
$\quad method = classical[\,foreuler\,],$
$\quad output = array([0,.1,.2,.3,.4,.5,.6,.7,.8,.9,1,1.1,1.2,1.3,1.4,1.5,$
$\quad\quad 1.6,1.7,1.8,1.9,2,2.1,2.2]), stepsize = 0.1;$

$eulersol := \left[\begin{bmatrix} t & x(t) & y(t) \end{bmatrix}\right], [$

$$\begin{bmatrix}
0. & 3. & 6. \\
0.1 & 2.70000000000000016 & 5.09999999999999964 \\
0.2 & 2.49000000000000022 & 4.41000000000000014 \\
0.3 & 2.35500000000000042 & 3.89100000000000044 \\
0.4 & 2.28330000000000056 & 3.51210000000000022 \\
0.5 & 2.26587000000000050 & 3.24891000000000042 \\
0.6 & 2.29584900000000046 & 3.08228100000000050 \\
0.7 & 2.36814750000000052 & 2.99729310000000070 \\
0.8 & 2.47913313000000058 & 2.98244961000000064 \\
0.9 & 2.62638314700000075 & 3.02903633100000080 \\
1. & 2.80849082490000113 & 3.13061337210000090 \\
1.1 & 3.02491539795000097 & 3.28261343571000142 \\
1.2 & 3.27586733019300080 & 3.48202576040100098 \\
1.3 & 3.56222237717070068 & 3.72714912133710108 \\
1.4 & 3.88545926605448954 & 4.01740066138760987 \\
1.5 & 4.24761691359331550 & 4.35317002985981106 \\
1.6 & 4.65126798169934742 & 4.73571047471254403 \\
1.7 & 5.09950628126664274 & 5.16706027567719950 \\
1.8 & 5.59594611051119450 & 5.64998930603964044 \\
1.9 & 6.14473208245662317 & 6.18796663887938080 \\
2. & 6.75055837941773440 & 6.78514602455594052 \\
2.1 & 7.41869668833186857 & 7.44636680444243292 \\
2.2 & 8.15503233394294114 & 8.17716842683139332
\end{bmatrix}]$$

The power of Euler's method is two-fold. First it is easy to use and second as a numerical method it can be used to estimate a solution to a system of differential equations that does not have a closed form solution.

Assume we have the following predator–prey system that does not have a closed from analytical solution:

$$\frac{dx}{dt} = 3x - xy$$
$$\frac{dy}{dt} = xy - 2y$$
$$x(0) = 1, y(0) = 2$$
$$t_0 = 1, \Delta t = .1$$

We will obtain an estimate *f* the solution using Euler's method (Figure 4.24).

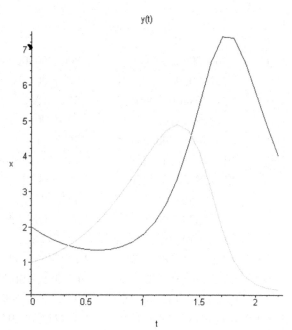

FIGURE 4.24
Plot of system of ODEs.

```
>with(LinearAlgebra):with(DEtools):
>ode1: = diff(x(t),t) = 3*x(t)-x(t)*y(t);
```

$$ode1 := \frac{d}{dt}x(t) = 3x(t) - x(t)y(t)$$

```
>ode2: = diff(y(t),t) = x(t)*y(t)-2*y(t);
```

$$ode2 := \frac{d}{dt}y(t) = x(t)y(t) - 2y(t)$$

```
>inits: = x(0) = 1,y(0) = 2;*102
```

$$inits := x(0) = 1, y(0) = 2$$

```
> eulersol: = dsolve({ode1,ode2,inits},numeric,
        method = classical[foreuler],
output = array([0,.1,.2,.3,.4,.5,.6,.7,.8,.9,1,1.1,1.2,1.3,1.4,1.5,1.6,1.7,1.8,1.9,2,
        2.1,2.2]),stepsize = 0.1);
```

$$eulersol :=
\begin{bmatrix}
& [t, \mathrm{x}(t), \mathrm{y}(t)] & \\
0. & 1. & 2. \\
0.1 & 1.10000000000000008 & 1.80000000000000004 \\
0.2 & 1.23200000000000021 & 1.63800000000000012 \\
0.3 & 1.39979840000000010 & 1.51220160000000026 \\
0.4 & 1.60806018198425638 & 1.42143901801574435 \\
0.5 & 1.86190228798054114 & 1.36572716301158748 \\
0.6 & 2.16618792141785876 & 1.34686678336611476 \\
0.7 & 2.52428764205455636 & 1.36925008248155188 \\
0.8 & 2.93593582846188727 & 1.44103817219427799 \\
0.9 & 3.39363701700781206 & 1.57591009774806334 \\
1. & 3.87692143779073328 & 1.79553476251787370 \\
1.1 & 4.34388314781755014 & 2.13254253132470328 \\
1.2 & 4.72069653578025861 & 2.63238558144231760 \\
1.3 & 4.89423614699907182 & 3.34857781466911942 \\
1.4 & 4.72363393293951717 & 4.31773530989456944 \\
1.5 & 4.10118401049446124 & 5.49372835024256556 \\
1.6 & 3.07846012684130698 & 6.64806176699554552 \\
1.7 & 1.95541885784630654 & 7.36502872064382874 \\
1.8 & 1.10187291030753887 & 7.33219458140772317 \\
1.9 & 0.624520125164112150 & 6.67367032336186838 \\
2. & 0.395092020148348210 & 5.75572040125449292 \\
2.1 & 0.286215706118782388 & 4.83198024107766244 \\
2.2 & 0.233781554289212323 & 4.00388305652733401
\end{bmatrix}$$

```
> with(plots):
> plot1: = odeplot(eulersol,[t,x(t)],0..10, color = green,title = `x(t)`):
> plot2: = odeplot(eulersol,[t,y(t)],0..10,color = blue,title = `y(t)`):
> display(plot1,plot2);
```

We experiment and find that when we plot x(t) versus y(t) we have an approximately a closed loop (Figure 4.25).

```
> odeplot(eulersol,[x(t),y(t)],0..22, color = green,title = `System`);
```

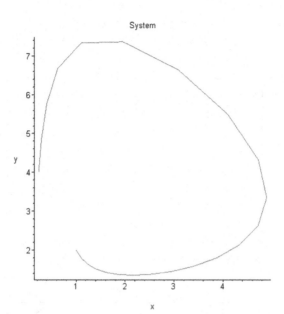

FIGURE 4.25
Plot of system of ODEs.

Another method for numerical estimates is the Runge-Kutta 4 applied to systems. We illustrate with the same predator–prey example (Figure 4.26 and Figure 4.27).

```
> restart;
> with(LinearAlgebra):with(DEtools):
> ode1: = diff(x(t),t) = 3*x(t)-x(t)*y(t);
```

$$ode1 := \frac{d}{dt}x(t) = 3x(t) - x(t)y(t)$$

```
> ode2: = diff(y(t),t) = x(t)*y(t)-2*y(t);
```

$$ode2 := \frac{d}{dt}y(t) = x(t)y(t) - 2y(t)$$

```
> inits: = x(0) = 1,y(0) = 2;
```

$$inits := x(0) = 1, y(0) = 2$$

```
> rk4sol: = dsolve({ode1,ode2,inits},numeric, method = classical[rk4],
```

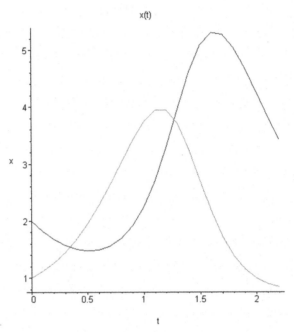

FIGURE 4.26
Plot of predator–prey system of ODEs.

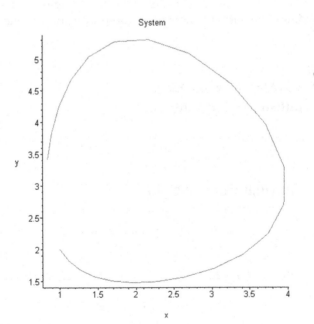

FIGURE 4.27
Plot of System of ODEs.

output = array([0,.1,.2,.3,.4,.5,.6,.7,.8,.9,1,1.1,1.2,1.3,1.4,1.5,1.6,1.7,1.8,1.9,2,2
.1,2.2]),stepsize = 0.1);

$$rk4sol = \begin{bmatrix} & & [t, x(t), y(t)] & \\ 0. & & 1. & & 2. \\ 0.1 & 1.11554071453956705 & 1.81968188493959970 \\ 0.2 & 1.26463746535620780 & 1.67761981669985926 \\ 0.3 & 1.45146389024689194 & 1.57278931221810914 \\ 0.4 & 1.68031037053259168 & 1.50542579227745321 \\ 0.5 & 1.95456986108324960 & 1.47762563817635396 \\ 0.6 & 2.27500034092209802 & 1.49412980191491540 \\ 0.7 & 2.63687047094120208 & 1.56336399876409616 \\ 0.8 & 3.02561346371586382 & 1.69866644464148696 \\ 0.9 & 3.41107833928229720 & 1.91916110241906601 \\ 1. & 3.74212348522216676 & 2.24852435676575446 \\ 1.1 & 3.94706090545572064 & 2.70772001337807611 \\ 1.2 & 3.95001172191016892 & 3.29658238745701324 \\ 1.3 & 3.70934802583802936 & 3.96638771088000476 \\ 1.4 & 3.25874674267819132 & 4.60727591934640213 \\ 1.5 & 2.70448435236999174 & 5.08419319694337712 \\ 1.6 & 2.16641586882149006 & 5.30768472071020270 \\ 1.7 & 1.71962027608752543 & 5.27274016506954890 \\ 1.8 & 1.38441109626295588 & 5.03717315896791806 \\ 1.9 & 1.14891557150367586 & 4.67751425643222784 \\ 2. & 0.99172673132949417 & 4.25984946993614156 \\ 2.1 & 0.893335577155144112 & 3.83076393278872685 \\ 2.2 & 0.839403883501402824 & 3.41909584046305914 \end{bmatrix}$$

```
>with(plots):
>plot1: = odeplot(rk4sol,[t,x(t)],0..10, color = green,title = `x(t)`):
>plot2: = odeplot(rk4sol,[t,y(t)],0..10,color = blue,title = `y(t)`):
>display(plot1,plot2);

>odeplot(rk4sol,[x(t),y(t)],0..22, color = green,title = `System`);
```

4.5.3 R for System of Ordinary Differential Equations

We provide R commands and code. R provides numerical results, plots, and phase portraits. You must install deSolve and phaseR from the CRAN library.

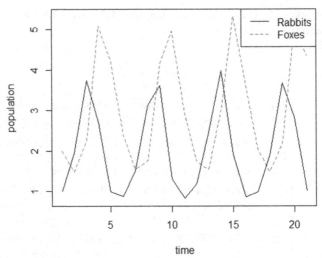

FIGURE 4.28
Screenshot of predator–prey system of ODEs in R.

Let's resolve our predator–prey model in R. We will obtain numerical value *a* and a portrait.

The ODE we will solve is

$$\frac{dx}{dt} = 3 * x(t) - x(t)y(t)$$

$$\frac{dy}{dt} = x(t)y(t) - 2y(t)$$

$$x(0) = 1, \; y(0) = 2$$

The R code commands to obtain plots of x(t) and y(t) are (Figure 4.28):

```
library(deSolve)

LotVmod<- function (Time, State, Pars) {
  with(as.list(c(State, Pars)), {
    dx = x*(alpha - beta*y)
    dy = -y*(gamma - delta*x)
    return(list(c(dx, dy)))
  })
}
```

```
Pars <- c(alpha = 3, beta = 1, gamma = 2, delta = 1)
State <- c(x = 1, y = 2)
Time <- seq(0, 10, by = 0.5)

out <- as.data.frame(ode(func = LotVmod, y = State, parms = Pars,
   times = Time))

matplot(out[,-1], type = "l", xlab = "time", ylab = "population")
legend("topright", c("Rabbits", "Foxes"), lty = c(1,2), col = c(1,2), box.
   lwd = 0)
matplot(out[,-1], type = "l", xlab = "time", ylab = "population")
   legend("topright", c("Prey", "Predator"), lty = c(1,2), col = c(1,2), box.
   lwd = 0)
}
```

R for developing Phase Portraits uses flow field.

```
lotkavolterra<-function(t,y,parameters) {
x<-y[1]
y<-y[2]
lambda<-parameters[1]
epsilon<-parameters[2]
eta<-parameters[3]
delta<-parameters[4]
dy<-numeric(2)
dy[1]<-lambda*x-epsilon*x*y
dy[2]<-eta*x*y-delta*y
list(dy)
}

lotkavolterra.flowField<-flowField(lotkavolterra,
xlim = c(0,5),ylim = c(0,10), parameters = c(3,1,2,1), points = 19, add = FALSE)
grid()
lotkavolterra.nullclines<-
nullclines(lotkavolterra, xlim = c(-1,5), ylim = c(-1,10),
parameters = c(3,1,2,1),points = 500)
y0<-matrix(c(1,2,2,2,3,4),ncol = 2,nrow = 3, byrow = TRUE)
lotkavolterra.trajectory<-trajectory(lotkavolterra, y0 = y0,t.end = 10),
parameters = c(3,1,2,1), colour = rep("black",3))
```

Phase Portrait plot from these commands (Figure 4.29).

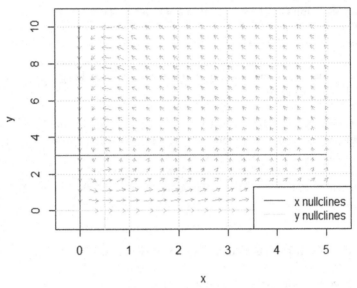

FIGURE 4.29
Screenshot of trajectories in phase portrait in R.

Following the trajectories, we see that the trajectories are elliptical.
We provide some additional R code and commands that allow you to see
the numerical values.

```
# parameters
pars <- c(alpha = 3, beta = 1, delta = 1, gamma = 2)
# initial state
init<- c(x = 1, y = 2)
# times
times <- seq(0, 100, by = 1)
```

Next, we need to define a function that computes the derivatives in the ODE
system at a given point in time.

```
deriv<- function(t, state, pars) {
with(as.list(c(state, pars)), {
   d_x<- alpha * x - beta * x * y
   d_y<- delta * x * y – gamma * y
   return(list(c(x = d_x, y = d_y)))
})
}
```

```
}
lv_results<- ode(init, times, deriv, pars)
```

The lv-results provide the numerical values over time.

```
lv_model<- function(pars, times = seq(0, 50, by = 1)) {
# initial state
state <- c(x = 1, y = 2)
# derivative
deriv<- function(t, state, pars) {
 with(as.list(c(state, pars)), {
  d_x<- alpha * x - beta * x * y
  d_y<- delta * beta * x * y - gamma * y
  return(list(c(x = d_x, y = d_y)))
})
}
# solve
ode(y = state, times = times, func = deriv, parms = pars)
}
lv_results<- lv_model(pars = pars, times = seq(0, 50, by = 0.25))
```

We provide some of the numerical outputs as well.

```
>lv_results
```

	time	x	y
1	0.00	1.0000000	2.000000
2	0.25	1.3530677	1.620569
3	0.50	1.9545794	1.477617
4	0.75	2.8293344	1.621674
5	1.00	3.7421685	2.248472
6	1.25	3.8602053	3.627084
7	1.50	2.7044130	5.084510
8	1.75	1.5381962	5.175234
9	2.00	0.9916079	4.259705
10	2.25	0.8259647	3.224968
11	2.50	0.8694643	2.409064
12	2.75	1.0854469	1.858638

13	3.00	1.5077400	1.551265
14	3.25	2.1955170	1.485853
15	3.50	3.1296117	1.747729
16	3.75	3.9144918	2.587278
17	4.00	3.6073011	4.144647
18	4.25	2.2865674	5.279742
19	4.50	1.3130843	4.951042

4.6 Exercises

1. Find and classify all the rest points. Then sketch a few trajectories to indicate the motion.

   (a) $dx/dt = x$, $dy/dt = y$.
   (b) $dx/dt = -x$, $dy/dt = 2y$.
   (c) $dx/dt = y$, $dy/dt = -2x$.
   (d) $dx/dt = -x + 1$, $dy/dt = -2y$.

2. Given the following system of Linear First-Order ODEs of species cooperation (symbiosis):

$$dx_1/dt = -0.5\, x_1 + x_2$$

$$dx_2/dt = 0.25\, x_1 - 0.5\, x_2$$

and $x_1(0) = 200$ and $x_2(0) = 500$.

   (a) Perform Euler's Method with step-size $h = 0.1$ to obtain graphs of numerical solutions for $x_1(t)$ and $x_2(t)$ versus t and for x_1 versus x_2. You can put both $x_1(t)$ and $x_2(t)$ versus t on one axis if you want.
   (b) From the graphs discuss the long-term behavior of the system (discuss stability).
   (c) Analytically using eigenvalues and eigenvectors solve the system of DEs to determine the population of each species for $t > 0$.
   (d) Determine if there is a steady state solution for this system.
   (e) Obtain real plots of $x_1(t)$ and $x_2(t)$ versus t and for $x_1(t)$ versus $x_2(t)$. Compare to the numerical plots. Briefly discuss.

3. Given a competitive hunter model defined by the system:

$$dx/dt = 15x - x^2 - 2xy = x\,(15 - x - 2y)$$

$$dy/dt = 12y - y^2 - 1.5xy = y\,(12 - y - 1.5x)$$

(a) Perform a graphical analysis of this competitive hunter model in the *x–y* plane.
(b) Identify all equilibrium points and classify their stability.
(c) Find the numerical solutions using Euler's Method with step size h = 0.05. Try it from two separate initial conditions: first, use $x(0) = 5$ and $y(0) = 4$, then use $x(0) = 3$, $y(0) = 9$. Obtain graphs of $x(t)$, $y(t)$ individually (or on the same axis) and then a plot of x versus y using your numerical approximations. Compare to your phase portrait analysis.

4. Since bass and trout both live in the same lake and eat the same food sources, they are competing for survival. The rate of growth for bass (dB/dt) and for trout (dT/dt) are estimated by the following equations:

$$dB/dt = (10 - B - T)B$$

$$dT/dt = (15 - B - 3T)T.$$

Coefficients and values are in thousands.

(a) Obtain a "qualitative" graphical solution of this system. Find all equilibrium points of the system and classify each as unstable, stable, or asymptotically stable.
(b) If the initial conditions are $B(0) = 5$ and $T(0) = 2$, determine the long-term behavior of the system from your graph in part (a). Sketch it out.
(c) Using Euler's Method, h = 0.1 and the same initial conditions as above, obtain estimates for B and T. Using these estimates determine a more accurate graph by plotting B versus T for the solution from $t = 0$ to $t = 7$.
(d) Euler's Method:

$$x_{n+1} = x_n + h f(x_n, y_n) \ \& \ y_{n+1} = y_n + h g(x_n, y_n)$$

Compare the graph in part (c) to the possible solutions found in (a) & (b). Briefly comment.

6. Find the equilibrium values for the SIR model presented.
7. In the Predator–Prey Model, determine the outcomes with the following sets of parameters.

(a) Initial foxes are 200 and initial rabbits are 400.
(b) Initial foxes are 2,000 and initial rabbits are 10,000.
(c) Birth rate of rabbits increases to 0.1.

8. In the SIR model determine the outcome with the following parameters changed.

   (a) Initially 5 are sick and 10 the next week.
   (b) The flu last 1 week.
   (c) The flu last 4 weeks.
   (d) There are 4,000 students in the dorm and 5 are initially infected and 30 more the next week.

4.7 Projects

1. **Diffusion.** Diffusion through a membrane leads to a first-order system of ordinary linear differential equations. For example, consider the situation in which two solutions of substance are separated by a membrane of permeability P. Assume the amount of substance that passes through the membrane at any particular time is proportional to the difference between the concentrations of the two solutions. Therefore, if we let x_1 and x_2 represent the two concentrations, and V_1 and V_2 represent their corresponding volumes, then the system of differential equations is given by:

$$\frac{dx_1}{dt} = \frac{P}{V_1}(x_2 - x_1)$$
$$\frac{dx_2}{dt} = \frac{P}{V_2}(x_1 - x_2)$$

where the initial amounts of x_1 and x_2 are given.

Consider two salt concentrations of equal volume V separated by a membrane of permeability P. Given that $P = V$, determine the amount of salt in each concentration at time t if $x_1(0) = 2$ and $x_2(0) = 10$.

Write out the system of differential equations that models this behavior.

Using the methods described in Chapter 4, solve this system. Clearly indicate your eigenvalues and eigenvectors.

Plot the solutions for x_1 and x_2 on the same axis and label each. Comment about the plots.

Use a numerical method (Euler or Runge-Kutta 4) and iterate a numerical solution to predict $x_1(4)$, use a step size of 0.5. Obtain a plot of your numerical approach. Compare it to the analytical plot.

Comment about what you see in the plots.

Diffusion through a double-walled membrane, where the inner wall has permeability P_1 and the outer wall has permeability P_2 with $0<P_1<P_2$. Suppose the volume of the solution within the inner wall is V_1 and between the two walls is V_2. Let x represent the concentration of the solution within the inner wall and y, the concentration between the two walls. This leads to the following system:

$$\frac{dx}{dt} = \frac{P_1}{V_1}(y - x)$$

$$\frac{dy}{dt} = \frac{1}{V_2}(P_2(C - y) + P_1(x - y))$$

$$x(0) = 2, y(0) = 1, C = 10$$

Also assume the following:

$$P_1 = 3$$

$$P_2 = 8$$

$$V_1 = 2$$

$$V_2 = 10.$$

Set up the system of ODEs with all coefficients.

Use the method of Variation of parameter for systems,

$$X = X_c + \varphi(t)\int \varphi^{-1}(t)F(t)dt$$

to find both X_c and X_p.

Use the initial conditions to find the particular solution, find the coefficients for X_c in the solution $X_c + X_p$.

Plot the solutions for $x(t)$ and $y(t)$ on the same axis. Comment about the solution.

2. **An Electrical Network.** An electrical network containing more than one loop also gives rise to a system of differential equations. For instance, in the electrical network displayed below, there are two resistors and two inductors. At branch point B in the network, the current $i_1(t)$ splits in two directions. Thus,

$$i_1(t) = i_2(t) + i_3(t).$$

Kirchhoff's law applies to each loop in the network. For loop ABEF, we find that

$$E(t) = i_1 R_1 + L_1 di_2/dt.$$

The sum of the voltage drops across the loop ABCDEF is

$$E(t) = i_1 R_1 + L_2 di_3/dt + i_3 R_3.$$

Substituting, we find the following systems for equations:

$$\frac{di_1}{dt} = -\frac{(R_1 + R_2)}{L_1} i_1 + \frac{R_2}{L_2} i_2 + 0$$
$$\frac{di_2}{dt} = (\frac{R_2}{L_2} - \frac{1}{R_2 C}) i_2 - \frac{(R_1 + R_2)}{L_2} i_1 + \frac{E(t)}{L_2}$$
$$i_2(0) = 1, i_1(0) = 0$$

Initially, let $E(t) = 0$ Volts, $L_1 = 1$ henry, $L_2 = 1$ henry, $R_1 = 1$ omhs, $R_2 = 1$ omhs, $C = 3$.

Write out the system of differential equation that models this behavior.

Using the methods described in Chapter 4, solve this system. Clearly indicate your eigenvalues and eigenvectors.

Plot the solutions for x_1 and x_2 on the same axis and label each. Comment about the plots.

Use a numerical method (Euler or Runge-Kutta) and iterate a numerical solution to predict $x_i(4)$, use a step size of 0.5. Obtain a plot of your numerical approach. Compare it to the analytical plot. Comment about the plots.

Now, let $E(t) = 100*\sin(t)$

Set up the system of ODEs with all coefficients.

Use the method of Variation of parameter for systems,

$$X = X_c + \varphi(t) \int \varphi^{-1}(t) F(t) dt$$

to find both X_c and X_p.

Use the initial conditions to find the particular solution, find the coefficients for X_c in the solution $X_c + X_p$.

Plot the solutions for $x(t)$ and $y(t)$ on the same axis. Comment about the solution.

3. **Interacting Species.** Suppose $x(t)$ and $y(t)$ represent respective populations of two species over time, t. One model might be

$$X' = R_1 X, X(0) = X_0$$

$$Y' = R_2 Y, Y(0) = Y_0.$$

where R_1 and R_2 are intrinsic coefficients. Models involving competition between species or Predator–Prey Models most often include interaction terms between the variables. These interactions terms, if included, will preclude any analytical solution attempts so we will simplify these models for this project.

Let's model bass and trout attempting to coexist in a small pond in South Carolina.

$$B' = -0.5B + T + H$$
$$T' = 0.25B - 0.5T + K$$

$$B(0) = 2{,}000, \ T(0) = 5{,}000.$$

Initially, let $H = K = 0$.

Write out the system of differential equation that models this behavior.

Using the methods described in Chapter 8, solve this system. Clearly indicate your eigenvalues and eigenvectors.

Plot the solutions for x_1 and x_2 on the same axis and label each. Comment about the plots.

Use a numerical method (Euler or Runge-Kutta) and iterate a numerical solution to predict $x_i(10)$, use a step size of 0.5. Obtain a plot of your numerical approach. Compare it to the analytical plot, C.

Comment about the plots.

Now, let $H = 1{,}500, \ K = 1{,}000$.

Set up the system of ODEs with all coefficients.

Use the method of Variation of parameter for systems,

$$X = X_c + \varphi(t)\int \varphi^{-1}(t)F(t)dt$$

to find both X_c and X_p.

Use the initial conditions to find the particular solution, find the coefficients for X_c in the solution $X_c + X_p$.

Plot the solutions for $x(t)$ and $y(t)$ on the same axis. Comment about the solution.

Do these species coexist? Briefly explain. If any die out, determine when this happens?

4. **Trapezoidal Method.** The trapezoidal method is a more stable numerical method that is shown in Numerical Analysis textbooks (see Burden and Faires, *Numerical Analysis*, Brooks-Cole Publishers, pp. 344–346). Find the Trapezoidal algorithm and modify it for systems of ODEs. Write a Maple program to obtain the Trapezoidal estimates and compare these to both Euler and Runge-Kutta estimates.

4.8 References and Suggested Further Reading

Abell, M. L. and J. P. Braselton (2000). *Differential Equation with Maple V*. 2nd Edition. Academic Press, San Diego, CA.

Allman, E. and J. Rhodes (2003). *Mathematical Models in Biology: An Introduction*. Cambridge University Press, Cambridge, UK.

Barrow, D., A. Belmonte, A. Boggess, J. Bryant, T. Kiffe, J. Morgan, M. Rahe, K. Smith, and M. Stecher (1998). *Solving Differential Equations with Maple V Release 4*. Brooks-Cole Publisher, Pacific Grove, CA.

Fox, W. (2013). *Mathematical Modeling with Maple, Brooks-Cole*. Cengage Publishers, Boston, MA.

Fox, W. and W. Bauldry (2020). *Problem Solving with Maple*. Taylor &Francis. CRC Press, Boca Raton, FL.

Giordano, F. R. and M. D. Weir. (1991) *Differential Equations: A Modeling Approach*. Addison-Wesley Publishers, Reading, MA.

Giordano, F. R., W.Fox, and S. Horton (2012). *A First Course in Mathematical Modeling*, 5th ed. Brooks-Cole, Boston, MA.

Zill, D. (2015). *A First Course in Differential Equations with Modeling Applications*, 10th ed. Brooks-Cole, Cengage Publishers, Boston, MA.

5

Regression and Advanced Regression Methods and Models

Objectives

1. Understand the concept of correlation and linearity
2. Build and interpret nonlinear regression models
3. Build and interpret logistics regression models
4. Build and interpret Poisson regression models

5.1 Introduction

We are analyzing a large data set of literacy and violence with the desire of predicting in the near future. We plot the data, see Figure 5.1, and draw a linear regression line through the data.

Should we use the linear model to predict? We introduce modeling fitting and the ordinary least squares method, Equation (5.1), that minimizes sum of squared error.

$$\text{Minimize} \, S = \sum_{i=1}^{n} (y_i - f(x_i))^2 \tag{5.1}$$

In the ordinary least squares method, linear regression, we are more concerned with the mathematical modeling and the use of the model for *explaining* or *predicting* the phenomena being studied or analyzed. We provide only a few diagnostic measures at this point such as: percent relative error and residual plots. Percent relative error is calculated by Equation (5.2).

$$\%RE = \frac{100(y_i - f(x_i))}{y_i} \tag{5.2}$$

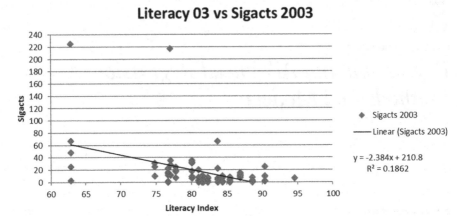

FIGURE 5.1
Plot of data literacy versus violent acts in 2003.

We usually provide a rule of thumb when looking at the magnitude of the percent relative errors and we want most of them to be less than 20% and those near where we need to predict less than 10%.

Secondly, we provide the visualization of the plot of residuals versus the model. We examine the plot for trends or patterns, such as shown in Figure 5.2. If a pattern is seen, we deem the model not adequate even though we might have to use it. If there is no visual pattern, then we may conclude the model is adequate.

Students have several misconceptions about *correlation*, many of which are supported by a poor definition. A good definition states that ***correlation is a measure of the <u>linear relationship</u> between variables***. The key term is *linear*. Some definitions for correlation actually state it is a measure of the relationship between two variables and they do not even mention linear. This is true in Excel. The Excel definition found in the Excel help menu is:

> Returns the correlation coefficient of the array1 and array2 cell ranges. Use the correlation coefficient to determine the relationship between two properties. For example, you can examine the relationship between a location's average temperature and the use of air conditioners.

It is no wonder students have misconceptions. Students, like the latter definition, often lose the term linear and state or think that correlation does measure the relationship between variables. As we will show, that is false thinking.

Some of the misconceptions that students have expressed are listed below:

- Correlation does everything.
- Correlation measures relationship.
- Correlation or relationships implies causation.

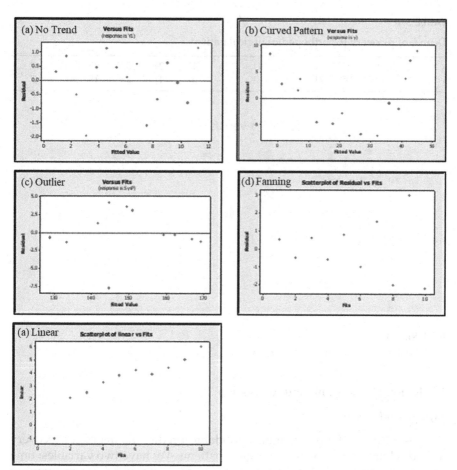

FIGURE 5.2

Patterns for residuals: (a) no pattern; (b) curved pattern; (c) outliers; (d) fanning pattern; (e) linear trend (adapted from Afifi and Azen,1979, p. 143).

- If it looks linear (visual) and the correlation value is large, then the model found will be useful.
- If the computer lets me do it, it must be right.
- Model diagnostics are not useful if the correlation value is large.
- I have to use the *regression* package taught from class.

We provide examples to illustrate some of the misconceptions surrounding correlations and the possible corrections that we illustrate later in the chapter. Additionally when we address the advanced regression techniques we will present the technique and the mathematics and then provide examples with appropriate technology.

TABLE 5.1

Degree of recovery for orthopedic surgical patients

t	2	4	6	7	9	11	13	15	21	33	45	51	62
y	87	80	74	71	66	61	56	52	41	25	15	12	7

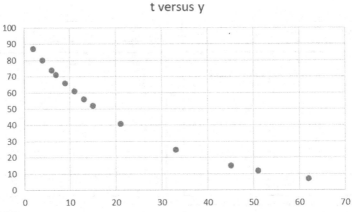

FIGURE 5.3
Scatterplot of the data with negative trend.

Modeling, Correlation, and Regression

Exponential Decay

We desire to build a mathematical model to predict the degree of recovery after discharge for orthopedic surgical patients. We have two variables: time in days in the hospital, t, and a medical prognostic index for recovery, y, where a large value of this index indicates a good prognosis.

We strongly believe the first step in any analysis effort is to plot the data and comment on the trends that they observe. We provide the scatterplot, Figure 5.3, showing the negative trend.

We need to obtain the correlation coefficient, ρ. Using the Excel command, $= correl(array\ 1, array\ 2)$, to obtain the correlation coefficient of -0.94105. We present two rules of thumb for correlation from the literature. First, from Devore (1995), for math, science, and engineering data we have the following:

$0.8 < |\rho| \leq 1.0$ Strong linear relationship
$0.5 < |\rho| \leq 0.8$ Moderate linear relationship
$|\rho| \leq 0.5$ Weak linear relationship

According to Johnson (2012) for non-math, non-science and non-engineering data we find a more liberal interpretation:

$0.5 < |\rho| \leq 1.0$ Strong linear relationship

$.5 < |\rho| \leq 0.3$ Moderate linear relationship

$0.1 < |\rho| \leq 0.3$ Weak linear relationship

$|\rho| \leq 0.1$ No linear relationship.

Further, we emphasize the interpretation of $|\rho| \approx 0$. This can be interpreted as either no linear relationship or the existence of a nonlinear relationship. Most students fail to pick up on the importance of the nonlinear relationship aspect of the interpretation.

Using either rule of thumb the correlation coefficient, $|\rho| = 0.96675$, indicates a strong linear relationship. We obtain this value, look at Figure 5.3, and think we will have an excellent regression model.

We obtain the following model, $y = 78.3222 - 1.3304*t$. The sum of squared error is 583.70. The correlation is, as we stated, -0.96675, and R^2, the coefficient of determination, is 0.9346. These are all indicators of a "good" model. Figure 5.4 gives this output.

Next, we examine both the percent relative error and the residual plot. The percent relative errors are:

Although some are small others are quite large with 4 of the 13 over 20% in error. The last one is over 159%. How much confidence would you have in predicting? The residual plot, Figure 5.5, clearly shows a curved pattern.

Furthermore, assume we need to predict the index when time was 100 days. Using our regression model, we would predict the index as -54.72. A negative value is clearly unacceptable and makes no common sense. He model does not pass the commonsense test. So, with a strong correlation of $-.94105$

SUMMARY OUTPUT

Regression Statistics	
Multiple R	0.966752
R Square	0.934609
Adjusted R Squa	0.928664
Standard Error	7.284503
Observations	13

ANOVA

	df	SS	MS	F	Significance F
Regression	1	8342.60388	8342.604	157.2178	7.39817E-08
Residual	11	583.7038125	53.06398		
Total	12	8926.307692			

	Coefficients	Standard Error	t Stat	P-value	Lower 95%	Upper 95%	Lower 95.0%	Upper 95.0%
Intercept	78.32227	3.044255338	25.72789	3.53E-11	71.6219071	85.02263	71.62191	85.02263
t	-1.33043	0.106106153	-12.5387	7.4E-08	-1.563966323	-1.09689	-1.56397	-1.09689

FIGURE 5.4
Screenshot of regression output in Excel.

TABLE 5.2

Residuals and percent error of regression model

t	y	Predicted y	Residuals	%Error
2	87	75.66	11.34	13.03
4	80	73.00	7.00	8.75
6	74	70.34	3.66	4.95
7	71	69.01	1.99	2.80
9	66	66.35	-0.35	-0.53
11	61	63.69	-2.69	-4.41
13	56	61.03	-5.03	-8.98
15	52	58.37	-6.37	-12.24
21	41	50.38	-9.38	-22.89
33	25	34.42	-9.42	-37.67
45	15	18.45	-3.45	-23.02
51	12	10.47	1.53	12.75
62	7	-4.16	11.16	159.49

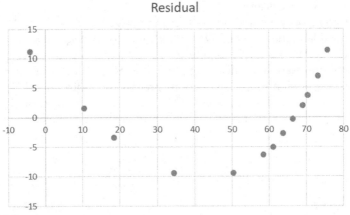

Residual

FIGURE 5.5

Residual plot showing a curved pattern.

what went wrong? The residual plot diagnostic shows a curved pattern. In regression analysis this suggests a nonlinear term is missing from the model.

If we try a parabolic model, $y = b_0 + b_1 x + b_2 x^2$, we get similar results to the linear model. We run the regression of y versus x and x^2 with an intercept. The model is

$$y = 89.73806 - 2.77{*}x + 0.023847{*}x^2.$$

The correlation is now 0.998371 and the R^2 is 0.9967. The sum of squared error is now 29.04. The output is seen in Figure 5.7. The residual plot appears to have removed only some of the curved pattern, see Figure 5.6.

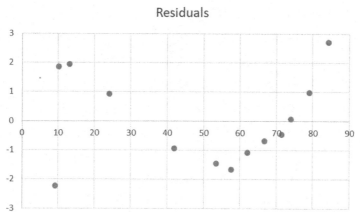

FIGURE 5.6
Residual plot of quadratic model.

SUMMARY OUTPUT						
Regression Statistics						
Multiple R	0.99837148					
R Square	0.99674561					
Adjusted R Square	0.99609473					
Standard Error	1.70439674					
Observations	13					
ANOVA						
	df	*SS*	*MS*	*F*	*Significance F*	
Regression	2	8897.25801	4448.629	1531.386	3.65046E-13	
Residual	10	29.04968245	2.904968			
Total	12	8926.307692				
	Coefficients	*Standard Error*	*t Stat*	*P-value*	*Lower 95%*	*Upper 95%*
Intercept	89.7380574	1.090820781	82.26655	1.72E-15	87.30755727	92.16855759
t	-2.77700286	0.107592274	-25.8104	1.75E-10	-3.016733388	-2.537272335
t2	0.02384722	0.001725827	13.81785	7.67E-08	0.020001833	0.027692597

FIGURE 5.7
Excel screenshot regression output for the parabolic model.

If we use the model to predict at $x = 100$ days we find $y = 50.5091$. The answer is now positive but does not pass the commonsense test. The quadratic is now curving upwards toward positive infinity, see Figure 5.8a.

Thus our previous models have not worked well. In this chapter we desire to introduce some additional regression models: nonlinear regression of power models, logistics regression, and Poisson regression.

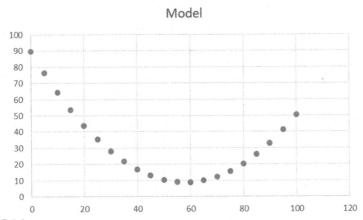

FIGURE 5.8a
Quadratic model, $y = 89.73806 - 2.77*x + 0.0847*x^2$.

Counter	Enter X and Y			
	X	y	Model	Error
1	2	87	87.09609	0.096086481
2	4	80	80.36299	0.362989696
3	6	74	74.15041	0.150405304
4	7	71	71.22661	0.226606138
5	9	66	65.72032	-0.279676086
6	11	61	60.63971	-0.360286703
7	13	56	55.95187	-0.048132909
8	15	52	51.62642	-0.373577962
9	21	41	40.555	-0.44500432
10	33	25	25.02586	0.025860749
11	45	15	15.44307	0.443071703
12	51	12	12.13126	0.131263053
13	62	7	7.79331	0.793310186

FIGURE 5.8b
Excel screenshot hospitalization data.

5.2 Nonlinear Regression

Let's try a nonlinear regression model. Here we limit ourselves to only certain nonlinear regression forms such as ax^b, ae^{bx}, and $a\sin(bx + c)$. We return to our previous example.

Example 5.1 Exponential Decay

We desire the model, $y = ae^{bx}$. The model we need is to minimize the function, $\sum_{i=1}^{n}(y_i - a(\exp(bx_i)))^2$. To do this we take the partial derivatives with respect to the unknown parameters, a and b.

$$f = \text{Minimize} \ \sum_{i=1}^{n}\left(y(i) - \left(\sum_{i=1}^{n} a e^{bx(i)} \right) \right)$$

$$\frac{\partial f}{\partial a} = -(n+1)\left(\sum_{i=1}^{n} e^{b\,x(i)} \right) + \sum_{i=1}^{n} e^{bx(i)} = 0$$

$$\frac{\partial f}{\partial b} = (n+1)\left(\sum_{i=1}^{n} a\,x(i)e^{bx(i)} \right) + \sum_{i=1}^{n} a\,x(i)e^{b\,x(i)} = 0$$

There exist no closed form solutions to these partial derivatives set equal to zero. Therefore, numerical methods must be used to solve for the parameters a and b. Numerical methods will be covered in later chapters. In this chapter we will present solution using technology.

5.3 Technology Examples for Regression

5.3.1 Excel for Regression

Excel does not have a nonlinear regression package, so we use the solver to obtain a solution. A clear understanding of Method 1 from Chapter 3 will be quite helpful and we suggest reviewing it and the use of the Solver to minimize sum of squared error.

Here we use the Solver in Excel to minimize the function, $\sum_{i=1}^{n}(y_i - a(\exp(bx_i)))^2$. We will use the Solver to do least squares; *we develop the necessary equations.*

Since we have previously used the data (Figure 5.8b), we move directly to the nonlinear model. We initial our two parameters as $a = 1$ and $b = 1$. We then generate a column labeled model that computes $a*exp(b*x)$ using our x data and our initial parameters. Next we, compute the difference, errors, between y and the model's prediction at each x and label as error ($error = y_i - f(x_i)$). Next we square the errors and then we sum the column of squared error. We highlight the sum and open the Solver. We want to minimize the sum of squared errors while changing our initial values. We choose GRG and obtain our output (Figure 5.9).

We click Solve.

Counter	Enter X and Y		Model	Error	Error^2
	x	y			
1	2.0000	87.0000	0.0000	-87.0000	7569.0000
2	4.0000	80.0000	0.0000	-80.0000	6400.0000
3	6.0000	74.0000	0.0000	-74.0000	5476.0000
4	7.0000	71.0000	0.0000	-71.0000	5041.0000
5	9.0000	66.0000	0.0000	-66.0000	4356.0000
6	11.0000	61.0000	0.0000	-61.0000	3721.0000
7	13.0000	56.0000	0.0000	-56.0000	3136.0000
8	15.0000	52.0000	0.0000	-52.0000	2704.0000
9	21.0000	41.0000	0.0000	-41.0000	1681.0000
10	33.0000	25.0000	0.0000	-25.0000	625.0000
11	45.0000	15.0000	0.0000	-15.0000	225.0000
12	51.0000	12.0000	0.0000	-12.0000	144.0000
13	62.0000	7.0000	0.0000	-7.0000	49.0000
					(41127.0000)

This value is SSE. WE enter into the Solver to minimize this value. Our decision cells are in M4 & M5.

FIGURE 5.9
Excel screenshot Solver.

Counter	Enter X and Y		Model	Error	Error^2				
	x	y							
1	2.0000	87.0000	87.0966	0.0966	0.0093				
2	4.0000	80.0000	80.3633	0.3633	0.1320			b0	94.39395
3	6.0000	74.0000	74.1506	0.1506	0.0227			b1	-0.04023
4	7.0000	71.0000	71.2267	0.2267	0.0514				
5	9.0000	66.0000	65.7203	-0.2797	0.0782				
6	11.0000	61.0000	60.6397	-0.3603	0.1299				
7	13.0000	56.0000	55.9517	-0.0483	0.0023				
8	15.0000	52.0000	51.6262	-0.3738	0.1397				
9	21.0000	41.0000	40.5547	-0.4453	0.1983				
10	33.0000	25.0000	25.0254	0.0254	0.0006				
11	45.0000	15.0000	15.4427	0.4427	0.1960				
12	51.0000	12.0000	12.1309	0.1309	0.0171				
13	62.0000	7.0000	7.7930	0.7930	0.6289				
					1.6065				

FIGURE 5.10
Excel screenshot of model parameters.

We obtain the model, $y = 94.39395\, e^{-0.04023}$ (Figure 5.9). As a nonlinear model, correlation has no meaning nor does R^2. We computed the sum of squared error as that was our objective function. Our *SSE* is now 1.6065. This value is substantially smaller than the 583.7 obtained in the linear model. The model and residual plot are shown in Figures 5.11 and 5.12.

The percent relative errors are also much improved with only four being greater than 20% and none larger than the absolute value of 35.838%.

If we use this model to predict at $x = 100$ we find by substitution that $y = 1.689$. This answer passes the common sense test.

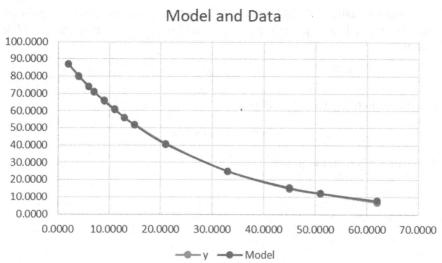

FIGURE 5.11

Plot of the model, $y = 94.390395*\exp(-0.04023*x)$.

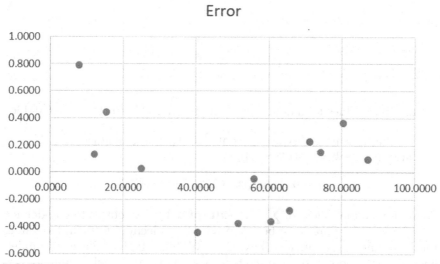

FIGURE 5.12

Residual plot from nonlinear model with no pattern.

5.3.2 Maple for Regression

Maple has nonlinear regression commands. We point out the need to a good starting point or Maple might not obtain the solution desired. We recommend for this problem doing a *ln–ln* transformation and getting approximate parameter values for a and b before performing the nonlinear regression model.

We suggest using a *ln–ln* transformation of the data and then a regression model so we will get a good approximation of the parameters to use as initial values.

$$y = ae^{bx}.$$

Taking the ln of each side yields

$$ln\,(y) = ln(a) + b{*}x.$$

So we use this to get a model,

Our initial values for a nonlinear fit model should be 96.445 and −0.04136.

>c1: = NonlinearFit(a*exp(b*x), Xdata, Ydata, x, initialvalues
= [a = 96.4458, b = −0.41389e−1]);

$$c1: = 94.3934403029403\; e^{(-0.0402288170687175\, x)}$$

We minimize the function, $\sum_{i=1}^{n}(y_i - a(\exp(bx_i)))^2$. We obtain the model, $y = 94.39344\, e^{-0.040239t}$. Care must be taken with the selection of the initial values for the unknown parameters (Fox, 2011). We find Maple yields good models based upon "good" input parameters. As a nonlinear model, correlation has no meaning nor does R^2. We computed the sum of squared error as that was our objective function. Our *SSE* was 2.32. This value is substantially smaller than the 583.7 obtained in the linear model. The model appears reasonable and he residual plot showing no pattern are seen in Figures 5.13 and 5.14. The new R^2 is approximately 0.9997.

$$rsq: = 1{-}SSE/SST;$$

$$rsq: = 0.9997395685.$$

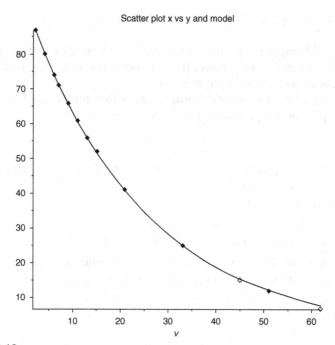

FIGURE 5.13
Exponential model with data.

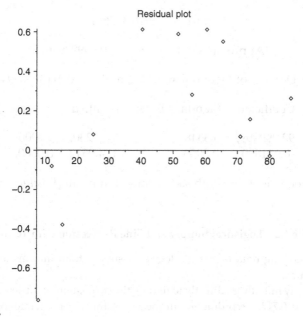

FIGURE 5.14
Residual plot.

>*pointplot([predict, residual], title = 'Residual Plot');*
>

If we use this model to predict at $x = 100$ we will find, by substitution, that $y = 1.6958$. This new result passes the commonsense test. This model would be our recommended model for this data.

Further we created a statistical output procedure for the nonlinear regression in Maple and we produced the following output.

>

$STAToutput := \mathbf{proc}(n, a1, b1, SE1, SE2, t1, t2, PV1, PV2, SSE, SST, MSE, rsq)$
  $print("SSE=", SSE);$
  $print("SST=", SST);$
  $print("Approximate R\text{-}square=", rsq)$
  $* \, print("Degrees of Freedom and MSE are", n+(-2), MSE)$
  $* \, print("Coefficient Standard Error T\text{-}Statistic P\text{-}Value\backslash n\backslash n");$
  $prinff("\%12.6f\%10.4f,\%10.4f,\%10.4f\backslash n", a1, SE1, t1, PV1);$
  $prinff("\%12.6f\%10.4f,\%10.4f,\%10.4f\backslash n", b1, SE2, t2, PV2)$
  $\mathbf{end \; proc}$

>*STAToutput(n, a1,b1,SE1,SE2,t1,t2,PV1,PV2,SSE,SST,MSE,rsq);*

"SSE = ",, 2.324692033

"SST = ", 8926.307691

"Approximate R-square = ", 0.9997395685

"Degrees of Freedom and MSE are", 11, 0.2113356394

Coefficient	Standard Error	T-Statistic	P-Value
93.999999,	0.3348,	280.7230,	0.0000
-0.041890,	0.0003,	-124.0542,	0.0000.

We see the coefficients are both statistically significant (*P-values* less than 0.5).

Example 5.2   Logistical Supply and Sine Regression with Technology

The following data represents logistical supply train information over 20 months.

We may initially go directly to finding the correlation coefficient. In this case, it is 0.712. According to our measures, this is a moderate to strong value. So, is the model to use linear? We plot the data to observe the

TABLE 5.3

Logistics supply sage

Month	Usage(tons)	Month	Usage (tons)
1	20	11	19
2	15	12	25
3	10	13	32
4	18	14	26
5	28	15	21
6	18	16	29
7	13	17	35
8	21	18	28
9	28	19	22
10	22	20	32

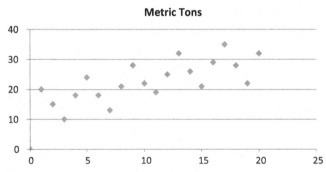

FIGURE 5.15

Scatterplot of usage in metric tons over 20 months.

trends, see Figure 5.15. We see what appears to be an oscillating pattern with a slight linear trend. Since this oscillating pattern is repeating five times within the data, it is more likely for the trend to be oscillating rather than linear.

We connect the dots to help see if there is a clear pattern, see Figure 5.16.

Using the Excel Solver using the same column as before accept we have initial parameter a,b,c,d,e and our model is $a*\sin(b*time + c) + d*time + e$. We obtain the model,

Usage = 6.316*sin(1.574*months + 0.0835) + 0.876*months + 13.6929.

The sum of squared error is only 11.58 that is quite a bit smaller than with the linear model. Our model is overlaid with the data in Figure 5.17, visually representing a good model.

Clearly the sine regression does a much better job in predicting the trends than just using a simple linear regression.

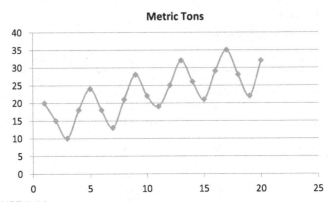

FIGURE 5.16

Scatterplot of usage in metric tons over 20 months with dots connected.

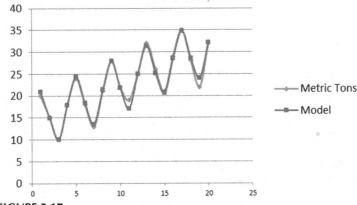

FIGURE 5.17

Overlay of model and data.

For illustrative purposes a linear model is found, Usage = 14.58 + 0.792*months. The sum of squared error is 406.16 with an R^2 of about 0.506944. Let's assume the fit for the data needs to represent oscillations with a slight line trend. Perhaps develop the model for the following model:

$$Usage = a*\sin(b*time + c) + d*time + e.$$

We find we have a good fit only if we do a good job in approximating the parameters. Using Maple initially with Maple's default initial values we obtain a very poor model. We show this to emphasize the importance of good initial values.

```
> msupply := NonlinearFit(a + b · x5 + c · sin(x5 · d + e), months,
    demand, x5);
```

$$msupply := 15.2811288067385 + 0.755027150679343\, x5$$
$$-0.619677627271433\sin\left(0.753071218688569\, x5 + 5.85956792055896\right)$$

>*p1: = plot(msupply, x5 = 0.25):*

>*p2: = pointplot([months, demand], title = 'Supply'):*

>*display ({p1,p2});*

$> p1 := plot\left(\boldsymbol{msupply}, x5 = 0..25\right):$

$> p2 := pointplot\left(\left[\boldsymbol{months, demand}\right], \boldsymbol{title} = \boldsymbol{'Supply'}\right) :> display\left(\left\{\boldsymbol{p}1, \boldsymbol{p}2\right\}\right);$

$> p1 := plot\left(msupply, x5 = 0..25\right):$
$> p2 := pointplot\left(\left[months, demand\right], title = 'Supply'\right):$
$> dispaly\left(\left\{p1, p2\right\}\right);$

The fit is not very good a shown in Figure 5.18. The trend is not captured.

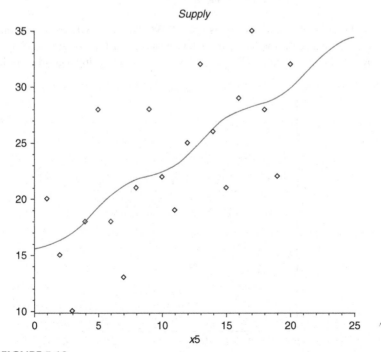

FIGURE 5.18
Not a good fit.

So, we go back to the data and estimate the parameters of intercept, slope, amplitude, phase shift, from trigonometry. We use those in the model and obtain a much better fit.

> $msupply := NonlinearFit(a + b \cdot x5 + c \cdot \sin(x5 \cdot d + e), months,$
> $demand, x5, initialvalues = [a = 16, b = .9, c = 6, d = 1.6, e = 1]);$

$msupply := 14.1865330075917 + 0.847951259234693x5$
        $+ 6.68918275750584 \sin(1.57350123298938x5 + 0.0826250652440048)$

> $p1 := plot(msupply, x5 = 0 .. 25):$

> $p2 := pointplot([months, demand], title = 'Supply'):$

> **$display(\{p1, p2\}); dispaly(\{p1, p2\});$**

This model, shown in Figure 5.19, does capture the trend.

$$msupply1 := 14.1865330075917 + 0.847951259234693x5$$
$$+ 6.68918275750584 \sin(1.57350123298938x5$$
$$+ 0.0826250652440048)$$

The sum of squared error is only 11.58. The new SSE is quite a bit smaller than with the linear model. Our nonlinear (oscillating) model is overlaid with the data in Figure 5.19, visually representing a good model.

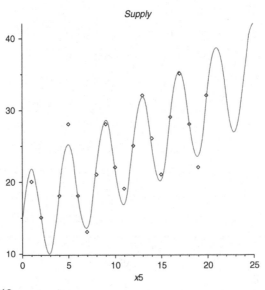

FIGURE 5.19
Sine regression model and original data as a better fit.

Clearly the sine regression does a much better job in predicting the trends than just using a simple linear regression.

5.3.3 R for Regression

We provide the R code for all regression analysis here.

```
###########################
##
## Setting up workspace
##
###########################


## Clear previous workspace, if any
rm(list = ls())

## Set working directory
os_detect<- Sys.info()['sysname']
if (os_detect == 'Darwin'){
setwd('/Users/localadmin/Dropbox/Research/StatsChapter')
}

## Load packages for analysis
pacman::p_load(
data.table, tidyverse, ggplot2, stargazer, easynls, pscl, pander
)


###########################
##
## Reading in the data sets used in this chapter
##
###########################


## Read in spring data
spring_data<- read_csv('./Data/01_correlation.csv')

## Parsed with column specification:
## cols(
```

```
## x = col_integer(),
## y = col_double()
##)

sigacts_data<- read_csv('./Data/06_poisson_sigacts.csv')

## Parsed with column specification:
## cols(
## sigacts_2008 = col_integer(),
## ggi_2008 = col_double(),
## literacy = col_double(),
## poverty = col_double()
##)

recovery_data<- read_csv('./Data/02_exponential_decay.csv')

## Parsed with column specification:
## cols(
## T = col_integer(),
## Y = col_integer()
##)

shipping_data<- read_csv('./Data/03_sine_regression_shipping.csv')

## Parsed with column specification:
## cols(
## Month = col_integer(),
## UsageTons = col_integer()
##)

afghan_data<- read_csv('./Data/04_sine_regression_casualties.csv')

## Parsed with column specification:
## cols(
## Year = col_integer(),
## Month = col_integer(),
```

```
## Casualties = col_integer()
##)


war_data<- read_csv('./Data/05_bin_logit_conflict.csv')

## Parsed with column specification:
## cols(
## side_a = col_integer(),
## cd_pct = col_double()
##)


sigacts_data<- read_csv('./Data/06_poisson_sigacts.csv')

## Parsed with column specification:
## cols(
## sigacts_2008 = col_integer(),
## ggi_2008 = col_double(),
## literacy = col_double(),
## poverty = col_double()
##)


alliance_data<- read_csv('./Data/07_bin_logit_alliance.csv')

## Parsed with column specification:
## cols(
## statea = col_character(),
## stateb = col_character(),
## alliance_present = col_integer(),
## igo_overlap = col_integer()
##)


## Format and subset casualties data
afghan_data<- mutate(
  afghan_data
  , Date = as.Date(paste0(Year, '-', Month, '-', '01 '), format = '%Y-%m-%d')
) %>% filter(
```

```
Date > = as.Date('2006-01-01 ')
 , Date < = as.Date('2008-12-01')
) %>% mutate(
 DateIndex = 1:36
)
```

```
## Print data as a tibble
print(spring_data)
```

```
## # A tibble: 11× 2
##       x        y
##     <int>   <dbl>
## 1     50    0.1000
## 2    100    0.1875
## 3    150    0.2750
## 4    200    0.3250
## 5    250    0.4375
## 6    300    0.4875
## 7    350    0.5675
## 8    400    0.6500
## 9    450    0.7250
## 10   500    0.8000
## 11   550    0.8750
```

Correlation in R

Using the *cor()* command in *R* on the data table:

```
## Calculate and print correlation matrix
print(cor(spring_data))
```

```
##         x           y
## x 1.0000000   0.9992718
## y 0.9992718   1.0000000
```

Plotting in R

```
## Generate a plot visualizing the data
spring_cor_plot<- ggplot(
aes(x = x, y = y)
, data = spring_data) +
geom_point() +
annotate(
'text'
, x = 100
, y = 0.75
, label = 'Correlation coefficient:\n 0.999272'
, hjust = 0) +
ggtitle('Spring data scatterplot ') +
theme_bw()

## Print the plot to console
plot(spring_cor_plot)
```

Fitting an Ordinary Least-Squares (OLS) Model with Form $y|x + \epsilon$ to the Spring Data in R

```
## Fit OLS model to the data
spring_model<- lm(
y ~ x
, data = spring_data
)
```

Correlation Matrix in R

```
## Calculate and print correlation matrix
print(cor(recovery_data))

##          T          Y
## T 1.0000000  -0.9410528
## Y -0.9410528  1.0000000
```

Quadratic Regression of Hospital Recovery Data

```
## Generate model
recovery_model2 <- lm(Y ~ T + I(T^2), data = recovery_data)
```

```
## # A tibble: 15 × 6
##    T   Y index predicted residuals pct_relative_error
##    <int><int> <int> TextInd   <dbl>      <dbl>         <dbl>
## 1     2    54     1    52.460836   1.5391644     2.8503045
## 2     5    50     2    47.640993   2.3590072     4.7180144
## 3     7    45     3    44.575834   0.4241663     0.9425917
## 4    10    37     4    40.200199  -3.2001992    -8.6491871
## 5    14    35     5    34.780614   0.2193857     0.6268164
## 6    19    25     6    28.672445  -3.6724455   -14.6897820
## 7    26    20     7    21.364792  -1.3647924    -6.8239618
## 8    31    16     8    17.033457  -1.0334567    -6.4591042
## 9    34    18     9    14.790022   3.2099781    17.8332119
## 10   38    13    10    12.213370   0.7866302     6.0510012
## 11   45     8    11     8.844363  -0.8443634   -10.5545422
## 12   52    11    12     6.926437   4.0735627    37.0323886
## 13   53     8    13     6.770903   1.2290967    15.3637082
## 14   60     4    14     6.511355  -2.5113548   -62.7838691
## 15   65     6    15     7.214379  -1.2143795   -20.2396576
```

Prediction

```
## Create a set of hypothetical patient observations with days in the hos-
   pital from 1 to 120
patient_days = tibble(T = 1:120)
```

```
## Feed the new data to the model to generate predicted recovery
   index values
predicted_values = predict(
  recovery_model2
  , newdata = patient_days
  )
```

Nonlinear Regression:
Exponential Decay Modeling of Hospital Recovery Data

```
## Fit NLS model to the data
## Generate model
recovery_model3 <- nls(
  Y ~ a * (exp(b * T))
  , data = recovery_data
  , start = c(
  a = 1
  , b = 0.05
  )
  , trace = T
)
```

Sinusoidal Regression

The functional form for the sinusoidal model we use here can be written as:

$$Usage = a * sin(b * time + c) + d * time + e$$

This function can be expanded out trigonometrically as:

$$Usage = a * time + b * sin(c * time) + d * cos(c(time)) + e$$

```
## Generate model
shipping_model2 <- nls(
UsageTons ~ a * Month + b*sin(c*Month) + d*cos(c*Month) + e
, data = shipping_data
, start = c(
  a = 5
  , b = 10
  , c = 1
  , d = 1
  , e = 10
  )
  , trace = T
)
```

Sinusoidal Regression of Afghanistan Casualties

Visualizing data on casualties in Afghanistan between 2006 and 2008 shows an increasing trend overall, and significant seasonal oscillation. Once again, we want to fit a nonlinear model that accounts for the oscillation present in the data. We use the same sinusoidal functional form

$$Casualties = a * sin(b * time + c) + d * time + e$$

which as before can be expressed as

$$Casualties = a * time + b * sin(c * time) + d * cos(c * time) + e$$

The logistic model in R is treated as one case of a broader range of generalized linear models (GLM), and can be accessed via the conveniently named *glm()* function. Note that because *glm()* implements a wide range of generalized linear models based on the inputs provided, it is necessary for the user to specify both the family of model (binomial) and the link function (logit).

```
## Generate model
war_model<- glm(
  side_a ~ cd_pct
  , data = war_data
  , family = binomial(link = 'logit')
)
```

Poisson Regression

Visualizing count data in a histogram is a useful way of assessing how the data are distributed.

Histogram Plot

　　## `stat_bin()` using `bins = 30`. Pick better value with `binwidth`.

Poisson regression in R is also treated as a special case of GLMs, similar to the logistic regression covered in the previous section. As such, it can be implemented using the same *glm()* function, but now specifying the model family as 'Poisson', which tells R to implement a Poisson model. The model we use here can be specified as

$$Y = e^{\beta_0 + \beta_1 GGI + \beta_2 Literacy + \beta_3 Poverty}$$

```
## Generate model
sigacts_model<- glm(
sigacts_2008 ~ ggi_2008 + literacy + poverty
 , data = sigacts_data
 , family = poisson
 )
```

5.4 Logistics Regression Models

Introduction

In data analysis, **logistic regression** (sometimes called the **logistic model** or logit **model**) is a type of regression used for predicting the outcome of a binary dependent variable (a variable which can take only two possible outcomes, e.g. "yes" vs. "no" or "success" vs. "failure") based on **one or more** predictor variables. Logistic regression attempts to model the probability of a "yes/success" outcome using a linear function of the predictors. Specifically, the log-odds of success (the logit of the probability) is fit to the predictors using linear regression. Logistic regression is one type of discrete choice model, which in general predict categorical dependent variables – either binary or multi-way.

Like other forms of regression, logistic regression makes use of one or more predictor variables that may be either continuous or categorical. Also, like other linear regression models, the expected value (average value) of the response variable is fit to the predictors – the expected value of a Bernoulli distribution is simply the probability of success. Unlike ordinary linear regression, however, logistic regression is used for predicting binary outcomes (Bernoulli trials) rather than continuous outcomes, and models a transformation of the expected value as a linear function of the predictors, rather than the expected value itself.

For example, logistic regression might be used to predict whether a patient has a given disease (e.g. diabetes), based on observed characteristics of the patient (age, gender, body mass index, results of various blood tests, etc.). Another example might be to predict whether a voter will vote Democratic or Republican, based on age, income, gender, race, state of residence, votes in previous elections, etc. Logistic regression is used extensively in numerous disciplines: the medical and social sciences fields, natural language processing, marketing applications such as prediction of a customer's propensity to purchase a product or cease a subscription, etc.

The model for just one predictor is shown in Equation (5.3):

$$Y_i = \frac{B_0}{1 + B_1 e^{B_2 X_i}} + \varepsilon_i \qquad (5.3)$$

where the error term are independent and identically distributed (*iid*) as normal random variables with constant variance.

For more than one predictor we use the model shown as Equation (5.4).

$$Y_i = \frac{e^{(b_0 + \Sigma b_i x_i)}}{1 + e^{(b_0 + \Sigma b_i x_i)}}. \qquad (5.4)$$

What is Logistic Regression?

Logistic Regression calculates the probability of the event occurring, such as the purchase of a product. In general, the object being predicted in a regression equation is represented by the dependent variable or output variable and is usually labeled as the Y variable in the Regression equation. In the case of Logistic Regression, this "Y" is binary. In other words, the output or dependent variable can only take the values of 1 or 0. The predicted event either occurs or it doesn't occur – your prospect either will buy or won't buy. Occasionally this type of output variable also referred to as a Dummy Dependent Variable.

An Example of Logistic Regression

To simplify the analysis in Excel, we create a maximum *ln* likelihood function for the logit expression:

$$\text{Ln } L(B_i) = \Sigma Y_i (B0 + \Sigma B_i X_i) - \Sigma \ln(1 + \exp(B0 + \Sigma B_i X_i))$$

I know this looks intimidating, but it is not. We can build the functions and optimize in Excel using this function.

Example 5.3 Binary Logistic Regression

We have the following data (Table 5.4) where the response, Y, is a binomial from Bernoulli trials – like yes or no. In this case it is a success, 1, or a failure, 0.

TABLE 5.4

Example 5.3. Binary logistics regression data

item	Status	Number	Difference
1	1	4	19.2
2	1	2	24.1
3	0	4	-7.1
4	1	3	3.9
5	0	9	4.5
6	0	6	10.6
7	0	2	-3
8	0	11	16.2
9	1	6	72.8
10	0	7	28.7
11	1	3	11.5
12	1	2	56.3
13	0	5	-0.5
14	0	3	-1.3
15	0	3	12.9
16	0	8	34.1
17	0	10	6.6
18	1	5	-2.5
19	0	13	24.2
20	0	7	2.3
21	1	3	36.9
22	0	4	-11.7
23	1	2	2.1
24	1	3	10.4
25	0	2	9.1
26	0	5	2
27	0	6	12.6
28	1	5	18
29	0	3	1.5
30	1	4	27.3
31	0	10	-8.4

The model we want is

$$Y_i = \frac{e^{(b_o + b_1 \text{number} + b_2 \text{difference})}}{1 + e^{(b_o + b_1 \text{number} + b_2 \text{difference})}}$$

This is how we proceed via technology.

5.4.1 Excel for Logistics Regression

A template is created using the Solver in Excel to do logistics regression and provide the necessary output to interpret the results.

We entered the data.

Create heading and initial values for our model's coefficients: B_0, B_1, and B_2. Usually we set them at 0.

1. Create the functions we need (I do this in two parts). Column P1 uses

2. $Yi*(B0 + B1*number + B2*difference)$ and Column P uses $ln (1 + \exp(B0 + B1*number + B2*difference))$.

3. Sum columns P1 and P2.

4. In an unused cell take the difference of P1–P2 – this is the objective function.

5. Open the Solver and Maximize this cell containing P1–P2, by changing cells with B0, B1, and B2. Insure to uncheck the non-negativity box.

6. Solve.

7. Obtain your model and use it as needed.

8. Repeat steps 3–8 for the model with intercept only.

We have the data entered in to three columns.

Next we create columns for $Y*X'B$ and $ln(1 + \exp(X'B))$ using initial values for $b0$, $b1$, and $b2$.

We sum these two columns separately and in another cell we take the difference in the sums (Figure 5.20).

This is our objective function that we maximize by changing the cells for $b0$, $b1$, and $b2$. By doing so we get the results below.

Bo	B1	B2
1.421207	-0.75534	0.112205

We obtain our model,

$$Y = \exp(1.4212 - 0.7553*number + 0.1122*difference)/$$
$$(1 + \exp(1.4212 - 0.7553*number + 0.1122*difference))$$

We can plot this as shown in Figure 5.21.

Before we accept this model, we require a minimum of a few diagnostics. We want to (1) examine the significance of each estimated coefficient $\{b0, b1, b2\}$ and (2) compare this full model to an intercept only model and one term model to measure the chi-square differences.

We start with the estimates of our full model's coefficients $\{b0 = 1.421207, b1 = -0.75534, \text{ and } b2 = 0.112205\}$. We need the following (a) estimates of the standard errors of these estimates, (b) $t*$ which equals the *estimates/se*, and (c) P-value for $t*$.

P1	Yi(B0+B1Number+B2*difference)	P2	LN(1+exp(Bo+B1*number+B2*difference))
0.554174			1.008141
2.614663			2.685301
0			0.087101
-0.40722			0.510124
0			0.007629
0			0.136619
0			0.502626
0			0.006264
5.057677			5.064017
0			0.421461
0.445538			0.940526
6.227665			6.229637
0			0.085876
0			0.315775
0			1.039183
0			0.372547
0			0.004544
-2.63602			0.069196
0			0.003398
0			0.026742
3.295545			3.331923
0			0.052891
2.390253			2.477904
0.322112			0.867117
0			1.263713
0			0.112174
0			0.16824
-0.33581			0.53927
0			0.411041
1.463034			1.671294
0			0.000846
18.99161			30.41312
-11.4215			

FIGURE 5.20
Excel screenshot logistics regression model.

We know that the estimates for the Variance-Covariance matrix are the inverse of the Hessian matrix evaluated at the estimates of $\{b0, b1,$ and $b2\}$. In a logistics equation the number of terms in the regression model affects the Hessian matrix. To obtain all this we will need the Hessian matrix, $H(X)$, so that we can find the inverse, $H(X)^{-1}$, and then $-H(X)^{-1}$ that is the variance-covariance matrix when evaluated at our final coefficient estimates. The main diagonal of this matrix are our Variances for each coefficient. If we take the square root of these coefficients, then we get the *se* of our coefficients.

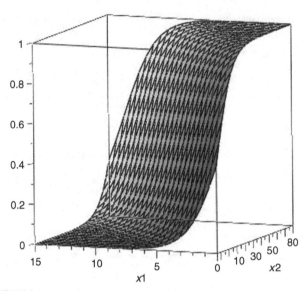

FIGURE 5.21
Plot of the logistic function, in this case a 3D CDF.

$$\left[\left[\left(-\left(\sum_{i=1}^{n}\left(\frac{e^{b0+b1x(i)}}{1+e^{b0+b1x(i)}}-\frac{\left(e^{b0+b1\,x(i)}\right)^2}{\left(1+e^{b0+b1\,x(i)}\right)^2}\right)\right)\right),-\right.\right.$$
$$\left.\left(\sum_{i=1}^{n}\left(\frac{x(i)e^{b0+b1x(i)}}{1+e^{b0+b1x(i)}}-\frac{\left(e^{b0+b1\,x(i)}\right)^2x(i)}{\left(1+e^{b0+b1x(i)}\right)^2}\right)\right)\right],$$
$$\left[\left(-\left(\sum_{i=1}^{n}\left(\frac{x(i)e^{b0+b1\,x(i)}}{1+e^{\,b0+b1\,x(i)}}-\frac{\left(e^{b0+b1\,x(i)}\right)^2x(i)}{\left(1+e^{b0+b1\,x(i)}\right)^2}\right)\right)\right),\right.$$
$$\left.\left.\left.\left(-\left(\sum_{i=1}^{n}\left(\frac{x(i)^2\,e^{b0+b1x(i)}}{1+e^{\,b0+b1\,x(i)}}-\frac{x(i)^2\left(e^{b0+b1\,x(i)}\right)^2}{\left(1+e^{b0+b1\,x(i)}\right)^2}\right)\right)\right)\right]\right]\right]\ \scriptstyle e$$

We would like to use pattern recognition and a simplification step to better see what is happening here and let $\pi = exp(b_0 + b_1x_1 + b_2x_2 + \ldots + b_nx_n)$.

Let $P = -\sum_{i=1}^{n}\left(\frac{\pi}{1+\pi}\right)-\left(\frac{\pi^2}{(1+\pi)^2}\right)$. Then we can more easily write the Hessian matrix for n terms and take its inverse.

We take the square root of the entries on the main diagonal as our estimates of the *se* for $\{b_0, b_1, \ldots, b_n\}$.

In our example we compute H and H^{-1}. We compute H using the sums of the columns in the matrix H. To obtain H^{-1}, we use the = MINVERSE command in Excel.

H

3.711818969	15.00133	44.21611
15.00132821	72.48339	218.9457
44.21611161	218.9457	1041.944

H^{-1}

1.655778953	-0.35711	0.004774
-0.35710513	0.114788	-0.00897
0.004774266	-0.00897	0.002641

We take the square root of the main diagonal entries {1.655778953, 0.114788, 0.002641} as our standard error, *se*, estimates for b0, b1, and b2 respectively. We find that our *se* estimates are

$$\{1.286770746, 0.338803, 0.051393\}.$$

We can enter and fill in the rest of our table.

Analysis of regression coefficients

We see from the results in Table 5.5 that the coefficients for b_1 and b_2 are significant at $\alpha = 0.05$.

Let's calculate the deviances for our model. We define the **deviances** as

$$dev_i = \pm\left[-2\left[Y_i \ln(\pi 1) + (1 - Y_i)\ln(1 - \pi 1)\right]\right]^{1/2},$$

where the sign is positive when $Y_i \geq \pi 1_i$ and negative when $Y_i < \pi 1_i$ and we define $\pi 1$ as $(1+\pi)^{-1}$ with π as we defined earlier.

TABLE 5.5

Statistical analysis of logistics regression analysis

Coefficient	Estimate from the Solver	Se from the square root of the V-C Matrix	Z-statistic = estimate/se	P-Value from P(Z> | Z-statistic |)
b_0	1.4212	1.2868	1.1044	0.2694
b_1	-0.7553	0.3388	-2.2296	0.0258
b_2	0.1122	0.0514	2.1620	0.0306

TABLE 5.6

Analysis of deviations

Model	ln likelihood	Deviance	df	Chi-square	P-value
Full Model	-11.4215	22.842	3		
Constant Model	-20.6904	41.3808	1		
Difference	-9.26887	18.53774	2	18.5377	$9.43x10^{-5}$

We find the difference is significant at $\alpha = 0.05$ so we choose the full model over the constant model.

Odds-Ratios

Interpretation of B parameters in the logistic model.

$$\pi^* = B_0 + B_1 x_1 + \ldots + B_n x_n$$

where

π^*

π^*

B_1 = Change in log-odds π^* for every 1 unit increase in x_1 holding all other x's fixed.

$e^{B_i}-1$ = Percentage change in odds ratio $\pi/(1-\pi)$ for every 1 unit increase in x_1 holding all other x's fixed.

So, $B_1 = -0.7553$, $e^{B1} = 0.47$, $e^{B1}-1 = -0.53$. For each unit of x_1, we estimate the odds of a fixed contract to decrease by 53% holding x_2 fixed.

For $B_2 = 0.1122$, $e^{B2} = 1.12$, $e^{B2}-1 = 0.12$. For each unit of x_2, we estimate the odds of a fixed contract to increase by 12% holding x_1 fixed.

We have presented a "how to" approach to logistic regression in Excel. We have given formulas and suggested how to put them into a table so that good analysis can be made.

5.4.2 Maple for Logistics Regression

Example 5.4 Damages versus Time

We allow damage, y, as a binary variable where 1 means damage and 0 means no damage as a function of flight time in hours, $x1$. Our data is

> $y := [1,1,0,1,0,0,0,0,1,0,1,1,0,0,0,0,0,1,0,0,1,0,1,1,0,0,0,1,$
  $0,1,0];$

$y := [1,1,0,1,0,0,0,0,1,0,1,1,0,0,0,0,0,1,0,0,1,0,1,1,0,0,0,1,0,1,0,]$

> $x1 := [4,2,4,3,9,6,2,11,6,7,3,2,5,3,3,8,10,5,13,7,3,4,2,3,2,$
  $5,6,6,4,10];$

$x1 := [4,2,4,3,9,6,2,11,6,7,3,2,5,3,3,8,10,5,13,7,3,4,2,3,2,5,6,6,4,10]$

> $model := \dfrac{\exp(4 + b1.t1)}{1 + \exp(a + b1.t1)}$

$$model := \frac{e^{a+b1\,t1}}{1 + e^{a+b1+t1}}$$

> $modelb := Nonlinear\,Fit(model, x1, y, t1, initialvalues = [a = 1,5, b1 = -.5]); 11$

$$modelb := \frac{e^{1.44319130879736 - 0.391897392001439\ t1}}{1 + e^{1.44319130879739 - 0.39189739200149\ t1}}$$

> $plot(modelb, t1 = 0..10);$

The model is displayed in Figure 5.22. It is up to the user to decide over what intervals of x, we call the y probability a 1 or a 0.

Example 5.5 Damages versus Time Differentials

> $x2 := [19.2, 24.1, -7.1, 3.9, 4.5, 10.6, -3, 16.2, 72.8, 28.7, 11.5, 56.3, -.5$
  $-1.3, 12.9, 34.1, 6.6, -2.5, 24.2, 2.3, 36.9, -11.7, 2.1, 10.4, 9.1, 2,$
  $12.6, 18, 1.5, 27.3, -8.4];$

$x2 := [19.2, 24.1, -7.1, 3.9, 4.5, 10.6, -3, 16.2, 72.8, 28.7, 11.5, 56.3,$
  $-0.5 - 1.3, 12.9, 34.1, 6.6, -2.5, 24.2, 2.3, 36.9, -11.7, 2.1, 10.4,$
  $9.1, 2, 12.6, 18, 1.5, 27.3, -8.4]$

> $y,$

$[1,1,0,1,0,0,0,0,1,0,1,1,0,0,0,0,0,1,0,0,1,0,1,1,0,0,0,1,0,1,0]$

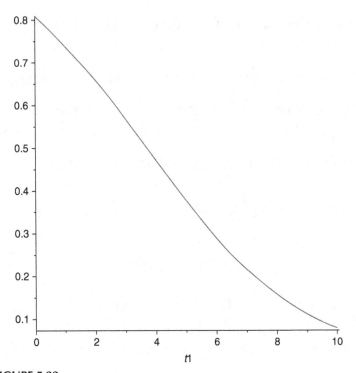

FIGURE 5.22
Logistics regression model for damages and time.

> $model := \dfrac{\exp(a + b1 \cdot t1)}{1 + \exp(a + b1 \cdot t1)};$

$$modal := \dfrac{e^{a+b1+t1}}{1+e^{a+b1+t1}}$$

> $modelb2 := NonlinearFit(model, x2, y, t1, initialvalues = [a = .5, b1 = 1])$

$$modelb2 := \dfrac{e^{-1.19966702762947 + 0.0545525473058010\, t1}}{1 + e^{-1.19966702762947 + 0.0545525473058010\, t1}}$$

> $plot(modelb2, t1 = 0..400);$

The model is displayed in Figure 5.23. It is up to the user to decide over what intervals of x, we call the y probability either a 1 or a 0.

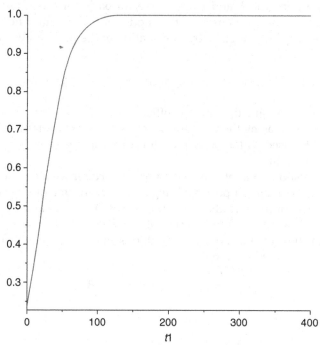

FIGURE 5.23
Logistics regression model for damages and time differences.

Poisson Regression Models

Many times students encounter situations where the outcome variable is not only numeric but also in the form of discrete counts. Often, it is a count of rare events such as the number of new cases of terrorist activities occurring within a population over a certain period of time, the number of certain types of IEDS encountered, or the number of significant acts of violence within a region. The goal of regression analysis in such instances is to model the dependent variable, y, as the estimate of outcome using some or all of the explanatory variables (in mathematical terminology estimating the outcome as a function of some explanatory or predictor variables).

Normality Assumption Lost

According to Neter et al. (1996) and Montgomery et al. (2006), in the case of logistic and Poisson regression the fact that probability lies between 0 and 1 imposes a constraint. We lose both the normality assumption of multiple linear regression and the assumption of constant variance. Without these assumptions the F and t tests have no basis for the analysis. When this happens, we must transform the model and the data. The new solution involves using the logistic transformation of the probability p or logit **p**, such that

$$ln(p/1-p) = \beta_0 + \beta_1 X_1 + \beta_2 X_2 \ldots\ldots + \beta_n X_n.$$

They go on to explain that the β coefficients could now be interpreted as increasing or decreasing the log odds of an event, and $\exp(\beta)$ (the odds multiplier) could be used as the odds ratio for a unit increase or decrease in the explanatory variable.

When the response variable is in the form of a *count* we face a yet different constraint. Counts are all positive integers and stand for rare events. Thus, the Poisson distribution (rather than the Normal) is more appropriate since the Poisson has a mean > 0 and our counts are all positive counting numbers. So, the logarithm of the response variable is linked to a linear function of explanatory variables such that

$$ln\,(Y) = \beta_0 + \beta_1 X_1 + \beta_2 X_2 \ldots + \beta_n X_n$$

and

$$Y = (e^{\beta_0})\,(e^{\beta_1 X_1})\,(e^{\beta_2 X_2}) \ldots (e^{\beta_n X_n}).$$

In other words, the typical Poisson regression model expresses the log outcome rate as a linear function of a set of predictors.

Assumptions in Poisson Regression

There are several key assumptions in Poisson regression that are different than the assumptions in the simple linear regression model. These include that the logarithm of the dependent variable changes linearly with equal incremental increases in the exposure variable. For example, if we measure risk in exposure per unit time and one group is counts per month and another is count per years, we can convert all exposures to strictly counts. We find that changes in the rate from combined effects of different exposures or risk factors are multiplicative. We find for each level of the covariates; the number of cases has variance equal to the mean which makes it follow a Poisson distribution. Further, we assume the observations are independent.

We use diagnostic methods to identify violations of the assumption to determine whether variances are too large or too small include plots of residuals versus the *mean* at different levels of the predictor variable. Recall that in the case of normal linear regression, diagnostics of the model used plots of residuals against fits (fitted values). This implies that some of the same diagnostics can be used in the case of Poisson Regression. We will use the residual plot, deviations versus the model to look for patterns as our main diagnostic method.

In Poisson regression we start with the basic model shown in Equation (5.5),

$$Y_i = E[Y_i] + \varepsilon_i \text{ for } i = 1, 2, \ldots, n. \tag{5.5}$$

$$Y_i = E[Y_i] + \varepsilon_i$$

The ith case mean response is denoted by u_i, where uican be one of many defined functions (Neter et al., 1996) but we elect to use only the form shown in Equation (5.6),

$$u_i = u(X_i B) = exp(X_i' B) \text{ where } u_i \geq 0. \tag{5.6}$$

We assume that the variable, Y_i, are independent Poisson random variables with expected value u_i.

In order to apply regression techniques, we will use the likelihood function (Neter et al., 1996; Montgomery et al., 2006). The likelihood function, L, is given in Equation (5.7).

$$L = \prod_{i=1}^{n} f_i(Y_i) = \prod_{i=1}^{n} \frac{(u(X_i, B)^{Y_i} \exp[-u(X_i, B)]}{Y_i!} \tag{5.7}$$

Most texts explain that maximizing this function is quite difficult so they use the logarithm of the likelihood function shown in Equation (5.8):

$$\ln(L) = \sum_{i=1}^{n} Y_i \ln(u_i) - \sum_{i=1}^{n} u_i - \sum_{i=1}^{n} \ln(Y_i!) \tag{5.8}$$

where u_i is the fitted model.

We maximize this function to obtain the best estimates for the coefficients of the model. Numerical search techniques are used to obtain these estimates. We mention here that "good" starting points are required to possibly obtain convergence (Fox, 2012).

Within the model development we are concerned about the deviations or residuals as we previously mentioned. In Poisson regression, the deviance is modeled as shown in Equation (5.9):

$$Dev = 2\left[\sum_{i=1}^{n} Y_i ln\left(\frac{Y_i}{u_i}\right) - \sum_{i=1}^{n}(Y_i - u_i)\right] \tag{5.9}$$

where u_i is the fitted model. We note that because of term $ln(Y_i/u_i)$ that if $Y_i = 0$ we must set the $ln(Y_i/u_i) = 0$.

Inferences for the coefficients are carried out in the same fashion as with logistics regression. To estimate the variance-covariance matrix we require the use the Hessian matrix. We define the Hessian, $H(X)$, as the matrix of second partial derivatives of the $ln(L)$ function. The variance-covariance matrix, $VC(X,B)$, is minus the inverse of this Hessian matrix evaluated with the final estimates of the coefficients, B.

$$VC(X,B) = -H(X)^{-1}$$

The main diagonal of the matrix are the estimates for the variance. Since we need the estimated standard deviations, se_B, we take the square root of each main diagonal entry to obtain this estimate.

We may then perform hypothesis tests of the coefficients using the t-test.

We use the logarithm of the likelihood function, Equation (5.9). The Hessian is defined as the matrix of second partial derivatives. We will illustrate two Hessian modeling examples and then we will make a useful observation.

Assume that our model is, $y_i = exp(b_0 + b_1 {}^* x_i)$. Putting this model into Equation (5.9) we have:

$$\ln(L) = \sum_{i=1}^{n} Y_i \ln\left(\exp(b0 + b1xi)\right) - \sum_{i=1}^{n} \exp(bo + b1xi) - \sum_{i=1}^{n} Y_i!$$

We define the second partials as follows:

$$g_{ij} = \frac{\partial^2 \left(\ln(L)\right)}{\partial b_i \partial b_{ij}} \text{ for all i and j} \tag{5.10}$$

The estimates for the variance-covariance matrix are defined and displayed in Equation (5.11):

$$s^2(b) = [(-g_{ij})_{B=b}]^{-1} \tag{5.11}$$

We take the partial derivatives and set up the Hessian matrix, g_{ij} as:

$$\begin{bmatrix} -\left(\sum_{i=1}^{n} e^{b0+b1x(i)}\right) & -\left(\sum_{i=1}^{n} x(i)e^{b0+b1x(i)}\right) \\ -\left(\sum_{i=1}^{n} x(i)e^{b0+b1x(i)}\right) & -\left(\sum_{i=1}^{n} x(i)^2 e^{b0+b1x(i)}\right) \end{bmatrix}.$$

If our model were, $y_i = exp(b_0 + b_1 x_{1i} + b_2 x_{2i})$ then we find the Hessian matrix, g_{ij} as

$$
-\left(\sum_{i=1}^{n} e^{b0+b1\,x1(i)+b2\,x2(i)}\right) \quad -\left(\sum_{i=1}^{n} x1(i) e^{b0+b1x(i)+b2x2(i)}\right) - \left(\sum_{i=1}^{n} x2(i) e^{b0+b1x(i)+b2x2(i)}\right)
$$

$$
-\left(\sum_{i=1}^{n} x1(i) e^{b0+b1x(i)+b2x2(i)}\right) \quad -\left(\sum_{i=1}^{n} x1(i)^2 e^{b0+b1x(i)+b2x2(i)}\right) - \left(\sum_{i=1}^{n} x1x2(i) e^{b0+b1x(i)+b2x2(i)}\right)
$$

$$
-\left(\sum_{i=1}^{n} x2(i) e^{b0+b1x(i)+b2x2(i)}\right) \quad -\left(\sum_{i=1}^{n} x1x2(i) e^{b0+b1x(i)+b2x2(i)}\right) - \left(\sum_{i=1}^{n} x2(i)^2 e^{b0+b1x(i)+b2x2(i)}\right)
$$

Now, we see the pattern and we can extend the pattern to easily obtain the Hessian for the model with n independent variables,

$$
y_i = exp(b_0 + b_1 x_{1i} + b_2 x_{2i} + \ldots + b_n x_{ni}) \text{ and the } \Sigma\, exp(b_0 + b_1 x_{1i} + b_2 x_{2i} + \ldots + b_n x_{ni})
$$

is a common term, we call this sum P. This gives us a generic Hessian matrix for Poisson regression with our choice of model.

Assume that our model is $y_i = exp(b_0 + b_1 {}^* x_i)$. Putting this model into Equation (5.8) we have,

$$
\ln(L) = \sum_{i-1}^{n} Y_i \ln\left(\exp(b_0 + b_1 x_i)\right) - \sum_{i=1}^{n} \exp(b_0 + b_1 x_i) - \sum_{i=1}^{n} Y_i\,!
$$

We define the second partial derivatives as follows:

$$
g_{ij} = \frac{\partial^2 (\ln(L))}{\partial b_i \partial b_{ij}} \text{ for all } i \text{ and } j
$$

The estimates for the variance-covariance matrix are defined and are displayed as:

$$
s^2(b) = [(-g_{ij})_{B=b}]^{-1}.
$$

We take these partial derivatives and set up the Hessian matrix, g_{ij} as shown in the matrix below:

$$
g_{ij} = \begin{bmatrix} -\left(\sum_{i=1}^{n} e^{b_0+b_1 x_i}\right) & -\left(\sum_{i=1}^{n} x_i e^{b_0+b_1 x_i}\right) \\ -\left(\sum_{i=1}^{n} x_i e^{b_0+b_1 x_i}\right) & -\left(\sum_{i=1}^{n} x_i^2 e^{b_0+b_1 x_i}\right) \end{bmatrix}.
$$

When our model slightly differs, such as $y_i = exp(b_0 + b_1 x_{1i} + b_2 x_{2i})$, then we find the Hessian matrix, g_{ij}. We note the similarities between the last two Hessian matrices.

$$g_{ij} = \begin{bmatrix} -\left(\displaystyle\sum_{i=1}^{n} e^{b_0+b_1x_1i+b_2x_2i}\right) & -\left(\displaystyle\sum_{i=1}^{n} x_{1i}e^{b_0+b_1x_1i+b_2x_2i}\right) & -\left(\displaystyle\sum_{i=1}^{n} x_{2i}e^{b_0+b_1x_1i+b_2x_2i}\right) \\ -\left(\displaystyle\sum_{i=1}^{n} x_{1i}e^{b_0+b_1x_1i+b_2x_2i}\right) & -\left(\displaystyle\sum_{i=1}^{n} x_{1i}^2 e^{b_0+b_1x_1i+b_2x_2i}\right) & -\left(\displaystyle\sum_{i=1}^{n} x_{1i}x_{2i}e^{b_0+b_1x_1i+b_2x_2i}\right) \\ -\left(\displaystyle\sum_{i=1}^{n} x_2 e^{b_0+b_1x_1i+b_2x_2i}\right) & -\left(\displaystyle\sum_{i=1}^{n} x_{1i}x_{2i}e^{b_0+b_1x_1i+b_2x_2i}\right) & -\left(\displaystyle\sum_{i=1}^{n} x_{2i}^2 e^{b_0+b_1x_1i+b_2x_2i}\right) \end{bmatrix}.$$

We see the pattern in the matrix of partial derivatives and we can extend the pattern to easily obtain the Hessian for a model when we have n independent variables,

$y_i = exp(b_0 + b_1 x_{1i} + b_2 x_{2i} + ... + b_n x_{ni})$ and we identify the common term in the matrix as the summation, $\Sigma exp(b_0 + b_1 x_{1i} + b_2 x_{2i} + ... + b_n x_{ni})$. We call this summation P. This gives us a generic Hessian matrix for Poisson regression to use with our choice of the model from $y_i = exp(b_0 + b_1 x_{1i} + b_2 x_{2i} + ... + b_n x_{ni})$ depending on the number of independent variables.

$$g_{ij} = -\begin{bmatrix} P & \sum x_{1i}P & \sum x_{2i}P & \sum x_{3i}P & \cdots & \sum x_{ni}P \\ \sum x_{1i}P & \sum x_{1i}^2 P & \sum x_{1i}x_{2i}P & \sum x_{1i}x_{3i}P & \cdots & \sum x_{1i}x_{ni}P \\ \sum x_{2i}P & \sum x_{1i}x_{2i}P & \sum x_{2i}^2 P & \sum x_{2i}x_{3i}P & \cdots & \sum x_{2i}x_{ni}P \\ \cdot & \cdot & \cdot & \cdot & \cdot & \cdot \\ \cdot & \cdot & \cdot & \cdot & \cdot & \cdot \\ \sum x_{ni}P & \sum x_{1i}x_{ni}P & \sum x_{2i}x_{ni}P & \sum x_{3i}x_{ni}P & & \sum x_{ni}^2 P \end{bmatrix}.$$

This is the generic Hessian matrix so we need to replace the formulas with numerical values and compute the inverse of the negative of this matrix. Once we replace the variables with their respective values we should have a non-singular square matrix that we can take the inverse.

The main diagonal entries of this matrix inverse are the estimates for the variances of the coefficients to the estimates of **b**. The square root of the entries of the main diagonal are the estimates of the *se* of the coefficients of **b** to be used in the hypothesis testing for each coefficient, **b** as:

$$t^* = bi/se(b_i).$$

Once we replace the variables with their respective values we should have a non-singular square matrix that we can take the inverse.

We now have all the equations that we need to build the tables of outputs for Poisson regression that are similar to packaged outputs.

TABLE 5.7

Poisson regression analysis

	Degrees of freedom (df)	Deviance	Meandeviance	Ratio
Regression	m	$D_{reg} = D_t - D_{res}$	$MDev(reg) = D_{reg}/m$	$|MDev(reg)|$
Residual	k-l-m	D_{res} = Result from equation (5) using the full model with predictors.	MDEV(res) = Dres/ (k-l-m)	
Total	k-l	D_t = Result from equation (5) using only y = exp(bo) as the best model	MDev(t) = Dt/(k-l)	

Estimates of Regression Coefficients

We use one for the constant plus one for every predictor variable in the model being examined for the number of coefficients. Estimates are the final values (that converged) for the numerical search method to maximize the $ln(L)$ equation. The values of se are the square roots of the main diagonal of the inverse of (-) the hessian matrix. The values of $t^* = $ *(final coefficient estimate)/se* and the p-value are displayed, where the p-value is the probability associated with the $|t^*|$ from $P(T>|t^*|)$.

Coefficients	Estimate	se	t^*	P-value

In our summary of Poisson regression analysis, let m = number of variables in the model, let k = number of data elements of the dependent variable, Y.

5.5 Technology Examples for Poisson Regression

5.5.1 Excel for Poisson Regression

The first example will be explained in more detail than the second example for illustrative purposes to show how we used the equations and Excel to perform Poisson Regression. We note that a prerequisite for using Poisson regression is that data for the dependent variable, Y, must be discrete count data. We have chosen two data sets (www.oxfordjournals.org/our_journals/tropej/online/ma_chap13.pdf) that have published solutions in the literature to be our examples.

Example 5.6 Caesarian Births

The data (Table 5.8):

Csec = number of C-sections performed
Hosp = type of hospital public or private (0 or 1)
Birth = number of births at the hospital.

   This data was obtained through the record at 4 private hospitals and 16 public hospitals. We desired to build a model to predict the number of *c-section* births as a function of the type of hospital and number of births. Since the y variable are counts of *c-section* births we might use Poisson regression.
   We present an Excel spreadsheet and illustrate how we do this. We list the steps required to obtain the model.

Step 1. Calculate the baseline constant model, $y = exp(constant)$, using the Solver to obtain the value of the constant that minimizes deviations, Equation (5.7). We minimize Equation (5.7) by varying the value cell of the constant.

Step 2. Repeat Step 1 for the full model, $y = exp(b_0 + b_1 x_1 + b_2 x_2 + \ldots + b_n x_n)$, using the Solver to obtain the values of the parameters $\{b_0, b_1, b_2, \ldots, b_n\}$ that minimize the deviations, Equation (5.7).

TABLE 5.8

Caesarian birth data

Csec	Hosp	Birth
8	0	236
16	1	739
15	1	970
23	1	2,371
5	1	309
13	1	679
4	0	26
19	1	1,272
33	1	3,246
19	1	1,904
10	1	357
16	1	1,080
22	1	1,027
2	0	28
22	1	2,507
2	0	138
18	1	502
21	1	1,501
24	1	2,750
9	1	192

Step 3. Using the pattern recognition format explained concerning Equation (5.9) compute the standard errors for the parameter estimates.

Step 4. Compute the individual deviations and plot the indexed deviations.

Step 5. Compute the p-values as appropriate.

Step 6. Put all values into tables.

We illustrate these steps.

Step 1. We have entered the data so first we calculate the information for the model $y = exp(constant)$ (Figure 5.24)

Model 1	$y = exp(bo)$
bo	
2.711377991	(initially 0)

Step 2. We repeat for the full model, $y = exp(b_0 + b_1*hosp + b_2*birth)$ in Figure 5.25.

Model 1	y*ln(exp(M1))	ln(Fact(Y))		Deviation eq	
15.05	21.69102393	10.6046029		-5.05549	-7.05
15.05	43.38204786	30.6718601		0.979372	0.950000002
15.05	40.67066987	27.8992714		-0.04992	-0.05
15.05	62.36169379	51.6066756		9.754673	7.950000002
15.05	13.55688996	4.78749174		-5.5097	-10.05
15.05	35.24791388	22.5521639		-1.90357	-2.05
15.05	10.84551196	3.17805383		-5.30033	-11.05
15.05	51.51618183	39.3398842		4.428159	3.950000002
15.05	89.47547371	85.054467		25.90928	17.95
15.05	51.51618183	39.3398842		4.428159	3.950000002
15.05	27.11377991	15.1044126		-4.08793	-5.05
15.05	43.38204786	30.6718601		0.979372	0.950000002
15.05	59.6503158	48.4711814		8.352618	6.950000002
15.05	5.422755982	0.69314718		-4.03646	-13.05
15.05	59.6503158	48.4711814		8.352618	6.950000002
15.05	5.422755982	0.69314718		-4.03646	-13.05
15.05	48.80480384	36.3954452		3.221888	2.950000002
15.05	56.93893781	45.3801389		6.996033	5.950000002
15.05	65.07307179	54.7847294		11.20022	8.950000002
15.05	24.40240192	12.8018275		-4.62738	-6.05
301	816.1247753	608.501426	Sums	49.99514	3.70058E-08
OBJ Func	-93.37665016				
				99.99028	Total Deviations

FIGURE 5.24
Excel screenshot of Poisson regression example.

Model 2	y*ln(exp(m2))	ln(fact(Y))	Dev Eq	y-u
4.17015	11.42361596	10.6046	5.211916	3.82985
13.9727	42.19368496	30.67186	2.167735	2.027302
15.06581	40.68642	27.89927	-0.06567	-0.06581
23.78976	72.89287012	51.60668	-0.7765	-0.78976
12.14472	12.48447121	4.787492	-4.43728	-7.14472
13.70199	34.02803267	22.55216	-0.68369	-0.70199
3.894155	5.437907292	3.178054	0.10727	0.105845
16.6249	53.40713457	39.33988	2.537206	2.375096
31.64463	114.0007586	85.05447	1.383991	1.355371
20.42949	57.32261015	39.33988	-1.37827	-1.42949
12.33629	25.1254571	15.10441	-2.09961	-2.33629
15.616	43.97273516	30.67186	0.388684	0.384001
15.34844	60.0823106	48.47118	7.920623	6.651556
3.896696	2.720257935	0.693147	-1.33396	-1.8967
24.86848	70.69922307	48.47118	-2.69629	-2.86848
4.038999	2.79199383	0.693147	-1.4057	-2.039
12.93357	46.07687136	36.39545	5.94982	5.066431
17.91382	60.59701966	45.38014	3.337952	3.086182
26.91911	79.02807853	54.78473	-2.75479	-2.91911
11.69012	22.1286941	12.80183	-2.35367	-2.69012
300.9998	857.1001469	608.5014	9.019768	0.000167
OBJ Func	-52.40111134		Dev due to residuals	
			18.0392	

FIGURE 5.25

Excel screenshot of updated Poisson regression example.

TABLE 5.9

Analysis of deviance

	Degrees of freedom (df)	Deviance	Meandeviance	Ratio
Regression	2 (two variables in model-hosp, births)	(99.9902-18.0392) = 81.951077	81.951077/2 = 40.9755	|40.9755| = 40.98
Residual	17 (Y-1-2)	18.0392	18.0392/17 = 1.06112	
Total	19 (Y-1)	99.9902	99.9902/19 = 5/26264	

Model 2	$y = exp(b0 + b1 + b2)$		
b0	b1	b2	
1.350998944	1.045138972	0.00032607	(all initially 0)

Step 3. We build the Hessian matrix (Figure 5.26) from knowing the pattern and then we take the inverse of (-) the Hessian matrix. After we take the inverse we take the square root of the values along the main diagonal as the standard errors.

-Hessian'		
300.9998327	284.9998324	467947.3576
284.9998324	284.9998324	466195.4648
467947.3576	466195.4648	1037620324
Inverse		
0.06254362	-0.06189194	-3.98392E-07
-0.06189194	0.074484805	-5.55339E-06
-3.9839E-07	-5.55339E-06	3.63851E-09

FIGURE 5.26a

Excel screenshot of Hessian matrix for Poisson regression example.

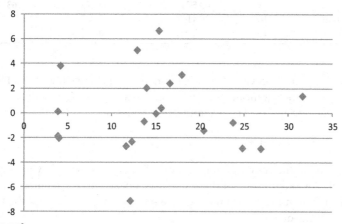

FIGURE 5.26b

Residual plot of deviations versus model for example.

We take the square root of the values along the main diagonal to obtain the standard errors of the estimates,

se(b0)	se(b1)	se(b2)
0.250087225	0.272919045	6.03201E-05

Step 4. We compute the deviations and obtain a plot, as shown in Figure 5.26.

Step 5 & 6. Now, we place the information into the tables noting that there are 20 data elements in Y, the full model has two variables plus a constant. Where appropriate we compute the p-values.

TABLE 5.10

Analysis of regression coefficients

	df	Coefficient	se	t*	P-values	Significant
b_0	2	1.351	0.2505367	5.3924	0.01	*
b_1	17	1.0451	0.2617086	3.9934	0.00047	*
b_2	19	0.000326	0.00000604	5.397	0.0000165	*

TABLE 5.11

Analysis of deviance check

	df	Deviance	Mean deviance	Ratio
Regression	1	63.575	63.575	63.575
Residual	18	36.414	2.023	
Total	19	99.9902	5.263	

To compute the p-values we use $t.dist.2t(x,df)$. First, we find that all the coefficients are significant at a 0.05 level. We accept the full model to use for predictive analysis. For example, our model is

$$Y = e^{(1.350993605 + 1.045142687 * hosp + 0.000326073 * births)}$$

We examine the residual plot, Figure 5.27, for the residuals versus the model. We look for one of these patterns: linear, curved, fanning (in or out), or for randomness of the plot. We see no pattern and accept the full model as adequate.

Since the model is adequate, we use the model for predictions and interpolations. If we know we have a private hospital with 363 births then our estimate for the number of caesarians births is

$$Y = e^{(1.350993605 + 1.04514268 + 0.000326073 * 363)} = 12.36 \text{ or approximately 12.}$$

Checking Our Analysis Decisions

We can always checking our analysis to see if a smaller model was more adequate. We could also build an intermediate model:

$$Y = e^{(b0 + hosp * b1)}$$

We would use all the same equations as before and we could have built the following two tables:

How do we know which equation to use?

$$Y = e^{(b0 + b1 * hosp)} \text{ or}$$

$$Y = e^{(b0 + b1 * hosp + b2 * births)}$$

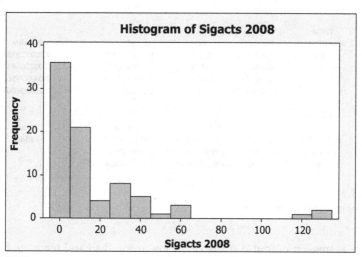

FIGURE 5.27
Histogram of SigActs 2008.

Deviance serves the purpose of comparing models. Model I, the one variable predictor model, has a regression deviance of 63.575 (or about 64% is explained by the model) where Model II, the full model, has a regression deviance of 81.95 (or about 82% is explained by the model). This is a difference of 18.375 with a change of *df* of 1. We can use aχ^2 at 1 degree of freedom to find it is significant at beyond 0.005. Thus, the model with more *df* and smaller residual deviance is better.

Example 5.7 Terrorist Actions in the Philippines

Recall Figure 5.1: the figure is representative of many similar figures trying to build linear models to gain insights into causes of terrorist actions. Since the number of violence acts, SIGACTS, are counts. We examine the histogram in Figure 5.28, noticing it appears to follow a Poisson distribution. A goodness of fit test for the Poisson distribution confirms that it follows a Poisson distribution ($\chi^2 = 933.11$, $p = 0.000$).

A Poisson regression was run yielding the model,

$$y = exp(7.828 - 0.034x_1 - 0.01799x_2 + 0.00400x_3)$$

where

   y = counts of violent activities
   X_1 = government satisfaction level
   X_2 = literacy level
   X_3 = poverty level.

Table 1				
Coefficients	Estimates	se	t*	P-Value
bo	7.8281417	0.535	14.63204056	0.002319158
b1	-0.0341572	0.002	-17.07862043	0.001705446
b2	-0.0179907	0.006	-2.998447403	0.047775567
b3	0.00400015	0.003	1.333383431	0.156996007

Table 2				
	df	Deviance	Mean Dev	Ratio
Regression	3	2214.064499	738.0214997	738.02
Residual	77	169.8891451	2.206352534	
Total	80	2383.953644	29.79942055	

FIGURE 5.28
Excel screenshot of Poisson regression.

The regression output is displayed below using the Excel template for Poisson regression.

We interpret the odds ratio for the coefficients to help explain our case.

- $\text{Exp}(-.0342) = 0.96647$. This means for 1 more value of Government Satisfaction that Violence goes down slightly by about 3.4%.

- $\text{Exp}(-0.01799) = 0.9812$. This means for every 1 unit increase in literacy that Violence goes down slightly, by 1.2%.

- $\text{Exp}(0.004) = 1.004$. This means that as Poverty increases Violence increases slightly, by 0.400%.

Based on the findings of this research, poverty, good governance, and literacy are the ones that are most strongly related to violent conflict in the Philippines. For governance, the poor delivery of basic social services, corruption and inefficiency in the bureaucracy, and poor implementation of laws instigate frustration that could eventually lead to aggression. A calibrated level of authority or political control is required to attain social and political stability. For literacy, the findings correspond to the claim that conflict and aggression are influenced by literacy.

Consistency in quantitative analysis among separate datasets indicates that good governance and literacy may be causal factors of conflict. Although the presence of correlations or relationship is present with the variables, this does not necessarily prove a causal link. Conflict is rarely caused by a single factor. It is usually caused by the interplay of long-term structural conditions with short-term proximate issues.

Insurgency has been an enduring problem in the Philippines. The government's Whole of the Nation Approach should be appropriately implemented with the full cooperation of the stakeholders. Insurgency is mainly driven by structural problems in the Philippine society that is

TABLE 5.12

Analysis of regression coefficients

	Coefficient	Estimated SE	t*	Significant at 0.05
b_0	2.132	0.102	20.95	*
b_1	0.0004405	.0000540	8.17	*

beyond the scope of the military. The local governments units should take the lead in resolving the conflict with the military and police as support. The military and police should only handle security concerns while the local government units address the socio-economic factors of the conflict.

Among the variables, good governance is considered to be the primordial factor in the resolution of conflict. Failure in governance can lead to the escalation of conflict that could further result in the breakdown in delivery of critical political goods such as security, rule of law, and social services. Good governance would eventually have a causal effect that leads to the eradication of poverty through the creation of jobs and improved social benefits; enhancement of literacy through the establishment of an efficient education program; and recognition and appreciation of ethnic diversity through formulation and implementation of laws that would protect culture, identity, and ancestral domain of ethnic groups.

Poisson Regression in R for This Example
Visualizing count data in a histogram is a useful way of assessing how the data are distributed.

Histogram plot

`stat_bin()` using `bins = 30`. Pick better value with `binwidth`.

Poisson regression in R is also treated as a special case of GLMs, similar to the logistic regression covered in the previous section. As such, it can be implemented using the same *glm()* function, but now specifying the model family as 'Poisson', which tells R to implement a Poisson model. The model we use here can be specified as

$$Y = e^{\wedge}(\beta\_0 + \beta\_1\,GGI + \beta\_2\,Literacy + \beta\_3\,Poverty)$$

```
## Generate model
sigacts_model<- glm(
  sigacts_2008 ~ ggi_2008 + literacy + poverty
  , data = sigacts_data
  , family = poisson
)
```

5.5.2 Maple for Poisson Regression

Poisson Regression with Three Parameters in Maple

We revisit the birth example with an additional predictor being the type of hospital: rural or urban coded as a {0,1}.

The data:

$$
data := \begin{bmatrix}
8 & 0 & 236 \\
16 & 1 & 739 \\
15 & 1 & 970 \\
23 & 1 & 2371 \\
5 & 1 & 309 \\
13 & 1 & 679 \\
4 & 0 & 26 \\
19 & 1 & 1272 \\
33 & 1 & 3246 \\
19 & 1 & 1904 \\
10 & 1 & 357 \\
16 & 1 & 1080 \\
22 & 1 & 1027 \\
2 & 0 & 28 \\
22 & 1 & 2507 \\
2 & 0 & 138 \\
18 & 1 & 502 \\
21 & 1 & 1501 \\
24 & 1 & 2750 \\
9 & 1 & 192
\end{bmatrix}
$$

We ran the Maple procedures in our worksheet and obtained final statistical analysis:

>*STAToutput(nobs,xx1,xx2,xx3,SE1,SE2,SE3,t1,t2,t3,PV1,PV2, PV3,DF1,DF2,RD,ResD,SST,MDR,OddsRation1,OddsRatio2);*

"Degrees of Freedom and Regression Deviations = ", 2, 81.95107805

"Degrees of Freedom and Residuals Deviations = ", 17, 18.0392

"Total = ", 99.9902780547385532

"Mean Deviation = ", 1.061129412

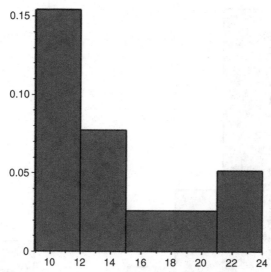

FIGURE 5.29
Histogram of violence in the Philippines.

Coefficient	Standard Error	T-Statistic	P-Value	Odds Ratio
1.350999	0.2501	5.4021	0.0163	
1.045138	0.2729	3.8295	0.0310	2.8438
0.000326	0.0001	5.4063	0.0163	1.0003

The model is $y = exp(1.35099 + 1.045138x_1 + 0.000326{*}x_2)$. The P-values are significant, so we tend to accept the model.

We return to our Philippines example briefly described earlier in this chapter.

Since the number of violence acts, SIGACTS, are counts. We examine the histogram in Figure 5.29, noticing it appears to follow a Poisson distribution. A goodness of fit test for the Poisson distribution confirms it follows a Poisson distribution ($\chi^2 = 933.11$, p = 0.000).

>*Histogram(yhc, binbounds = [9, 12, 15, 21, 24]);*

We do a chi-square test in Maple to determine whether the data follows a Poisson distribution (Figure 5.30).

```
> with(Statistics):
> infolevel[Statistics]: = 1:
```

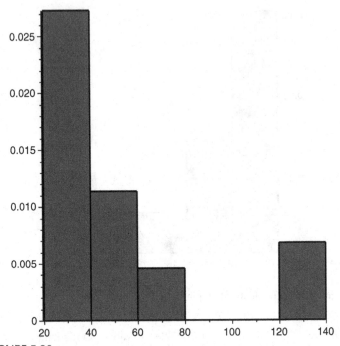

FIGURE 5.30
Histogram of SigActs of violence.

Specify the matrices of categorized data values.

> Ob: = Array([7,0,1,3,1,8]):
> Ex: = Array([2.3,2.9,3.9,3.9,3.1,3.6]):

Perform the goodness-of-fit test upon this sample.

>ChiSquareGoodnessOfFitTest(Ob, Ex, level = .05);
Chi-Square Test for Goodness-of-Fit

Null Hypothesis:
Observed sample does not differ from expected sample

Alt. Hypothesis:
Observed sample differs from expected sample

Categories: 6
Distribution: ChiSquare(5)
Computed statistic: 21.6688
Computed pvalue: 0.000605193
Critical value: 11.0704974062099

Result: [Rejected]
There exists statistical evidence against the null hypothesis

$hupothesis = false, criticalvalue = 11.0704674062099, distribution$
$\quad = ChiSquare(5), pvalue = 0.000605192984603264, statistic$
$\quad = 21.66880882$

> $violencedata := \left[seq(violence[i, 2], i = 1..80) \right];$

$violencedata := \left[seq(violence[i, 2], i = 1..80) \right];$

$violencedata := [122, 44, 2, 42, 31, 28, 64, 10, 1, 12, 57, 18, 4, 4,$
$\qquad 29, 23, 52, 5, 35, 8, 33, 26, 4, 0, 2, 5, 27, 8, 26, 2, 11,$
$\qquad 0, 8, 0, 125, 0, 14, 7, 0, 64, 0, 0, 15, 2, 0, 3, 126, 0, 0, 0, 4, 0, 0, 7, 0, 0]$

> $Histogram(violencedata, binbounds = [20, 40, 60, 80, 100, 120, 140]);$

A Poisson regression was run and yields the following outputs.
Model is $y = exp(9.5599 - 0.06398x)$.
A Poisson regression was run yielding the following model (Figure 5.31),

> $pmodel := exp(b0 + b1 x);$

$pmodel := e^{b0 + b1\, x}$

> $NonlinearFit(pmodel, literacyratedata, violencedata, x);$

$y = e^{8.58852448202650\, - 0.0526327458474074\, x}$

where

y = Counts of violent activities

x = Literacy level

> $p1 := plot\left(e^{9.5599 - 0.06398x}, x = 75 ..190, title \right.$
$\qquad = "Violence\ versus\ Literacy");$

> $p2 := pointplot\left(\left[literacyratedata, violencedata \right] \right):$

> $display(\{p1, p2\});$

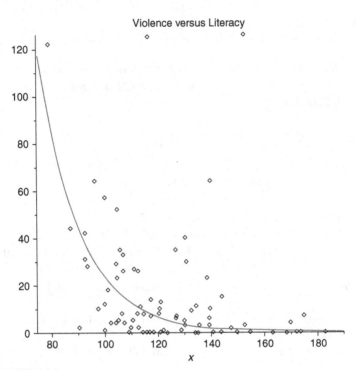

FIGURE 5.31

Plot of Poisson regression of SigActs of violence.

Plots (Figure 5.32)

> $p1 := pointplot([xhc, yhc], title = "Plots"):$

> $p2 := plot(exp(xx2x + xx1), x = 0\ 180, thickness = 3, color = BLUE):$

>

> $display(p1, p2);$

We summarize our statistics:

> $STAToutput(nobx, xx1, xx2, SE1, SE2, t1, t2, PV1, PV2, DF1, DF2, RD, ResD, SST);$

"Degrees of Freedom and Regression Deviations = ", 1, 2065.9443

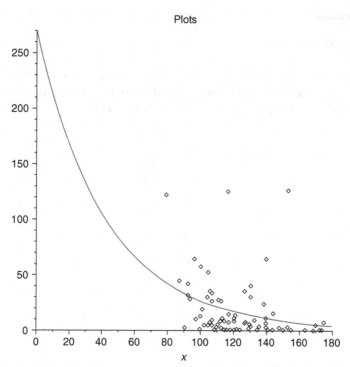

FIGURE 5.32
Poisson regression model.

"Degrees of Freedom and Residuals Deviations = ", 78, 309.389073 *200
"Total = ", 2375.33337348472469
"Mean Deviation = ", 3.966526577

Coefficient	Standard Error	T-Statistic	P-Value	Odds Ratio
5.599020	0.1653	33.8755		0.0004
−0.023550	0.0014	−16.4345	0.0018	0.9767

Again, we accept that the fit looks pretty good. We interpret the odds ratio for the coefficients to help explain our results, 0.976. This means that for every 1 unit increase in literacy Violence goes down slightly, 2.4%. This suggests improving literacy within the country.

Interpretation: Based on the findings of this research literacy affects violent conflict in the Philippines. Thus, considering literacy, the findings correspond to the claim that conflict and aggression is influenced by literacy.

5.6   Exercises

1. The following data represents changes in growth where x = body weight and y = normalized metabolic rate for 13 animals. Build a reasonable nonlinear regression model for $y = axb$

x	y
110	198
115	173
120	174
230	149
235	124
240	115
360	130
362	102
363	95
500	122
505	112
510	98
515	96

2. We have data on the average tread on a radial racing tire over time. The variable x will be the hours in heavy racing use and the variable y will be the average tread thickness in centimeters (cm). Plot the data and discuss trends and build a reasonable nonlinear regression model for y = axb.

Number	Hours(x)	Tread (cm) (y)
1	2	5.4
2	5	5.0
3	7	4.5
4	10	3.7
5	14	3.5
6	19	3.5
7	26	2.0
8	31	1.6
9	34	1.8
10	38	1.3
11	45	0.8
12	52	1.1
13	53	0.8
14	60	0.4
15	65	0.6

3. Let's assume our suspected data is nonlinear. Use nonlinear regression techniques to estimate the final parameters a, b, and c.

ROW	x	y	z
1	101	15	0.788
2	73	3	304.149
3	122	5	98.245
4	56	20	0.051
5	107	20	0.27
6	77	5	30.485
7	140	15	1.653
8	66	16	0.192
9	109	5	159.918
10	103	14	1.109
11	93	3	699.447
12	98	4	281.184
13	76	14	0.476
14	83	5	54.468
15	113	12	2.81
16	167	6	144.923
17	82	5	79.733
18	85	6	21.821
19	103	20	0.223
20	86	11	1.899
21	67	8	5.18
22	104	13	1.334
23	114	5	110.378
24	118	21	0.274
25	94	5	81.304

5.7 Projects

1. We need more advanced fitting techniques. The method chosen for illustrative purpose is Gauss-Newton, a generalized nonlinear least squares program. The gradient of numerical partial derivatives is used in the solution process. The least squares estimates are found by iterating the matrix algebra equations:

$$b = (D'D)^{-1} DY$$

$g_1 = g_0 + b$, where
  g is the matrix of refined estimates
  b is the change in the estimates
  Y is the matrix of errors, $(y - f(x))$, and
  D is the matrix of the partial derivatives with respect to the unknown parameters.

The procedure iterates until the absolute difference between the sum of squared errors of successive iterations is less than some user-defined tolerance. Resolve our tire tread problem using this technique.

5.8 References and Suggested Further Reading

Afifi, A. and S. Azen. (1979). *Statistical Analysis: A Computer Oriented Approach*, 2nd ed. Academic Press, NY, pp. 143–144.

Devore, J. (1995). *Probability and Statistics for Engineering and the Sciences*, 4th ed. Wadsworth Publishing, Belmont, pp. 474–509.

Fox, W. P. (2011). Using the Excel Solver for nonlinear regression, *Computers in Education Journal (COED)*. October–December, 2(4): 77–86.

Fox, W. P. (2012). Issues and importance of "good" starting points for nonlinear regression for mathematical modeling with Maple: Basic model fitting to make predictions with oscillating data, *JCMST* 31(1): 1–16.

Giordano, F., W. Fox, and S. Horton (2013). *A First Course in Mathematical Modeling*. Cengage Publishing, Boston: MA.

Johnston, I. (2000). *I'll Give You a Definite Maybe: An Introductory Handbook on Probability, Statistics, and Excel Section 4: Correlations*. http://johnstoi.web.viu.ca/maybe/maybe/title.htm

Mendenhall, W. and T. Sincich (1996). *A Second Course in Statistics Regression Analysis*, 5th ed. Prentice Hall, Upper Saddle River, NJ, pp. 476–485.

Montgomery, D., E. Peck, and G. Vining (2006). *Introduction to Linear Regression Analysis*, 4th ed. John Wiley and Sons. Hoboken, NJ, pp. 428–448.

Neter, J., M. Kutner, C. Nachtsheim, and W. Wasserman (1996). *Applied Linear Statistical Models*, 4th ed. Irwin Press, Boston, pp. 609–614. www.oxfordjournals.org/our_journals/tropej/online/ma_chap13.pdf (April 10, 2012).

6

Linear, Integer, and Mixed Integer Programming

Objectives

1. Formulate a linear programming problem.
2. Solve and interpret a linear programming problem.
3. Perform sensitivity analysis and interpret that analysis.

6.1 Introduction

Consider the Emergency Service Coordinator (ESC) for a county that is interested in locating the county's three ambulances to maximize the residents that can be reached within 8 minutes in emergency situations. The county is divided into 6 zones and the average time required to travel from one region to the next under semi-perfect conditions are summarized in Table 6.1.

The population in zones 1, 2, 3, 4, 5, and 6 are given in Table 6.2.

Problem Identification: Determine the location for placement of the ambulances to maximize coverage within the allotted time.

Assumptions: Time travel between zones is negligible. Times in the data are averages under ideal circumstances.

Here we further assume that employing an optimization technique would be worthwhile. We begin by assuming a linear model and then we will enhance the model with integer programming.

Perhaps, consider planning the shipment of needed items from the warehouses where they are manufactured and stored to the distribution centers where they are needed.

There are three warehouses at different cities: Detroit, Pittsburgh, and Buffalo. They have 250, 130, and 235 tons of paper accordingly. There are four publishers in Boston, New York, Chicago, and Indianapolis. They ordered 75,

TABLE 6.1

Average travel times from zone *i* to zone *j* in perfect conditions

	1	2	3	4	5	6
1	1	8	12	14	10	16
2	8	1	6	18	16	16
3	12	18	1.5	12	6	4
4	16	14	4	1	16	12
5	18	16	10	4	2	2
6	16	18	4	12	2	2

TABLE 6.2

Populations in each zone

1	50,000
2	80,000
3	30,000
4	55,000
5	35,000
6	20,000
Total	270,000

TABLE 6.3

Distribution costs

From \ To	Boston (BS)	New York (NY)	Chicago (CH)	Indianapolis (IN)
Detroit (DT)	15	20	16	21
Pittsburgh (PT)	25	13	5	11
Buffalo (BF)	15	15	7	17

230, 240, and 70 tons of paper to publish new books. The costs in dollars of transportation of one ton of paper are given in Table 6.3.

Management wants you to minimize the shipping costs while meeting demand. This problem involves the allocation of resources and can be modeled as a linear programming problem as we will discuss.

In engineering management the ability to optimize results in a constrained environment is crucial to success. Additionally, the ability to perform critical sensitivity analysis, or "what if analysis" is extremely important for decision-making. Consider starting a new diet which needs to be healthy. You go to a nutritionist that gives you lots of information on foods. They recommend sticking to six different foods: Bread, Milk, Cheese, Fish, Potato, and Yogurt: and provide a table of information (Table 6.4),including the average cost of the items.

TABLE 6.4

Average cost of nutrition items

	Bread	Milk	Cheese	Potato	Fish	Yogurt
Cost, $	2.0	3.5	8.0	1.5	11.0	1.0
Protein, g	4.0	8.0	7.0	1.3	8.0	9.2
Fat, g	1.0	5.0	9.0	0.1	7.0	1.0
Carbohydrates, g	15.0	11.7	0.4	22.6	0.0	17.0
Calories, Cal	90	120	106	97	130	180

We go to a nutritionist and she recommends that our diet contains not less than **150** calories, not more than **10g** of protein, not less than **10g** of carbohydrates and not less than 8g of fat. Also, we decide that our diet should have **minimal cost**. In addition we conclude that our diet should include at least **0.5g** of fish and not more than **1** cups of milk. Again this is an allocation of resources problem where we want the optimal diet at minimum cost. We have six unknown variables that define weight of the food. There is a lower bound for Fish as 0.5 g. There is an upper bound for Milk as 1 cup. To model and solve this problem, we can use linear programming.

Modern linear programming was the result of a research project undertaken by the US Department of Air Force under the title of Project SCOOP (Scientific Computation of Optimum Programs). As the number of fronts in the Second World War increased, it became more and more difficult to coordinate troop supplies effectively. Mathematicians looked for ways to use the new computers being developed to perform calculations quickly. One of the SCOOP team members, George Dantzig, developed the simplex algorithm for solving simultaneous linear programming problems. The simplex method has several advantageous properties: it is very efficient, allowing its use for solving problems with many variables; it uses methods from linear algebra, which are readily solvable.

To provide a framework for our discussions, we offer the following basic model:

Maximize (or minimize) $f(X)$

Subject to

$$g_i(X) \begin{Bmatrix} \geq \\ = \\ \leq \end{Bmatrix} b_i \text{ for all } i.$$

Now let us explain this notation. The various components of the vector X are called the decision variables of the model. These are the variables that can be controlled or manipulated. The function, $f(X)$, is called the objective function. By subject to, we connote that there are certain side conditions, resource requirement, or resource limitations that must be met. These conditions are called constraints. The constant b_i represents the level of the associated constraint $g(Xi)$ and is called the right-hand side in the model.

Linear programming is a method for solving linear problems, which occur very frequently in almost every modern industry. In fact, areas using linear programming are as diverse as defense, health, transportation, manufacturing, advertising, and telecommunications. The reason for this is that in most situations, the classic economic problem exists – you want to maximize output, but you are competing for limited resources. The 'Linear' in Linear Programming means that in the case of production, the quantity produced is proportional to the resources used and also the revenue generated. The coefficients are constants and no products of variables are allowed.

In order to use this technique, the company must identify a number of constraints that will limit the production or transportation of their goods; these may include factors such as labor hours, energy, and raw materials. Each constraint must be quantified in terms of one unit of output, as the problem-solving method relies on the constraints being used.

An optimization problem that satisfies the following five properties is said to be a linear programming problem.

- There is a unique objective function, $f(X)$.
- Whenever a decision variable, X, appears in either the objective function or a constraint function, it must appear with an exponent of 1, possibly multiplied by a constant.
- No terms contain products of decision variables.
- All coefficients of decision variables are constants.
- Decision variables are permitted to assume fractional as well as integer values.

Linear problems, by the nature of the many unknowns, are very hard to solve by human inspection, but methods have been developed to use the power of computers to do the hard work.

6.2 Formulating Linear Programming Problems

A linear programming problem is a problem that requires an objective function to be maximized or minimized subject to resource constraints. The key to formulating a linear programming problem is recognizing the decision variables. The objective function and all constraints are written in terms of these decision variables.

The conditions for a mathematical model to be a linear program (LP) are:

- all variables continuous (i.e. can take fractional values),
- a single objective (minimize or maximize),
- the objective and constraints are linear, i.e. any term is either a constant or a constant multiplied by an unknown,
- the decision variables must be non-negative.

LPs are important because:

- many practical problems can be formulated as LPs,
- there exists an algorithm (called the *simplex* algorithm) that enables us to solve LPs numerically relatively easily.

We will return later to the simplex algorithm for solving LPs but for the moment we will concentrate upon formulating LPs.

Some of the major application areas to which LP can be applied are:

- Blending
- Production planning
- Oil refinery management
- Distribution
- Financial and economic planning
- Manpower planning
- Blast furnace burdening
- Farm planning

We consider below some specific examples of the types of problem that can be formulated as LPs. Note here that the key to formulating LPs is *practice*. However a useful hint is that common objectives for LPs are *minimize cost/maximize profit*.

Example 6.1 Manufacturing

Consider the following problem statement: A company wants to can two new different drinks for the holiday season. It takes 2 hours to can one gross of Drink A, and it takes 1 hour to label the cans. It takes 3 hours to can one gross of Drink B, and it takes 4 hours to label the cans. The company makes $10 profit on one gross of Drink A and a $20 profit of one gross of Drink B. Given that we have 20 hours to devote to canning the drinks and 15 hours to devote to labeling cans per week, how many cans of each type drink should the company package to maximize profits?

<u>Problem Identification</u>: Maximize the profit of selling these new drinks.
<u>Define variables</u>:

X_1 = the number of gross cans produced for Drink A per week
X_2 = the number of gross cans produced for Drink B per week

<u>Objective Function</u>:

$$Z = 10X_1 + 20X_2$$

<u>Constraints</u>:

(1)   Canning with only 20 hours available per week

$$2\,X_1 + 3\,X_2 \le 20$$

(2)   Labeling with only 15 hours available per week

$$X_1 + 4\,X_2 \le 15$$

(3)   Non-negativity restrictions

$X_1 \ge 0$ (non-negativity of the production items)
$X_2 \ge 0$ (non-negativity of the production items)

<u>The Complete Formulation</u>:

$$\text{Maximize } Z = 10X_1 + 20X$$

subject to

$$2\,X_1 + 3\,X_2 \le 20$$
$$X_1 + 4\,X_2 \le 15$$
$$X_1 \ge 0$$
$$X_2 \ge 0$$

We will see in the next section how to solve these two-variable problems graphically.

Example 6.2 Financial Planning

A bank makes four kinds of loans to its personal customers and these loans yield the following annual interest rates to the bank:

- First mortgage 14%

- Second mortgage 20%

- Home improvement 20%

- Personal overdraft 10%

The bank has a maximum foreseeable lending capability of $250 million and is further constrained by the policies:

1. First mortgages must be at least 55% of all mortgages issued and at least 25% of all loans issued (in $ terms).

2. Second mortgages cannot exceed 25% of all loans issued (in $ terms).

3. To avoid public displeasure and the introduction of a new windfall tax, the average interest rate on all loans must not exceed 15%.

Formulate the bank's loan problem as an LP so as to maximize interest income while satisfying the policy limitations.

Note here that these policy conditions, while potentially limiting the profit that the bank can make, also limit its exposure to risk in a particular area. It is a fundamental principle of risk reduction that risk is reduced by spreading money (appropriately) across different areas.

Financial Planning Formulation

Note here that as in *all* formulation exercises, we are translating a verbal description of the problem into an *equivalent* mathematical description.

A useful tip when formulating LPs is to express the variables, constraints, and objective in words before attempting to express them in mathematics.

Variables

Essentially, we are interested in the amount (in dollars) the bank has loaned to customers in each of the four different areas (not in the actual number of such loans). Hence let x_i = amount loaned in area i in millions of dollars (where $I = 1$ corresponds to first mortgages, $i = 2$ to second mortgages, etc.) and note that each $x_i \geq 0$ ($I = 1,2,3,4$).

Note here that it is conventional in LPs to have all variables ≥ 0. Any variable (X, say) which can be positive *or* negative can be written as $X_1 - X_2$ (the difference between two new variables) where $X_1 \geq 0$ and $X_2 \geq 0$.

Constraints

(a) limit on amount lent

$$x_1 + x_2 + x_3 + x_4 \leq 250.$$

(b)   policy condition 1

$$x_1 \geq 0.55(x_1 + x_2)$$

i.e. first mortgages $> = 0.55$(total mortgage lending) and also

$$x_1 \geq 0.25(x_1 + x_2 + x_3 + x_4).$$

i.e. first mortgages ≥ 0.25(total loans)

(c)   policy condition 2

$$x_2 \leq 0.25(x_1 + x_2 + x_3 + x_4).$$

(d)   policy condition 3 – we know that the total annual interest is $0.14x_1$ + $0.20x_2$ + $0.20x_4$ + $0.10x_4$ on total loans of $(x_1 + x_2 + x_3 + x_4)$. Hence the constraint relating to policy condition (3) is $0.14x_1 + 0.20x_2 +$ $0.20x_3 + 0.10x_4 \leq 0.15(x_1 + x_2 + x_3 + x_4)$.

Objective Function
To maximize interest income (which is given above) i.e.

$$\text{Maximize } Z = 0.14x_1 + 0.20x_2 + 0.20x_3 + 0.10x_4.$$

Example 6.3 Blending Formulation

Consider the example of a manufacturer of animal feed who is producing feed mix for dairy cattle. In our simple example the feed mix contains two active ingredients. One kg of feed mix must contain a minimum quantity of each of four nutrients as below:

Nutrient	A	B	C	D
gram	90	50	20	2

The ingredients have the following nutrient values and cost

	A	B	C	D	Cost/kg
Ingredient 1 (gram/kg)	100	80	40	10	40
Ingredient 2 (gram/kg)	200	150	20	0	60

What should be the amounts of active ingredients in one kg of feed mix that minimizes cost?

Blending problem solution

Variables
In order to solve this problem, it is best to think in terms of one kilogram of feed mix. That kilogram is made up of two parts – ingredient 1 and ingredient 2:

x_1 = amount (kg) of ingredient 1 in one kg of feed mix
x_2 = amount (kg) of ingredient 2 in one kg of feed mix
where $x_1 \geq 0, x_2 \geq 0$.

Essentially these variables (x_1 and x_2) can be thought of as the recipe telling us how to make up one kilogram of feed mix.

Constraints

*   nutrient constraints

$$100x_1 + 200x_2 \geq 90 \ (nutrient A)$$
$$80x_1 + 150x_2 \geq 50 \ (nutrient \ B)$$
$$40x_1 + 20x_2 \geq 20 \ (nutrient \ C)$$
$$10x_1 \geq 2 \ (nutrient \ D).$$

*   balancing constraint (an *implicit* constraint due to the definition of the variables)

$$x_1 + x_2 = 1.$$

Objective Function

Presumably to minimize cost, i.e.

$$\text{Minimize } Z = 40x_1 + 60x_2.$$

This gives us our complete LP model for the blending problem.

Example 6.4 Production Planning Problem

A company manufactures four variants of the same table and in the final part of the manufacturing process there are assembly, polishing, and packing operations. For each variant the time required for these operations is shown below (in minutes) as is the profit per unit sold.

		Assembly	Polish	Pack	Profit ($)
Variant	1	2	3	2	1.50
	2	4	2	3	2.50
	3	3	3	2	3.00
	4	7	4	5	4.50

Given the current state of the labor force the company estimate that, each year, they have 100,000 minutes of assembly time, 50,000 minutes of polishing time, and 60,000 minutes of packing time available. How many of each variant should the company make per year and what is the associated profit?

Variables

Let:

x_i be the number of units of variant i ($i = 1, 2, 3, 4$) made per year where $x_i \geq 0$ I = 1, 2, 3, 4

Constraints

Resources for the operations of assembly, polishing, and packing

$$2x_1 + 4x_2 + 3x_3 + 7x_4 \leq 100{,}000 \ (assembly)$$
$$3x_1 + 2x_2 + 3x_3 + 4x_4 \leq 50{,}000 \ (polishing)$$
$$2x_1 + 3x_2 + 2x_3 + 5x_4 \leq 60{,}000 \ (packing)$$

Objective Function

$$\text{Maximize } Z = 1.5x_1 + 2.5x_2 + 3.0x_3 + 4.5x_4$$

Example 6.5 Shipping

Consider planning the shipment of needed items from the warehouses where they are manufactured and stored to the distribution centers where they are needed as shown in the introduction. There are three warehouses at different cities: Detroit, Pittsburgh, and Buffalo. They have 250, 130, and 235 tons of paper accordingly. There are four publishers in Boston, New York, Chicago, and Indianapolis. They ordered 75, 230, 240, and 70 tons of paper to publish new books.

There are the following costs in dollars of transportation of one ton of paper:

From \ To	Boston (BS)	New York (NY)	Chicago (CH)	Indianapolis (IN)
Detroit (DT)	15	20	16	21
Pittsburgh (PT)	25	13	5	11
Buffalo (BF)	15	15	7	17

Management wants you to minimize the shipping costs while meeting demand.

We define x_{ij} to be the travel from city i(1 is Detroit, 2 is Pittsburg, 3 is Buffalo) to city j (1 is Boston, 2 is New York, 3 is Chicago, and 4 is Indianapolis).

$$\text{Minimize } Z = 15x_{11} + 20x_{12} + 16x_{13} + 21x_{14} + 25x_{21} + 13x_{22} + 5x_{23} + 11x_{24} + 1\,5x_{31}$$
$$+ 15x_{32} + 7x_{33} + 17x_{34}$$

Subject to:

$$x_{11} + x_{12} + x_{13} + x_{14} \leq 250 \text{ (availability in Detroit)}$$
$$x_{21} + x_{22} + x_{23} + x_{24} \leq 130 \text{ (availability in Pittsburg)}$$
$$x_{31} + x_{32} + x_{33} + x_{34} \leq 235 \text{ (availability in Buffalo)}$$
$$x_{11} + x_{21} + x_{31} \geq 75 \text{ (demand Boston)}$$
$$x_{12} + x_{22} + x_{32} \geq 230 \text{ (demand New York)}$$
$$x_{13} + x_{23} + x_{334} \geq 240 \text{ (demand Chicago)}$$
$$x_{14} + x_{24} + x_{34} \geq 70 \text{ (demand Indianapolis)}$$
$$x_{ij} \geq 0.$$

Integer Programming

For Integer programming, we will take advantage of technology. We will not present the branch and bound technique but suggest a thorough review of the topic can be found in Winston or other similar math programming texts.

For mixed integer programming we show the formulations and the use of via technology.

Nonlinear Programming

It is not our plan to present material on how to formulate or solve nonlinear programs in this chapter. Often, we have nonlinear objective functions or constraints. Suffice it to say, we will recognize these and use technology to assist in the solution. Excellent nonlinear programming can be read for additional information.

6.3 Graphical Linear Programming

Many applications in business and economics involve a process called optimization. In optimization problems, you are asked to find the minimum or the maximum result. This section illustrates the strategy in graphical simplex of linear programming. We will restrict ourselves in this graphical context to two-dimensions. Variables in the simplex method are restricted to positive variables (for example $x \geq 0$).

A two-dimensional linear programming problem consists of a linear objective function and a system of linear inequalities called resource constraints. The objective function gives the linear quantity that is to be maximized (or minimized). The constraints determine the *set of feasible solutions*. Let's illustrate.

New Components to an iPhone XS

Let's start with a manufacturing example. Suppose a small business wants to know how many of two types of high-speed computer chips to manufacturer weekly to maximize their profits. First, we need to define our decision variables. Let

x_1 = number of logic board to produce weekly,

x_2 = number of camera to produce week.

The company reports a profit of \$140 for each new logic board and \$120 for each new camera sold.

The production line reports the following information:

	Chip A	Chip B	Quantity available
Assembly time (hours)	2	4	1400
Installation time (hours)	4	3	1500
Profit (per unit)	140	120	

The constraint information from the table becomes inequalities that are written mathematical as:

$$2x_1 + 4x_2 \leq 1{,}400 \text{ (assembly time)}$$
$$4x_1 + 3x_2 \leq 1{,}500 \text{ (installation time)}$$
$$x_1 \geq 0, x_2 \geq 0.$$

The profit equation is:

$$\text{Profit } Z = 140x_1 + 120x_2.$$

The Feasible Region

We use the constraints of the linear program,

$$2x_1 + 4x_2 \leq 1{,}400 \text{ (assembly time)}$$
$$4x_1 + 3x_2 \leq 1{,}500 \text{ (installation time)}$$
$$x_1 \geq 0, x_2 \geq 0.$$

The constraints of a linear program, which include any bounds on the decision variables, essentially shape the region in the x–y plane that will be the domain for the objective function prior to any optimization being performed.

Every inequality constraint that is part of the formulation divides the entire space defined by the decision variables into 2 parts: the portion of the space containing points that violate the constraint, and the portion of the space containing points that satisfy the constraint.

It is very easy to determine which portion will contribute to shaping the domain. We can simply substitute the value of some point in either *half-space* into the constraint. Any point will do, but the origin is particularly appealing. Since there's only one origin, if it satisfies the constraint, then the *half-space* containing the origin will contribute to the domain of the objective function.

When we do this for each of the constraints in the problem, the result is an area representing the intersection of all the *half-spaces* that satisfied the constraints individually. This intersection is the domain for the objective function for the optimization. Because it contains points that satisfy all the constraints simultaneously, these points are considered feasible to the problem. The common name for this domain is the *feasible region*.

Consider our constraints:

$$2x_1 + 4x_2 \leq 1{,}400 \text{ (assembly time)}$$
$$4x_1 + 3x_2 \leq 1{,}500 \text{ (installation time)}$$
$$x_1 \geq 0, x_2 \geq 0.$$

For our graphical work we use the constraints: $x_1 \geq 0, x_2 \geq 0$ to set the region. Here, we are strictly in the $x_1 - x_2$ plane (the first quadrant).

Let's first take constraint #1 (assembly time) in the first quadrant: $2x_1 + 4x_2 \leq 1{,}400$ shown in Figure 6.1

We graph each constraint as an equality, one at a time. We choose a point, usually the origin to test the validity of the inequality constraint. We shade all the areas where the validity holds. We repeat this process for all constraints to obtain Figure 6.2.

Figure 6.2 shows a plot of (1) the assembly hour's constraint and (2) the installation hour's constraint in the first quadrant. Along with the non-negativity restrictions on the decision variables, the intersection of the half-spaces defined by these constraints is the feasible region shown in red. This area represents the domain for the objective function optimization.

We region shaded in our feasible region.

Solving a Linear Programming Problem Graphically

We have decision variables defined and an objection function that is to be maximized or minimized. Although all points inside the feasible region provide feasible solutions the solution, if one exists, occurs according to the Fundamental Theorem of Linear Programming: *If the optimal solution exists, then it occurs at a corner point of the feasible region*

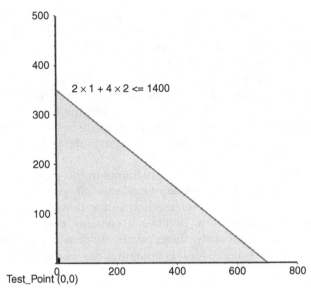

FIGURE 6.1
Shaded inequality.

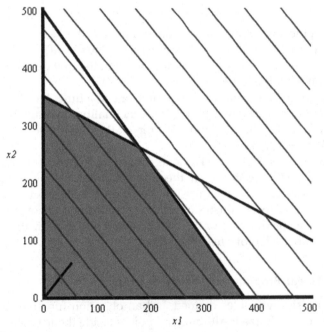

FIGURE 6.2
Plot of (1) the assembly hour's constraint and (2) the installation hour's constraint in the first quadrant.

Notice the various corners formed by the intersections of the constraints in example. These points are of great importance to us. There is a cool theorem (didn't know there were any of these, huh?) in linear optimization that states, "if an optimal solution exists, then an optimal corner point exists." The result of this is that any algorithm searching for the optimal solution to a linear program should have some mechanism of heading toward the corner point where the solution will occur. If the search procedure stays on the outside border of the feasible region while pursuing the optimal solution, it is called an *exterior point* method. If the search procedure cuts through the inside of the feasible region, it is called an *interior point* method.

Thus, in a linear programming problem, if there exists a solution, it must occur at a corner point of the set of feasible solutions (these are the vertices of the region). Note that in Figure 6.2 the corner points of the feasible region are the four coordinates and we might use algebra to find these: (0,0), (0,350) (375,0), and (180,260).

How did we get the point (180,260)?

This point is the intersection of the lines: $2x + 4y = 1,400$ and $4x + 3y = 1,500$. We use matrix algebra and solve for (x,y) from

$$\begin{bmatrix} 2 & 4 \\ 4 & 3 \end{bmatrix} \begin{bmatrix} x \\ y \end{bmatrix} = \begin{bmatrix} 1400 \\ 1500 \end{bmatrix}$$

Now, that we have all the possible solution coordinates for (x, y), we need to know which is the optimal solution. We evaluate the objective function at each point and choose the best solution.

Assume our objective function is to Maximize $Z = 2x + 2y$. We can set up a table of coordinates and corresponding Z-values as follows.

Coordinate of Corner Point	9. $Z = 140x + 120y$
(0,0)	$Z = 0$
(0,350)	$Z = 42,000$
(180,260)	$Z = 56,400$ *
(375,0)	$Z = 52,500$
Best solution is (180,260)	**$Z = 56,400$**

Graphically, we see the result by plotting the objective function line, $Z = 2x + 2y$, with the feasible region. Determine the parallel direction for the line to maximize (in this case) Z. Move the line parallel until it crosses the last point in the feasible set. That point is the solution. The line that goes through the origin at a slope of $-2/2$ is called the ISO-Profit line. We have provided this Figure 6.3 below:

We summarize the steps for solving a linear programming problem involving only two variables.

1. Sketch the region corresponding to the system of constraints. The points satisfying all constraints make up the feasible solution.

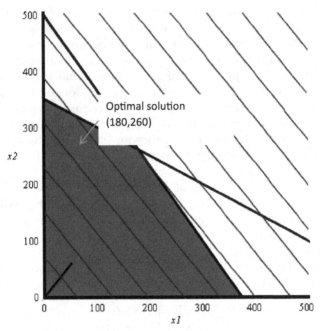

FIGURE 6.3
Iso-profit lines added.

2. Find all the corner points (or intersection points in the feasible region).
3. Test the objective function at each corner point and select the values of the variables that optimize the objective function. For bounded regions, both a maximum and a minimum will exist. For an unbounded region, if a solution exists, it will exist at a corner.

Minimization Problem

$$\text{Minimize } Z = 5x + 7y$$

Subject to: $2x + 3y \geq 6$
$3x - y \leq 15$
$-x + y \leq 4$
$2x + 5y \leq 27$
$x \geq 0$
$y \geq 0$

The corner points in Figure 6.4 are (0,2), (0,4,) (1,5), (6,3), (5,0), and (3,0). See if you can find all these corner points.

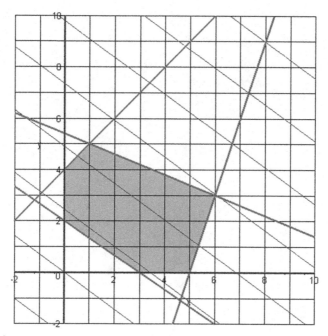

FIGURE 6.4
Solution zone for minimization problem iso-profit lines added.

If we evaluate $Z = 5x + 7y$ at each of these points, we find:

Corner point	$Z = 5x + 7y$ (Minimize)
(0,2)	$Z = 14$
(1,5)	$Z = 40$
(6,3)	$Z = 51$
(5,0)	$Z = 25$
(3,0)	$Z = 15$
(0,4)	$Z = 28$

The minimum value occurs at (0, 2) with a Z value of 14. Notice in our graph that the blue ISO-Profit line will last cross the point (0,2) as it move out of the feasible region in the direction that Minimizes Z.

Example 6.6   Unbounded Case

Let's examine the concept of an unbounded feasible region. Look at the constraints:

$$x + 2y \geq 4$$

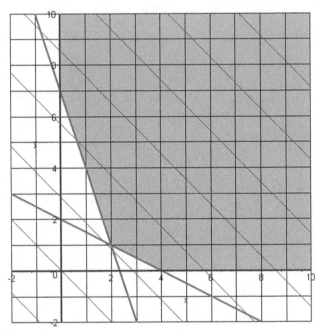

FIGURE 6.5
Solution space for unbounded case.

$$3x + y \geq 7$$

$$x \geq 0 \text{ and } y \geq 0$$

Note that the corner points are (0, 7), (2, 1) and (4, 0)and the region is unbounded (Figure 6.5).

If our solution is to Minimize $Z = x + y$ then our solution is: (2, 1) with $Z = 3$.

Determine why there is no solution to the LP to Maximize $Z = x + y$.

6.4 Technology Examples for Linear Programming

6.4.1 Excel for Linear Programming

Solver

Technology is critical to solving, analyzing, and performing sensitivity analysis on linear programing problems. Technology provides a suite of powerful, robust routines for solving optimization problems, including linear programs (LPs).

Consider the following example:

$$\text{Maximize } 2X1 + 6X2 + 5X3$$

Subject to:
$$X1 + X2 + X3 = 3$$
$$X1 + 2X2 + 3X3 \leq 10$$
$$2X1 + 6X2 + X3 \geq 5$$
$$X1, X2, X3 \geq 0$$

Using two variable models, we can graph it on paper and easily solve for the solution. With *more* than two variables (*X1, X2, X3*), LP must be used and this example will demonstrate how to use MS Excel in that capacity.

Step 1. Open Excel, and type in/set-up an LP that looks exactly like the one below (Figure 6.6):

Key: RHS = Right Hand Side (for constraints)

Const 1 = Constraint #1 (same for Const 2 and 3)

Dec vars = Decision Variables (the values for *X1*, *X2*, and *X3* we seek)

To this point, the Linear Program is all "just typing," with no blocks identified with formulas in them. A title is input in cell A1 – call it whatever you want, in this case "Linear Program Example" – and the spacings between cells are merely to make it an "easy to read" format up to this point. In future problems, you can label the constraints and Decision Variables any way you would like – Instead of "Const 1", you can use "Labor Hours", or "Amount of Material Available", etc… instead of Dec Vars, you can use "Number of fixtures to produce", etc.…. You can use any terminology you like, as long as

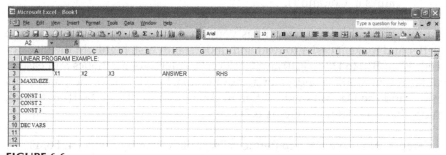

FIGURE 6.6
Excel worksheet maximize linear program (LP) example.

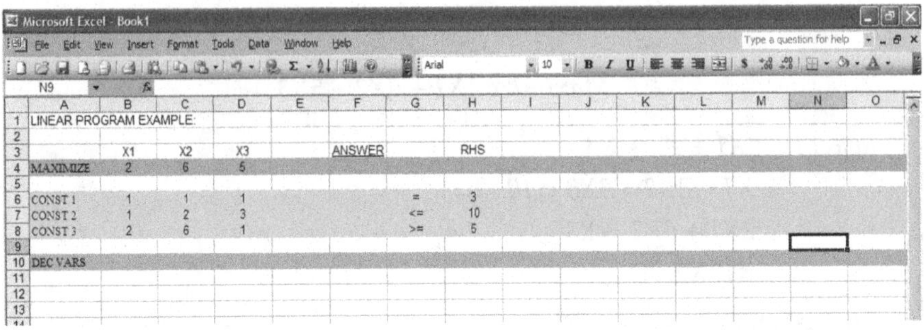

FIGURE 6.7a
Maximize LP example data.

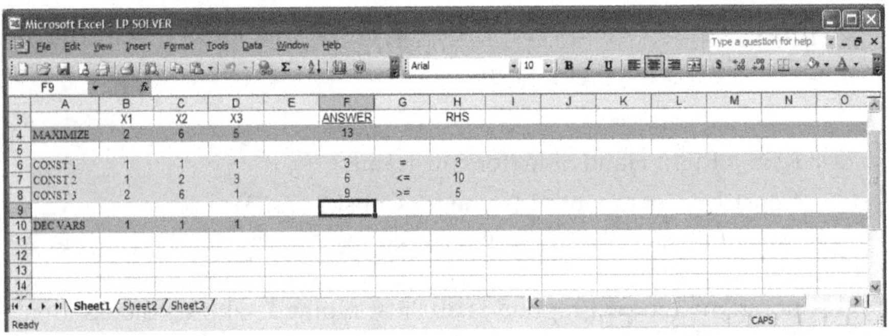

FIGURE 6.7b
Maximize LP defined cells.

you understand it – just remember that Dec Vars in Row 10 are the values of X1, X2, and X3 that you seek.

Step 2. Add in the numbers for your problem in the correct columns and rows. In Row G, add in the = and/or < = constraints, then the RHS values in Column H.

At this point, there are two things you can do to make your program look better and make it easier to read and understand: first, highlight the block of cells B3 to B10 to H3 to H10 and Click on "Center", for aesthetics and ease of data presentation; second, color in the Objective Function row, the Constraints rows, and the Dec Vars with different colors to offset them. Starting to look more like a model now (Figure 6.7a)!

Step 3. Now, we need to define cells. In this example, cells F4, F6, F7, F8 need to be defined. Also, in cells B10, C10, and D10, place the number "1" in each for reasons that will become clear later (briefly, they are to test that the answer cells F4, F6, F7 and F8 have been defined properly).

In cells F4, F6, F7 and F8, the "=" sign before the equation tells the cell that this is a mathematical formula to be computed.

Cell F4: This cell will contain the final answer to the Value of the Objective Function. Highlight the cell F4, and type in the following: = B4*B10 + C4*$C10$ + D4*D10, which signifies 2*X1 + 6*X2 + 5*X3.

After completion, the answer in F4 should be "13" at this point, since the Dec Vars are set to 1,1,1 ensuring your coding is correct. If your answer is not 13, re-check your coding in F4.

Cell F6: This cell contains the answer to the first constraint. Highlight cell F6, and type in the following:

= B6*B10 + C6 * C10 + D6* D10 which signifies 1*X1 + 1*X2 + 1*X3. Click on Enter -- the answer in F6 should be "3". Don't forget the "=" sign!

Cell F7: This cell contains the answer to the second constraint. Highlight cell F7, and type in the following:

= B7*B10 + C7 * C10 + D7* D10 which signifies 1*X1 + 2*X2 + 3*X3. Click on Enter: the answer in F7 should be "6".

Cell F8: This cell contains the answer to the third constraint. Highlight cell F8, and type in the following:

= B8*B10 + C8 * C10 + D8* D10 which signifies 2*X1 + 6*X2 + 1*X3. Click on Enter: the answer in F8 should be "9".

Now we are ready to use the Solver (Figure 6.7b). Save what you have, then go to "Tools" and then select "Solver". (If you cannot find Solver, it will be in Add-ins – extract it from there.)

1. Set Target Cell: Input F4.
2. Select "Max" button.
3. In the "By Changing Cells" area, merely highlight all three cells in B10, C10, and D10 with the mouse.
4. Now move to the constraints (Figure 6.8). Click on "Add".
   a. In the first constraint: Type in F6, select = , and type in the number 3, in the three areas. Click on OK.
   b. then Add ….
   c. In the second constraint: Type in F7, select ≤, type in 10. Click on OK, then Add…
   d. In the third constraint: Type in F8, select ≥, type in 5. Click on OK.

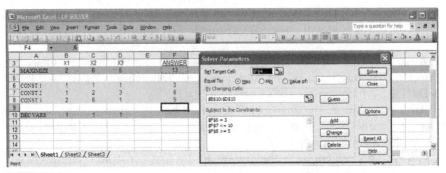

FIGURE 6.8
Maximize LP Solver parameter screen.

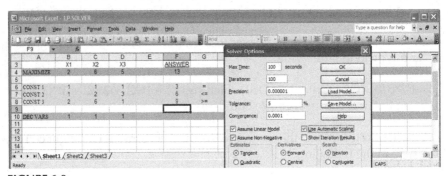

FIGURE 6.9
Maximize LP Solver options.

5. Now go to Options: Select "Assume Linear Model", "Assume Non-Negative", "Use Auto Scaling".
6. Click on OK, then Click on Solve (Figure 6.9).

The solution should now be revealed (Figure 6.10):

Objective Function = **18** in Cell F4

Constraints are Satisfied: 3 = 3

6 ≤ 10

18 ≥ 5

Decision Variables: **X1 = 0, X2 = 3, X3 = 0**

Thus, the problem is maximized by making three X2s only, with an objective value profit of 18. Now we present our previous example solved via each technology.

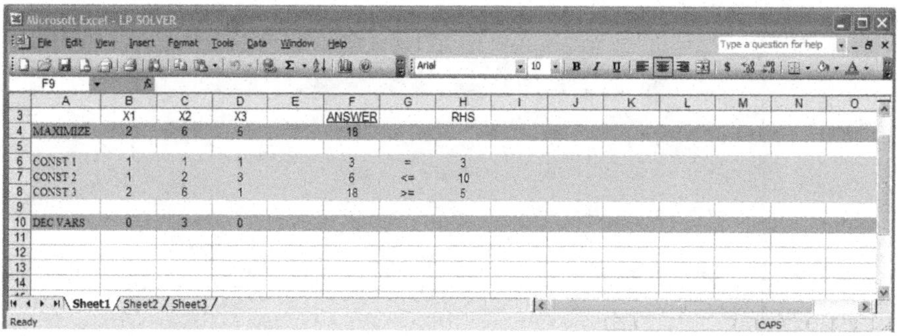

FIGURE 6.10
Maximize LP Solver solution.

$$\text{Maximize } Z = 25x_1 + 30x_2$$

Subject to:

$$20x_1 + 30x_2 \leq 690$$

$$5x_1 + 4x_2 \leq 120$$

$$x1, x2, \geq 0$$

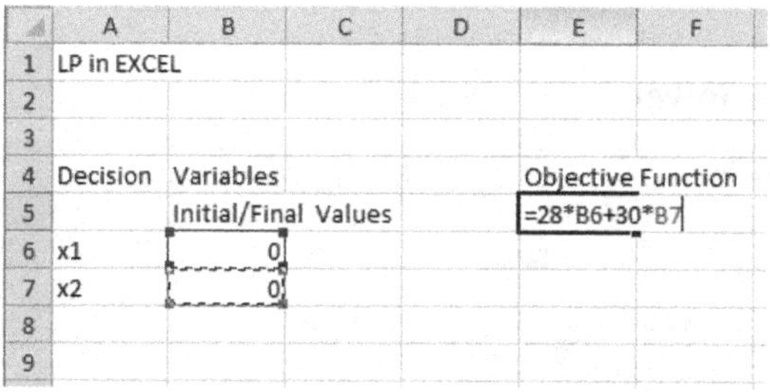

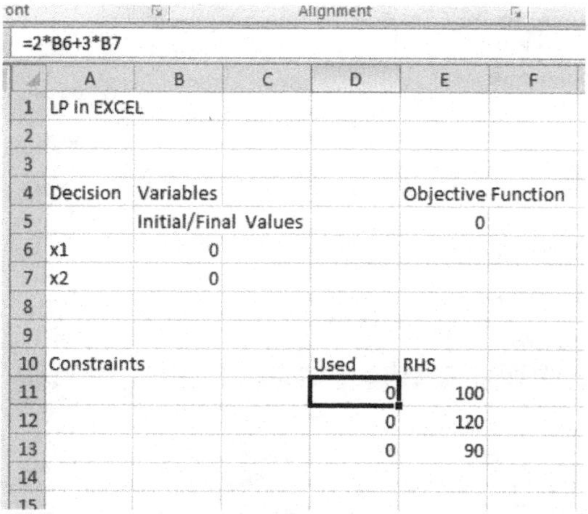

Solver

Constraints into Solver

We now have the Full Set UP. We can Click on Solve.

	A	B	C	D	E	F
1	LP in EXCEL					
2						
3						
4	Decision	Variables			Objective Function	
5		Initial/Final Values			972	
6	x1	9				
7	x2	24				
8						
9						
10	Constraints			Used	RHS	
11				90	100	
12				120	120	
13				90	90	
14						

Obtain the answers as $x1 = 9$, $x2 = 24$, $Z = 972$.

Additionally we can obtain reports from Excel. Two key reports are the answer report and the sensitivity report.

Answer Report

	A	B	C	D	E	F	G	H
1	Microsoft Excel 14.0 Answer Report							
2	Worksheet: [Book3]Sheet1							
3	Report Created: 8/28/2012 1:37:02 PM							
4	Result: Solver has converged to the current solution. All Constraints are satisfied.							
5	Solver Engine							
6	Engine: Simplex LP							
7	Solution Time: 0.015 Seconds.							
8	Iterations: 2 Subproblems: 0							
9	Solver Options							
10	Max Time Unlimited, Iterations Unlimited, Precision 0.000001, Use Automatic Scaling							
11	Max Subproblems Unlimited, Max Integer Sols Unlimited, Integer Tolerance 1%, Assume NonNegative							
12								
13								
14	Objective Cell (Max)							
15		Cell	Name	Original Value	Final Value			
16		$E$5	Initial/Final Values Objective Function	0	972			
17								
18								
19	Variable Cells							
20		Cell	Name	Original Value	Final Value	Integer		
21		$B$6	x1 Initial/Final Values	0	9	Contin		
22		$B$7	x2 Initial/Final Values	0	24	Contin		
23								
24								
25	Constraints							
26		Cell	Name	Cell Value	Formula	Status	Slack	
27		$D$11	Used	90	$D$11<=$E$11	Not Binding	10	
28		$D$12	Used	120	$D$12<=$E$12	Binding	0	
29		$D$13	Used	90	$D$13<=$E$13	Binding	0	

Sensitivity Report

	A B	C	D	E	F	G	H
1	Microsoft Excel 14.0 Sensitivity Report						
2	Worksheet: [Book3]Sheet1						
3	Report Created: 8/28/2012 1:37:03 PM						
4							
5							
6	Variable Cells						
7			Final	Reduced	Objective	Allowable	Allowable
8	Cell	Name	Value	Cost	Coefficient	Increase	Decrease
9	$B$6	x1 Initial/Final Values	9	0	28	6.285714286	8
10	$B$7	x2 Initial/Final Values	24	0	30	12	5.5
11							
12	Constraints						
13			Final	Shadow	Constraint	Allowable	Allowable
14	Cell	Name	Value	Price	R.H. Side	Increase	Decrease
15	$D$11	Used	90	0	100	1E+30	10
16	$D$12	Used	120	4.8	120	60	15
17	$D$13	Used	90	4.4	90	10	30
18							

We find our solution is $x_1 = 9$, $x_2 = 24$, P = \$972. From the standpoint of sensitivity analysis Excel is satisfactory in that it provides shadow prices.

Limitation: No tableaus are provided making it difficult to find alternate solutions.

Linear Programming in Excel using Matrices

We can use the revised simplex equations as matrices to create the tableau if desired. We define the following terms as matrices.

BV = matrix of basic variables,

B = right hand side column matrix,

a_j = column matrix for x_j in the jth constraints,

B = m x m matrix,

C_j = row matrix of costs,

C_{BV} = original costs of basic variables.

Summarizing formulas that we can use,

$$B^{-1}a_j$$

$$C_{BV}B^{-1}a_j - C_j$$

$$B^{-1}b$$

Objective Function

$$\text{Maximize } Z = 25x_1 + 30x_2$$

Subject to:

$$20x_1 + 30x_2 \le 690$$

$$5x_1 + 4x_2 \le 120$$

$$x1, x2, \ge 0$$

Set up the LP

Z	X1	X2	S1	S2	Rhs
1	−25	−30	0	0	0
0	20	30	1	0	690
0	5	4	0	1	120

Step 1. Place this matrix in Excel as shown below in Standard form.

B

Z	x1	x2	s1	s2	rhs
1	-25	-30	0	0	0
0	20	30	1	0	690
0	5	4	0	1	120

Step 2. Determine the entering variables as the cost in row 1 that is most
  negative. In this case, it is $x2$ at -30.
Step 3. Determines who leaves by performing the minimum positive ratio
  test. In this case we compare 690/30 and 120/4. We find 690/30 = 23
  is smaller so S1 departs.
Step 4. Write the new BV matrix replacing column 2 0,1,0 with the original
  column of $x2 - 30,30,4$.

New Bbv	1	-30	0
	0	30	0
	0	4	1

Step 5. Find the matrix inverse of the new Bbv matrix.

1	1	0
0	0.033333333	0
0	−0.133333333	1

Step 6. Multiple the new Bbv and the original B matrix to obtain the updated tableau.

z	x1	x2	s1	s2	rhs
1	-5	0	1	0	690
0	0.666666667	1	0.33333	0	23
0	2.333333333	0	-0.13333	1	28

Step 7. Check for optimality conditions, all Cj> = 0. If not return to Step 2 and repeat steps 2–7 again.
We are not optimal as the coefficient of x1 in the Cj row is –5.

We repeat steps 2–7.

B						
z	x1	x2	s1	s2	rhs
	1	-25	-30	0	0	0
	0	20	30	1	0	690
	0	5	4	0	1	120

We find we are now optimal as all cj> = 0. Our solution is $Z = 750$ when $x_1 = 12, x_2 = 15$.

6.4.2 Maple for Linear Programming

Maple provides a suite of powerful, robust routines for solving optimization problems, including linear programs (LPs). Using Maple's flexible mathematical programming language, it is simple to conduct thorough sensitivity studies on solutions to optimization problems. Maple has intrinsic built-in features that solve linear programming problems. We will present several of these features.

Method 1. With (simplex)

---
10. Command

---
with(simplex):
Maximize(objective function, constraints, NONNEGATIVE)
Subs(result, objective function)

Example 6.7 Maximize $Z = x + y$

Subject to:

$$x + 3y \le 9$$

$$2x + y \le 8$$

$$x \ge 0$$

$$y \ge 0$$

We evoke the (Simplex) library. Then we enter the set of constraints and call the set *cnsts*. Note we did not have to enter the non-negativity constraints. Then we enter the objective function and call it **obj**. We maximize the objective function with the constraints set and now tell Maple that the variables are non-negative. The results of the decision variables that satisfy the requirements are returned by Maple. We then substitute those results back into the objective function to get the optimal value for the problem.

```
> with(simplex):
>cnsts: = {x+3*y < = 9, 2*x+y< = 8}:
obj: = x +y:

> maximize(obj,cnsts,NONNEGATIVE);
{y = 2, x = 3}
> subs({y = 2, x = 3},obj);
5
```

The solution is $x = 3$, $y = 2$, and $Z = 5$.

Method 2. Optimization[LPSolve]

This method evokes the use of the new Optimization package in Maple and requires us only to enter the same inputs as in Method I but input them in the LP Solver. We present the LPSolve command.

Optimization[LPSolve] – solve a linear program

Calling Sequence

    LPSolve(**obj, constr, bd, opts**)

Parameters

    obj – **algebraic**; linear objective function
    constr– (optional) **set(relation)** or **list(relation)**; linear constraints

bd – (optional) sequence of **name = range**; bounds for one or more variables

opts – (optional) equation(s) of the form **option = value** where **option** is one of **assume, binaryvariables, depthlimit, feasibilitytolerance, infinitebound, initialpoint, integertolerance, integervariables, iterationlimit, maximize, nodelimit** or **output**; specify options for the **LPSolve** command

New Bbv	1	-30	0
	0	30	0
	0	4	1

Example 6.8 Revisit Example 6.7 with the New Commands

We enter the constraints and objective function as we did in Method 1.

> $cnsts := \{x + 3 * y \le 9, 2 * x + y \le 8\}; obj := x + y;$

$$cnsts := \{x + 3y \le 9, 2x + y \le 8\}$$

$$obj := x + y$$

> $with(Optimization):$

> $LPSolve(obj, cnsts, assume = nonnegative, \max imize = true);$

$$\left[5., \left[x = 3., y = 1.99999999999999978\right]\right]$$

>

The output provides solution as Z = 5, is x = 3, y = 1.99999999999999978. Note, we would round this value to 2.

1	1	0
0	0.033333333	0
0	-0.133333333	1

Method III. Linear Algebra and LPSolve Command

This method evokes the use of the Optimization package in Maple and requires the use and knowledge of linear algebra. The coefficient matrix, A, the coefficient of the constraints is entered as a matrix A. The right-hand side values b are entered as a column matrix B. The cost coefficients (or profit coefficients) of the objective function is entered as a column matrix C. We then use the LPSolve command.

> *restart*;

> *with*(*Optimization*) :

> *with*(*Linear Algebra*):

> $A := <<1, 2>|<3, 1>>;$

$$A := \begin{bmatrix} 1 & 3 \\ 2 & 1 \end{bmatrix}$$

> $B := <9,8>$

$$B := \begin{bmatrix} 9 \\ 8 \end{bmatrix}$$

> $C := <1,1>;$

$$C := \begin{bmatrix} 1 \\ 1 \end{bmatrix}$$

> . $mySol \; ; LPSolve\left(C,[A,B],\right.$
> $\left. assume = nonnegative , maximize\right);$

$$mySol := \begin{bmatrix} 5, & \begin{bmatrix} 3 \\ 1.999999999999999 \end{bmatrix} 7 \end{bmatrix}$$

Note the optimal answer is read as the value of the objective function is 5 when x = 3 and y = 1.9999999999999978.

6.4.3 R for Linear Programming

Here are the R commands to solve the following LP problem. We will install and load the LPSolve package to solve LP problems in R. We will resolve the following LP in R.

$$\text{Maximize } Z = x + y$$

Subject to:

$$x + 3y \leq 9$$

$$2x + y \leq 8$$

$$x \geq 0$$

$$y \geq 0$$

R Code:

```
library(lpSolve)
obj.fun<-c(1,1)
constr<-matrix(c(1,3,2,1), ncol = 2,byrow = TRUE)
constr.dir<-c("< = ","< = ")
rhs<-c(9,8)
prod.sol<-lp("max",obj.fun,constr,constr.dir,rhs,conoute.sens = TRUE)
#Assessing output
prod.sol$obj.val
prod.sol$solution
prod.sol$duals

>prod.sol$objval

[1] 5

>prod.sol$solution

[1] 3 2

>prod.sol$duals

[1] 0.2 0.4 0.0 0.0
```

6.5 Linear Programming Case Study

Case Study 6.1 Supply Chain Operations for Gasoline Distribution

In our case study, we present a linear programming model for supply chain design. We consider producing a new mixture of gasoline. We desire to maximize profit for distribution and sales of the new mixture. There is a supply chain involved with a product that must be modeled. The product is made up of components that are produced separately.

Crude oil type	Cost/barrel	Barrel avail (000 of barrels)
X10	$0.30	6,000
X20	$0.40	10,000
X30	0.48	12,000

Blending information is as follows:

Gasoline	Compound X10 (%)	Compound X20 (%)	Compound X30 (%)	Expected demand (000 of barrels)
Blend 1	≥ 0.30 X10	≤.50 X20	≥ 0.3 X30	4,000
Blend 2	≤.4 B	≥ 0.35 X20	≤ 0.40 X30	22,000

Each barrel of blend 1 can be sold for $1.10 and each barrel of blend 2 can sell for $1.20. Long-term contracts require at least 10,000 barrels of each blend to be produced.

Let i = crude type 1, 2, 3 (X10,X20,X30 respectively)

Let j = gasoline blend type 1,2 (Blend 1, Blend 2)

We define the following decision variables:

Gij = amount of crude i used to produce gasoline j

For example, G_{11} = amount of crude X10 used to produce Blend 1.

G_{12} = amount of crude type X20 used to produce Blend 1

G_{13} = amount of crude type X30 used to produce Blend 1

G_{12} = amount of crude type X10 used to produce Blend 2

G_{22} = amount of crude type X20 used to produce Blend 2

G_{32} = amount of crude type X30 used to produce Blend 2

Revenue = 1.1 *(G11 + G12 + G13) + 1.2(G21 + G22 + G23)
Cost = .3 (G11 + G21) + .4 (G12+G22) + 0.48 (G13 + G23)

LP formulation

$$\text{Maximize Profit} = \text{Revenue} - \text{Cost}$$

Subject to:
  Oil availability

  $G11 + G21 \leq 6{,}000$
  $G12 + G22 \leq 10{,}000$
  $G13 + G23 \leq 12{,}000$

Contracts

  $G11 + G12 + G13 \geq 10{,}000$
  $G21 + G22 + G23 \geq 10{,}000$

Composition or product mix in mixture format

  $G11/(G11+G12+G13) \geq 0.30$
  $G12/(G11+G12+G13) \leq 0.50$
  $G13/(G11+G12+G13) \geq 0.30$

  $G21/(G21+G22+G23) \leq 0.4$
  $G22/(G21+G22+G23) \geq 0.35$
  $G23/(G21+G22+G23) \leq 0.40.$

All decision variables are non-negative

z	x1	x2	s1	s2	rhs	
1	-5	0	1	0	690	
0	0.666666667	1	0.033333	0	23	
0	2.333333333	0	-0.13333	1	28	

6.5.1 Gasoline Distribution Case Study with Excel

G11	3000
G12	2200
G13	4800
G21	3000
G22	7800
G23	7200
OBJF	
	21040

CONSTRAINTS

USED	RHS	
6000	6000	
10000	10000	
12000	120000	
10000	10000	
18000	10000	
0	0	>
-2800	0	<
1800	0	>
-4200	0	<
1500	0	>
0	0<	

We examine the sensitivity report.

Microsoft Excel 16.0 Sensitivity Report
Worksheet:[Rook8]Sheet3
Report Created: 6/25/2020 2:11:19 PM

Variable cells

Cell	Name	Final value	Reduced cost	Objective coefficient	Allowable increase	Allowable decrease
$C$5	G11	3000	0	0.8	0	1E+30
$C$5	G12	2200	0	0.7	0.333333333	0
$C$7	G13	4800	0	0.62	0	1.2
$C$8	G21	3000	0	0.9	1E+30	
$C$9	G22	7600	0	0.8	0	0.333333333
$C$10	G23	7200	0	0.72	1E+30	0

Cell	Name	Final Value	Shadow price	Constraint RHside	Allowable increase	Allowable decrease
$C$18	USED	6000	0.9	6000	2000	3000
$C$19	USED	10000	0.8	10000	4.500	5503
$C$20	USED	12000	0.72	12000	3666.666667	3000
$C$21	USED	10000	-0.1	10000	8000	7333.333333
$C$22	USED	18000	0	10000	8000	1E+30
$C$23	USED	0	0	0	2200	1500
$C$24	USED	-2800	0	0	1E+30	2800
$C$25	USED	1800	0	0	1800	1E+30
$C$26	USED	-4200	0	0	1E+30	4200
$C$27	USED	1500	0	0	1500	1E+30
$C$28	USED	0	0	0	1500	2200

In our analysis of the sensitivity report, we noticed a possible alternate optimal solution. We obtain the second solution. Two solutions are found yielding a maximum profit of $21,040.00.

Decision variable	Z = $21040.00	Z = $21,040.00
G_{11}	3000	3414.41
G_{12}	2200	1483.99
G_{13}	4800	5101.60
G_{21}	3000	2585.59
G_{22}	7800	8516.01
G_{23}	7200	6898.40

Any oil available at a cost less than $0.90/ barrel would make an additional profit. Also if we could relax the long-term contract from 10,000 barrels to less for Blend 1, at a cost less than $0.72 then we could make an additional profit. The shadow price.

25	z	x1	x2	s1	s2	rhs
26	1	-5	0	1	0	690
27	0	0.666666667	1	0.033333	0	23
28	0	2.333333333	0	-0.13333	1	28
29						
30	x1 enters as -5<0					
31	ratio test	34.5				
32		12				
33	s2 leaves because 12<34.5					
34						
35		New BV	1	-30	-25	
36			0	30	20	
37			0	4	5	
38						
39		B^-1	1	0.714286	2.142857	
40			0	0.071429	-0.28571	
41			0	-0.05714	0.428571	
42						
43	1	-3.55271E-15	-3.55271E-15	0.714286	2.142857	750
44	0	0	1	0.071429	-0.28571	15
45	0	1	0	-0.05714	0.428571	12

6.5.2 Gasoline Distribution Case Study with Maple

```
>with(Optimization);
> Revenue: = 1.1*(G11+G12+G13)+1.2*(G21+G22+G23);
> Cost: = .3*(G11+G21)+.4*(G12+G22)+.48*(G13+G23);
```

Revenue: = 1.1 G11 + 1.1 G12 + 1.1 G13 + 1.2 G21 + 1.2 G22
            + 1.2 G23

Cost: = 0.3 G11 + 0.3 G21 + 0.4 G12 + 0.4 G22 + 0.48 G13
          + 0.48 G23

➢ Profit: = Revenue-Cost;

Profit: = 0.8 G11 + 0.7 G12 + 0.62 G13 + 0.9 G21 + 0.8 G22 + 0.72 G23

>constr: = {G11-.3*(G11+G12+G13) > = 0, G13-.3*(G11+G12+G13) > = 0, G22-.35*(G21+G22+G23) > = 0, G11+G12+G13 > = 10000, G21+G22+G23 > = 10000, G11+G21 < = 6000, G12+G22 < = 10000, G12-.5*(G11+G12+G13) < = 0, G13+G23 < = 12000, G21-.4*(G21+G22+G23) < = 0, G23-.4*(G21+G22+G23) < = 0};

>constr: = {0 < = 0.7 G11 - 0.3 G12 - 0.3 G13, 0 < = 0.7 G13 - 0.3 G11 - 0.3 G12, 0 < = 0.65 G22 - 0.35 G21 - 0.35 G23, 10000 < = G11 + G12 + G13, 10000 < = G21 + G22 + G23, G11 + G21 < = 6000, G12 + G22 < = 10000, G13 + G23 < = 12000, 0.5 G12 - 0.5 G11 - 0.5 G13 < = 0, 0.6 G21 - 0.4 G22 - 0.4 G23 < = 0, 0.6 G23 - 0.4 G21 - 0.4 G22 < = 0}

>LPSolve(Profit, constr, assume = nonnegative, maximize);

[21040., [G11 = 3000.00000000000, G12 = 2200., G13 = 4800., G21 = 3000.00000000000, G22 = 7800.00000000000, G23 = 7200.00000000000]]

G11	3000
G12	2200
G13	4800
G21	3000
G22	7800
G23	7200

OBJF	
	21040

CONSTRAINTS

USED	RHS	
6000	6000	
10000	10000	
12000	12000	
10000	10000	
18000	10000	
0	0	>
-2800	0	<
1800	0	>
-4200	0	<
1500	0	>
0	0	<

6.5.3 Gasoline Distribution Case Study with R

We use our model already formulated. Here is the R Code.
library(lpSolve)

```
obj.fun<-c(.8,.7,.62,.9,.8,.72)
```

```
constr<-matrix(c(1,0,0,1,0,0,0,0,1,0,0,1,0,0,0,0,1,0,0,1,1,1,1,0,0,0,0,0,0,0,1,1,1,.7
    ,-.7,-.7,0,0,0,
```

```
-.5,.5,-.5,0,0,0,-.3,-.3,.3,0,0,0,0,0,0,.6,.-.4,-.4,0,0,0,-.35,.65,-.35, 0,0,0, -.4, -.4,.6),
    ncol = 6, byrow = TRUE)
```

```
constr.dir<-c("< = ","< = ","< = ",">  = ",">  = ", "> = "."< = "."> = ",
    "< = ",">  = ","< = ")
```

```
rhs< = c(6000,10000,12000, 10000,10000,0,0,0,0,0,0)
```

```
prod.sol<-lp("max",obj.fun,constr,constr.dir,rhs,conoute.sens = TRUE)
#Assessing output
prod.sol$obj.val
prod.sol$solution
prod.sol$duals
```

```
>prod.sol$objval
[1] 21040
>prod.sol$solution
[1] 3000 2200 4800 3000 7800 7200
```

We could obtain the duals and notice that a dual variable for a non-basic variable is 0 indicating an alternate optimal solution,

Microsoft Excel 16.0 Sensitivity Report
Worksheet: [Book8]Sheet3
Report Created: 6/25/2020 2:11:19 PM

Variable Cells

Cell	Name	Final Value	Reduced Cost	Objective Coefficient	Allowable Increase	Allowable Decrease
$C$5	G11	3000	0	0.8	0	1E+30
$C$6	G12	2200	0	0.7	0.333333333	0
$C$7	G13	4800	0	0.62	0	1.2
$C$8	G21	3000	0	0.9	1E+30	0
$C$9	G22	7800	0	0.8	0	0.333333333
$C$10	G23	7200	0	0.72	1E+30	0

Constraints

Cell	Name	Final Value	Shadow Price	Constraint R.H. Side	Allowable Increase	Allowable Decrease
$C$18	USED	6000	0.9	6000	2000	3000
$C$19	USED	10000	0.8	10000	4500	5500
$C$20	USED	12000	0.72	12000	3666.666667	3000
$C$21	USED	10000	-0.1	10000	8000	7333.333333
$C$22	USED	18000	0	10000	8000	1E+30
$C$23	USED	0	0	0	2200	1500
$C$24	USED	-2800	0	0	1E+30	2800
$C$25	USED	1800	0	0	1800	1E+30
$C$26	USED	-4200	0	0	1E+30	4200
$C$27	USED	1500	0	0	1500	1E+30
$C$28	USED	0	0	0	1500	2200

Case Study 6.2 Recruiting Raleigh Office (modified from McGrath, 2007)

Although this is a simple model it was adopted by the US Army recruiting commend for operations. The model determines the optimal mix of prospecting strategies that a recruiter should use in a given week. The two prospecting strategies initially modeled and analyzed are phone and email prospecting. The data came from the Raleigh Recruiting Company United States Army Recruiting Command in 2006. On average each phone lead yields 0.041 enlistments and each email lead yields 0.142 enlistments. The 40 recruiters assigned to the Raleigh recruiting office prospected a combined 19,200 minutes of work per week via phone and email. The company's weekly budget is $60,000.

1		2	Phone (x1)	3	Email (x2)
4	Prospecting time (Minutes)	5	60 minutes per lead	6	1 minute per lead
7	Budget (dollars)	8	$10 per lead	9	$37 per lead.

The decision variables are:

$x1$ = Number of phone leads

$x2$ = Number of email leads

Maximize $Z = 0.041 x1 + 0.142 x2$.

Subject to

$60 x1 + 1 x2 < 19200$ (prospecting minutes available)

$10 x1 + 37 x2 < 60000$ (budget dollars available)

$x1, x2 > 0$ (non-negativity).

If we examine all the intersections point we find a sub-optimal point, $x1 = 294.29$, $x2 = 154.082$, achieving 231.04 recruitments.

We examine the sensitivity analysis report,

Microsoft Excel 14.0 Sensitivity Report
Worksheet: [Book4]Sheet1
Report Created: 5/5/2015 2:21:25 PM
Variable Cells

Cell	Name	Final Value	Reduced cost	Objective coefficient	Allowable increase	Allowable decrease
$B$3	x1	294.2986425	0	0.041	8.479	0.002621622
$B$4	x2	1542.081448	0	0.142	0.0097	0.141316667

Constraints

Cell	Name	Final value	Shadow price	Constraint RH side	Allowable increase	Allowable decrease
$C$10		19200	4.38914E-05	19200	340579	17518.64865
$C$11		60000	0.003836652	60000	648190	56763.16667
$C$12		294.2986425	0	1	293.2986425	1E+30
$C$13		1542.081448	0	1	1541.081448	1E+30

First, we maintain a mixed solution over a fairly large range of values for the coefficient of $x1$ and $x2$. Further the shadow prices provide additional information. A one unit increase in prospecting minutes available yields an increase of approximately 0.00004389 in recruits while an increase in budget of $1 yields an additional 0.003836652 recruits. At initial look, it appears as though we might be better off with an additional $1 in resource.

Let's assume that it costs only $0.01 for each additional prospecting minute. Thus we could get 100*0.00004389 or a 0.004389 increase in recruits for the same unit cost increase. In this case, we would be better off obtaining the additional prospecting minutes.

6.6 Sensitivity Analysis with Technology

6.6.1 Excel and Sensitivity Analysis

Let's revisit our iPhone example from Section 6.3 with a new management question. Additionally, after solving the problem the company wants to know whether they should increase hours of assembly or hours of installation to increase profits.

The linear program is:

$$\text{Maximize profit } Z = 140x_1 + 120x_2$$

$$2x_1 + 4x_2 \leq 1{,}400 \text{ (assembly time)}$$
$$4x_1 + 3x_2 \leq 1{,}500 \text{ (installation time)}$$
$$x_1 \geq 0, x_2 \geq 0$$

We solve again in Excel, Figure 6.11 and save the sensitivity report, Figure 6.12.

Decision Variables	
x1	180
x2	260
Z=	56400
Constraints	
Used	RHS
1400	1400
1500	1500

FIGURE 6.11
Screenshot of iPhone solution.

Microsoft Excel 16.0 Sensitivity Report
Worksheet: [Book11]Sheet2
Report Created: 6/28/2020 10:17:48 AM

Variable Cells

Cell	Name	Final Value	Reduced Cost	Objective Coefficient	Allowable Increase	Allowable Decrease
$C$6	x1	180	0	140	20	80
$C$7	x2	260	0	120	160	15

Constraints

Cell	Name	Final Value	Shadow Price	Constraint R.H. Side	Allowable Increase	Allowable Decrease
$B$12	Used	1400	6	1400	600	650
$B$13	Used	1500	32	1500	1300	450

FIGURE 6.12
Screenshot of sensitivity analysis report.

Currently, our solution is $x_1 = 180$, $x_2 = 260$ with a Profit of $Z = \$56,400$. If we can purchase either an hour increase in assembly or instillation which should we do? In our sensitivity analysis report we look at the shadow prices, 6 for assembly and 32 for instillation. If we assume the cost to increase is identical then buying an additional hour of instillation nets us an increase in profit of \$32 while an hour increase in assembly nets an increase of \$6. If the cost/hour is less than the net increase, then we should recommend the increase.

6.6.2 Maple and Sensitivity Analysis

We have looked at Maple to solve linear programming problems and to generate tableaus for further analysis on problems. In this section we explore the use of Mapleto generate sensitivity analysis. This sensitivity analysis is essential for good decision-making. We will present methods to obtain sensitivity analysis on parameters of (1) coefficient of non-basic variables (2) coefficient of basic variables, and (3) changes in the resources (the RHS values). We will examine these one at a time.

We must reduce the analysis to formula form using the following matrix notation for our analysis:

CBV: matrix of the original cost coefficients of the basic variables
   B^{-1}: the inverse of the matrix of the basic variables in the order in which they enter the basis.
   a_j: the original column of the resources coefficient of the jth variable.
   cj: the original column of the cost coefficient of the jth variable.
   b_j: = the resource limit of the jth constraint.

Let's start with the linear programming problem given by

$$\text{Maximize } Z = 60d + 30t + 20c.$$

Subject to:

$$8d + 6t + c \le 48$$

$$4d + 2t + 1.5c \le 20$$

$$2d + 1.5t + 0.5c < 8$$

$$d, t, c \ge 0$$

We solve this LP to obtain the following the solution and also to create the tableaus.

```
> .cbv: =≪ 0 >|< −20 >|< −60 ≫;
cbv: = [0 − 20 − 60]
.cbvl := convert(cbv, Matrix);
cbvl := [0 − 20 − 60]
.NewB :=≪ 1,0,0 >|< 1,1.5,.5 >|< 8,4,2 ≫;
```

$$NewB := \begin{bmatrix} 1 & 1 & 8 \\ 0 & 1.5 & 4 \\ 0 & 0.5 & 2 \end{bmatrix}$$

```
> .0NewBinv := MatrixInverse(NewB);
    NewBinv := [1., 2., − 8.], [0., 1.99999999999999956,
        − 3.99999999999999912], [0., − .4999999999999888,
        1.499999999999978]

> .CbvBinv := MatrixMatrixMultiply(−1 · cbvl, NewBinv)
CbvBinv := [0. 10. 10.]
> .a2:= ≪ 6, 2, 1.5 ≫ ;
```

$$a2 := \begin{bmatrix} 6 \\ 2 \\ 1.5 \end{bmatrix}$$

```
> .MatrixMatrixMultiply(CbvBinv, a2);
[.35]
> .c2 := − ≪ 30 + delta ≫ ;
c2 := [−30 − δ]
> .solve(35 − 30 − delta ≥ 0, delta);
RealRange(−∞, 5)
```

The solution is $Z = 280$ when $s_1 = 24$, $d = 2$, and $c = 8$. The importance of the initial and final tableau are that they provide information about the variables needed in the sensitivity analysis.

Coefficient of Non-basic variables.

We will reduce the analysis to formula form using the following matrix notation. We will use

$$\bar{c}_j = c_{BV} B^{-1} a_j - c_j$$

For the current solution basis to remain the optimal basis, then all non-basic variables coefficients must remain "greater than or equal to" zero in the final tableau (for a maximization problem). Thus, $c_j \geq 0$.

Change of Non-basic Variables Coefficient

```
> .cbv :=≪ 0 >|< −20 >|< −60 ≫;
cbv := [0  − 20  − 60]
.cbvl := convert(cbv, Matrix);
cbvl := [0  − 20  − 60]
.NewB :=≪ 1,0,0 >|< 1,1.5,.5 >|< 8,4,2 ≫;
```

$$NewB := \begin{bmatrix} 1 & 1 & 8 \\ 0 & 1.5 & 4 \\ 0 & 0.5 & 2 \end{bmatrix}$$

```
> .0NewBinv := MatrixInverse(NewB);
NewBinv := [1., 2., − 8.], [0., 1.99999999999999956,
−3.99999999999999912], [0., −.4999999999999888,
1.499999999999978]
> .CbvBinv := MatrixMatrixMultiply(−1·cbvl, NewBinv)
CbvBinv := [0.  10.  10.]
>.a2:= ≪ 6, 2, 1.5 ≫ ;
```

$$a2 := \begin{bmatrix} 6 \\ 2 \\ 1.5 \end{bmatrix}$$

```
> .MatrixMatrixMultiply(CbvBinv, a2);
[.35]
> .c2 := − ≪ 30 + delta ≫ ;
c2 := [−30 − δ]
> .solve(35 − 30 − delta ≥ 0, delta);
RealRange(−∞,  5)
```

This means that if the coefficient of table, t, that is currently at a value of 30 is less than 35 then tables will never become a basic variables.

Change in a Basic Variable Cost Coefficient

First, change coefficient for d and then the coefficient for c.
  Change in "d"

```
> .cbv :=≪ 0 >|< −20 >|< −60 − delta ≫;
cbv := [0  −20  −60  −′]
> .CbvBinv1 := [0., 10.00000000 − 0.499999999999999888 ′,
        10.00000000 + 1.49999999999999978 ′]
>.a1 :=≪ 8, 4, 2 ≫;
```

$$a1 := \begin{bmatrix} 8 \\ 4 \\ 2 \end{bmatrix}$$

> .a3 := ≪ 1, 1.5, .5 ≫;

$$a3 := \begin{bmatrix} 1 \\ 1.5 \\ 0.5 \end{bmatrix}$$

> .t1 := MatrixMatrixMultiply(CbvBinv1, a1);
t1 := [60.000000000 + 1.000000000 ´]
> .t2 := MatrixMatrixMultiply(CbvBinv1, a2);
t2 := [35.000000000 + 1.250000000 ´]
> .t3 := MatrixMatrixMultiply(CbvBinv1, a3);
t3 := [20.000000000]
solve({10 − .5 · delta ≥ 0, 10 + 1.5 · delta ≥ 0, 5 + 1.25 · delta ≥}, {delta});
>.
{−4. ≤ ´, ´ ≤ 20.}

Change in coefficient for "t"

> .cbv := ≪ 0 >|< −20 − delta> | < −60 ≫;
cbv := [0   −20   −´   −60]
> .CbvBinv1 := MatrixMatrixMultiply(−1 · cbv, NewBinv);
CbvBinv1 := [0., 10.000000000 + 1.99999999999999956 ´,
       10.00000000 − 3.99999999999999912 ´]
> .t1 := MatrixMatrixMultiply(CbvBinv1, a1);
t1 := [60.00000000]
> .t2 := MatrixMatrixMultiply(CbvBinv1, a2);
t2 := [35.00000000 − 2.000000000 ´]
> .t3 := MatrixMatrixMultiply(CbvBinv1, a3);
t3 := [20.00000000 + 1.00000000 ´]
solve({5 − 2 · delta ≥ 0, 10 + 2 · delta ≥ 0, 10 − 4 · delta ≥ 0}, {delta});
>.

$$\left\{ -5 \le ´, ´ \le \frac{5}{2} \right\}$$

Let's interpret these results. First, for the coefficient of "*d*" we find that if the value is between 56 and 80 that "*d*" will remain a basic variable. If the coefficient of "*c*" is between 15 and 22.5 then "*c*" remains a basic variable.

Change in RHS Values

Test with a change in b1, originally 48.

> .b := ≪ 48 + delta, 20, 8 ≫;

$$b := \begin{bmatrix} 48 + ´ \\ 20 \\ 8 \end{bmatrix}$$

> .*MatrixMatrixMultiply*(*NewBinv*,*b*);

$$\begin{bmatrix} 24.+1.' \\ 8.00000000 \\ 2.0000000 \end{bmatrix}$$

> . .*solve*(24 + delta ≥ 0, delta);

RealRange(−24, ∞)

This is interpreted as if the resource of constraint 1 currently at 48 units is at least 24 then "d" and 'c' remain as basic variables.

6.7 Exercises

1. With the rising cost of gasoline and increasing prices to consumers, the use of additives to enhance performance of gasoline is being considered. Consider two additives, Additive 1 and Additive 2. The following conditions must hold for the use of additives:

   - Harmful carburetor deposits must not exceed 1/2 lb per car's gasoline tank.
   - The quantity of Additive 2 plus twice the quantity of Additive 1 must be at least 1/2 lb per car's gasoline tank.
   - 1 lb of Additive 1 will add 10 octane units per tank, and 1 lb of Additive 2 will add 20 octane units per tank. The total number of octane units added must not be less than six (6).
   - Additives are expensive and cost $1.53/lb for Additive 1 and $4.00/lb for Additive 2.
   - We want to determine the quantity of each additive that will meet the above restrictions and will minimize their cost.

   Required:

   (a) List the decision variables and define them.
   (b) List the objective function.
   (c) List the resources that constrain this problem.
   (d) Graph the "feasible region".
   (e) Label all intersection points of the feasible region.
   (f) Plot the objective function in a different color (highlight the objective function line, if necessary) and label it the ISO-Profit line.

(g) Clearly indicate on the graph the point that is the optimal solution.

(h) List the coordinates of the optimal solution and the value of the objective function.

(i) Assume now that manufacturer of additives has the opportunity to sell you a nice TV special deal to deliver at least 0.5 lbs of Additive 1 and at least 0.3 lbs of Additive 2. Use graphical LP methods to help recommend whether you should buy this TV offer. Support your recommendation.

(j) Write a one-page cover letter to your boss of the company that summarizes the results that you found.

2. A farmer has 30 acres on which to grow tomatoes and corn. Each 100 bushels of tomatoes require 1,000 gallons of water and 5 acres of land. Each 100 bushels of corn requires 6,000 gallons of water and 2 1/2 acres of land. Labor costs are $1 per bushel for both corn and tomatoes. The farmer has available 30,000 gallons of water and $750 in capital. He knows that he cannot sell more than 500 bushels of tomatoes or 475 bushels of corn. He estimates a profit of $2 on each bushel of tomatoes and $3 of each bushel of corn. How many bushels of each should he raise to maximize profits?

Required:

(a) List the decision variables and define them.

(b) List the objective function.

(c) List the resources that constrain this problem.

(d) Graph the "feasible region."

(e) Label all intersection points of the feasible region.

(f) Plot the objective function in a different color (highlight the objective function line, if necessary) and label it the ISO-Profit line.

(g) Clearly indicate on the graph the point that is the optimal solution.

(h) List the coordinates of the optimal solution and the value of the objective function.

(i) Assume now that farmer has the opportunity to sign a nice contract with a grocery store to grow and deliver at least 300 bushels of tomatoes and at least 500 bushels of corn. Use graphical LP methods to help recommend a decision to the farmer. Support your recommendation.

(j) If the farmer can obtain an additional 10,000 gallons of water for a total cost of $50, is it worth it to obtain the additional water? Determine the new optimal solution caused by adding this level of resource.

(k) Write a one-page cover letter to your boss that summarizes the result that you found.

3. *Fire Stone Tires*, headquartered in Akron, Ohio, has a plant in Florence, SC, which manufactures two types of tires: SUV 225 Radials and SUV 205 Radials. Demand is high because of the recent recall of tires. Each hundred SUV 225 Radials requires 100 gallons of synthetic plastic and 5 lbs of rubber. Each hundred SUV 205 Radials require 60 gallons synthetic plastic and 2 1/2 lbs of rubber. Labor costs are $1 per tire for each type tire. The manufacturer has weekly quantities available of 660 gallons of synthetic plastic, $750 in capital, and 300 lbs of rubber. The company estimates a profit of $3 on each SUV 225 radial and $2 of each SUV 205 radial. How many of each type tire should the company manufacture in order to maximize their profits?

Required:

(a) List the decision variables and define them.
(b) List the objective function.
(c) List the resources that constrain this problem.
(d) Graph the "feasible region".
(e) Label all intersection points of the feasible region.
(f) Plot the objective function in a different color (highlight the objective function line, if necessary) and label it the ISO-Profit line.
(g) Clearly indicate on the graph the point that is the optimal solution.
(h) List the coordinates of the optimal solution and the value of the objective function.
(i) Assume now that manufacturer has the opportunity to sign a nice contract with a tire outlet store to deliver at least 500 SUV 225 Radial tires and at least 300 SUV 205 radial tires. Use graphical LP methods to help recommend a decision to the manufacturer. Support your recommendation.
(j) If the manufacturer can obtain an additional 1,000 gallons of synthetic plastic for a total cost of $50, is it worth it to obtain this amount? Determine the new optimal solution caused by adding this level of resource.
(k) If the manufacturer can obtain an additional 20 lbs of rubber for $50, should they obtain the rubber? Determine the new solution caused by adding this amount.
(l) Write a one-page cover letter to your boss of the company that summarizes the results that you found.

4. Consider a toy maker that carves wooden soldiers. The company specializes in two types: Confederate soldiers and Union soldiers. The estimated profit for each is $28 and $30, respectively. A Confederate soldier requires 2 units of lumber, 4 hours of carpentry, and 2 hours of finishing to complete. A Union soldier requires 3 units of lumber, 3.5 hours of carpentry, and 3 hours of finishing to complete. Each week the company has 100 units of lumber delivered. The workers can provide at most 120 hours of carpentry and 90 hours of finishing. Determine the number of each type wooden soldiers to produce to maximize weekly profits. Formulate and then solve this linear programming graphically.

5. A company wants to can two different drinks for the holiday season. It takes 3 hours to can one gross of Drink A, and it takes 2 hours to label the cans. It takes 2.5 hours to can one gross of Drink B, and it takes 2.5 hours to label the cans. The company makes $15 profit on one gross of Drink A and an $18 profit of one gross of Drink B. Given that we have 40 hours to devote to canning the drinks and 35 hours to devote to labeling cans per week, how many cans of each type drink should the company package to maximize profits?

6. The Mariners Toy Company wishes to make three models of ships to maximize their profits. They found that a model steamship takes the cutter one hour, the painter 2 hours, and the assembler 4 hours of work; it produces a profit of $6.00. The sailboat takes the cutter 3 hours, the painter 3 hours, and the assembler 2 hours. It produces a $3.00 profit. The submarine takes the cutter one hour, the painter 3 hours, and the assembler one hour. It produces a profit of $2.00. The cutter is only available for 45 hours per week, the painter for 50 hours, and the assembler for 60 hours. Assume that they sell all the ships that they make, formulate this LP to determine how many ships of each type that Mariners should produce.

7. In order to produce 1,000 tons of non-oxidizing steel for BMW engine valves, at least the following units of manganese, chromium, and molybdenum, will be needed weekly: 10 units of manganese, 12 units of chromium, and 14 units of molybdenum (1 unit is 10 lbs). These materials are obtained from a dealer who markets these metals in three sizes small (S), medium (M), and large (L). One S case costs $9 and contains 2 units of manganese, 2 units of chromium, and one unit of molybdenum. One M case costs $12 and contains 2 units of manganese, 3 units of chromium, and one unit of molybdenum. One L case costs $15 and contains 1 unit of manganese, 1 units of chromium, and 5 units of molybdenum. How many cases of each kind (S, M, L) should be purchased weekly so that we have enough manganese, chromium, and molybdenum at the smallest cost?

8. The Super bowl Advertising agency wishes to plan an advertising campaign in three different media– television, radio, and magazines. The purpose or goal is to reach as many potential customers as possible. Results of a marketing study are given below:

	Day time TV	Prime time, TV	Radio	Magazines
Cost of advertising Unit	$40,000	$75,000	$30,000	$15,000
Number of potential customers reached per unit	400,000	900,000	500,000	200,000
Number of woman customers reached per unit	300,000	400,000	200,000	100,000

The company does not want to spend more than $800,000 on advertising. It further requires (1) at least 2 million exposures take place among woman; (2) TV advertising be limited to $500,000; (3) at least 3 advertising units be bought on day time TV and 2 units on prime time TV, and (4) the number of radio and magazine advertisement units should each be between 5 and 10 units.

9. A tomato cannery has 5,000 pounds of grade A tomatoes and 10,000 pounds of grade B tomatoes, from which they will make whole canned tomatoes and tomato paste. Whole tomatoes must be composed of at least 80% grade A tomatoes, whereas tomato paste must be made with at least 10% grade A tomatoes. Whole tomatoes sell for $0.08 per pound and grade B tomatoes sell for $0.05 per pound. Maximize revenue of the tomatoes.

HINT: Let X_{wa} = pounds of grade A tomatoes used to make whole tomatoes, X_{wb} = pounds of grade B tomatoes used to make whole tomatoes. The total number of whole tomato cans produced is the sum of $X_{wa} + X_{wb}$ after each is found. Also remember a percent is a fraction of the whole times 100%.

10. The McCow Butchers is a large-scale distributor of dressed meats for Myrtle Beach restaurants and hotels. Ryan's order meat for meatloaf (mixed ground beef, pork, and veal) for 1,000 pounds according to the following specifications:

   (a) Ground beef is to be no less than 400 lbs and no more than 600 lbs.
   (b) The ground pork is to be between 200 and 300 lbs.
   (c) The ground veal must weigh between 100 and 400 lbs.
   (d) The weight of the ground pork must be no more than one and one half (3/2) times the weight of the veal.

The contract calls for Ryan's to pay $1,200 for the meat. The cost per pound for the meat is: $0.70 for hamburger, $0.60 for pork, and $0.80 for the veal. How can this be modeled?

11. **Portfolio Investments**
    A portfolio manager in charge of a bank wants to invest $10 million. The securities available for purchase, as well as their respective quality ratings, maturate, and yields, are shown below.

Bond name	Bond type	Moody's quality scale	Bank's quality scale	Years to maturity	Yield at maturity	After-tax yield
A	MUNICI-PAL	Aa	2	9	4.3%	4.3%
B	AGENCY	Aa	2	15	5.4%	2.7%
C	GOVT 1	Aaa	1	4	5%	2.5%
D	GOVT 2	Aaa	1	3	4.4%	2.2%
E	LOCAL	Ba	5	2	4.5%	4.5%

The bank places certain policy limitations on the portfolios manager's actions:

(a) Government and Agency Bonds must total at least $4 million.
(b) The average quality of the portfolios cannot exceed 1.4 on the Bank's quality scale. Note a low number means high quality.
(c) The average years to maturity must not exceed 5 years.

Assume the objective is to maximize after-tax earnings on the investment.

6.8 References and Suggested Further Reading

Apaiah, R. and E. Hendrix (2006). Linear programming for supply chain design: A case on novel protein foods. Ph.D. Thesis, Wageningen University, Netherlands.

Balakrishnan, N., B. Render, and R. Stair (2007). *Managerial Decision Making*, 2nd ed. Prentice Hall, Saddle River, NJ.

Bazarra, M., S. J. J. Jarvis, and H. D. Sheralli (1990). *Linear Programming and Network Flows*. John Wiley & Sons, New York.

Ecker, J. and M. Kupperschmid (1988). *Introduction to Operations Research*. John Wiley & Sons, New York.

Fox, W. (2013). *Mathematical Modeling with Maple*. Cengage Publishers, Boston: MA.

Fox, W. (2018). *Mathematical Modeling for Business Analytics*. Taylor & Francis Publishers, CRC Press, Boca Raton, FL.

Fox, W. and W. Bauldry (2020). *Problems Solving with Maple*. Taylor & Francis, CRC Press, Boca Raton, FL.

Giordano, F., W. Fox, and S. Horton (2013). *A First Course in Mathematical Modeling*, 5th ed. Cengage, Boston, MA, chapter 7.

Hiller, F. S. and G. J. Liberman (1990). *Introduction to Mathematical Programming*. McGraw Hill Publishing Company, New York.

McGrath, G. (2007). Email marketing for the U.S. Army and Special Operations Forces Recruiting. Master's Thesis, Naval Postgraduate School. December.

Winston, W. L. (2002). *Introduction to Mathematical Programming Applications and Algorithms*, 4th ed. Duxbury Press, Belmont, CA.

7

Nonlinear Optimization Methods

Objectives

1. Formulate a nonlinear optimization problem.
2. Solve and interpret the nonlinear optimization problem.
3. Perform sensitivity analysis and interpret that analysis.

7.1 Introduction

Consider an oil-drilling rig that is 9.5 miles off shore. The drilling rig is to be connected by underwater pipe to a pumping station as depicted in Figure 7.1. The pumping station is connected by land-based pipe to a refinery, which is 14.5 miles down the shoreline from the drilling rig (see Figure 7.1). The underwater pipe costs $31,575 per mile, and land-based pipe costs $13,342 per mile. You are to determine where to place the pumping station to minimize cost of the pipe.

We will discuss how to find an optimal solution (if one exists) for the following unconstrained nonlinear optimization problem:

In this chapter, we will discuss models that require calculus to solve. We will review the calculus concepts for optimization and then apply them to modeling applications. We will introduce numerical methods to solve single variable optimization problems. We will present Excel's Solvers ability to solve single variable nonlinear optimization problems using both unconstrained and constrained optimization.

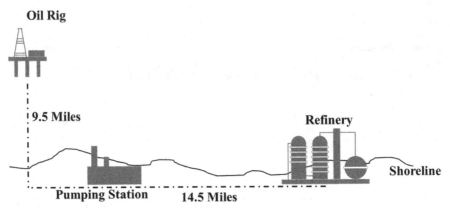

FIGURE 7.1
Oil drilling layout.

7.2 Unconstrained Single Variable Optimization and Basic Theory

We want to solve problems of the form:

$$\text{max (or min) } f(x)$$

$$x \in (a,b)$$

If $a = -\infty$ and $b = \infty$ then we are looking at R^2 in the xy plane. If either a, b, or both a and b are restricted then we must consider possible end points in our solution. We will examine three cases.

Case 1.   Points where $a < x < b$ and $f'(x) = 0$.

Case 2.   Points where $f'(x)$ does not exist.

Case 3.   End points a and b of the interval $[a, b]$.

Additionally, we will define points where $f'(x) = 0$ as critical points or stationary points. Here are some additional definitions and theorems from calculus that might be useful.

Definition: A function f has a maximum (global) at a point c, if $f(c) \geq f(x)$ for all x in the Domain of f and a function f has a minimum (global) at a point c, if $f(c) \leq f(x)$ for all x in the Domain of f.

Extreme Value Theorem: If f is continuous on a closed interval $[a, b]$, then f has both a global maximum and a global minimum over the interval.

We recall from your study of calculus the analysis of the first derivative test.

If $f'(x) > 0$ to the left of x^* and $f'(x) < 0$ to the right of x^*, then x^* is a local maximum and we find $f(x)$ increasing to the left and decreasing to the right of point x^*.

Also, recall the second derivative test. If $f'(x_0) = 0$, then we compute $f''(x_0)$.

If $f''(x_0) < 0$, then x_0 is a local maximum.

If $f''(x_0) > 0$, then x_0 is a local minimum.

If $f''(x_0) = 0$, then x_0 might be an inflection point.

Theorem 7.1: If $f'(x_0) = 0$ and $f''(x_0) < 0$, then x_0 is a local maximum.

If $f'(x_0) = 0$ and $f''(x_0) > 0$, then x_0 is a local minimum.

Theorem 7.2: If $f'(x_0) = 0$, and

1. if the first non-zero derivative at x_0 occurs at an odd-order
2. derivative then x_0 is neither a local maximum nor a local minimum
3. if the first non-zero derivative is positive and occurs at an
4. even-order derivative then x_0 is a local minimum
5. if the first non-zero derivative is negative and occurs at an even-order derivative then x_0 is a local maximum.

Case 2. Points where $f'(x)$ does not exist.

If $f(x)$ does not have a derivative at x_0, then x_0 might be a local maximum, a local minimum, or neither. In this case, we test points near x_0 and evaluate the function at those neighboring points where $x_1 < x_0 < x_2$.

Relationship between $f(x_0)$ and close neighbors	x_0 Classification
$f(x_0) > f(x_1), f(x_0) < f(x_2)$	Not a local extrema
$f(x_0) < f(x_1), f(x_0) > f(x_2)$	Not a local extrema
$f(x_0) \geq f(x_1), f(x_0) \geq f(x_2)$	Local maximum
$f(x_0) \leq f(x_1), f(x_0) \leq f(x_2)$	Local minimum

Case 3. Endpoints a and b.

From the figure below, we see the following:

1. If $f'(a) > 0$, then a is a local minimum.
2. If $f'(a) < 0$, then a is a local maximum.
3. If $f'(b) > 0$, then a is a local minimum.
4. If $f'(b) < 0$, then a is a local maximum.

If both $f'(a) = f'(b) = 0$, then draw a sketch and test neighboring points to determine if a and/or b is/are extrema.

Example 7.1 Minimize $1.5\, x^2$

Solution:

$$f(x) = 1.5\, x^2$$

$$\frac{df(x)}{dx} = 3x$$
$$3x = 0$$
$$x = 0$$
$$\frac{d^2 f(0)}{dx^2} = 3$$

The first derivative is equal to 0 at $x = 0$. The second derivative is positive at $x = 0$ so the critical point yields a minimal of the function.

Example 7.2 Minimize

Subject to. $-1 \le x \le 3$
  Solution: We find $f'(x) = 0$ at $x = 0$. The second derivative test yields $f''(x_0) = 0$ so we have an inflection point at x = 0. We also note the $f(0) = 0$. Next, we test the end points.

$$f(-1) = -1$$

$$f(3) = 27$$

We find a minimum at $x = -1$.

7.3 Models with Basic Applications of Max-Min Theory

Example 7.3 Chemical Company

A chemical manufacturing company sells sulfuric acid at a price of $100 per unit. If the daily total production cost in dollars for x units is:

$$C(x) = 100000 + 50x + 0.0025x^2$$

and the daily production is at most 7,000 units. How many units of sulfuric acid should the manufacturer produce to maximize daily profits?

Solution: Profit = Revenue– Cost

$$P = 100x - (100000 + 50x + 0.0025x^2)$$

$$P = -100000 + 50x - 0.0025x^2, \text{ for } 0 \le x \le 7000$$

$$\frac{dP}{dx} = 50 - .005x = 0$$

$$x = 10,000 \text{ units with } P(10,000) = \$150,000$$

$\dfrac{d^2P}{dx^2} = -.005$, since $P'' < 0$ we have a local maximum.

We must also check the end points.

$$x = 0, P = -\$100,000$$

$$x = 7,000, P = \$217,500$$

Our solution is $x = 7,000$ and $P = \$217,500$ at the end point since x* is outside of the domain of production even though its mathematical results are better: $x^* = 10,000$ units with $P(10,000) = \$240,000$.

Example 7.4 Manufacturing

A company wants to minimize their costs involved with manufacturing and storing items for future sales.

Let x = the number of batches of the item produced per year.
Let s = the storage costs of one unit for one year.
Let F = the fixed set up costs to produce the item (includes insurance, machines, labor, etc.).
Let v = the variable cost in producing one unit of the item.
Let T = the total number of units produced annually.

Total Production Cost, $M(x) = (F + \dfrac{vT}{x}) x$

Average Storage Cost, $s(x) = \dfrac{kT}{2x}$

Total Cost Function, $C(x) = \left(F + \dfrac{vT}{x}\right)x + \dfrac{kT}{2x}$

Solution: $C(x) = \left(F + \dfrac{vT}{x}\right)x + \dfrac{kT}{2x}$

$$\frac{dC(x)}{x} = F - \frac{kT}{2x^2} = 0$$

$$x^* = \sqrt{\frac{kT}{2F}}$$

$\dfrac{d^2C(x)}{dx^2} = \dfrac{kT}{x^3} > 0$ for real values of k, T, and x.

Since the 2nd derivative is greater than zero, $\dfrac{kT}{x^3} > 0$, the critical point,

$x^* = \sqrt{\dfrac{kT}{2x}}$, represents a minimum.

Example 7.5 SP6 Computer Development

A company spends $200 in variable costs to produce a SP6 computer, plus a fixed cost of $5,000 if any SP6 computers are produced. If the company spends x dollars on advertising their new SP6 computer, it can sell $x^{1/2}$ at $500 per computer. How many SP6 computers should the company produce to maximize profits?

Solution: *Maximize Profit = Revenue − Cost*
Cost = fixed + variable + advertising costs $= 5000 + 200 \cdot x^{1/2} + x$
Revenue $= 500 \cdot x^{1/2}$
Maximize P $= 500 \cdot x^{1/2} - (5{,}000 + 200 \cdot x^{1/2} + x)$

$$\frac{dP}{dx} = \frac{d(500 \cdot x^{1/2} - (5000 + 200 \cdot x^{1/2} + x))}{dx} = -\frac{-150 + \sqrt{x}}{\sqrt{x}} = \frac{150}{\sqrt{x}} - 1$$

Set $\dfrac{dP}{dx} = 0$, yields $\dfrac{150}{\sqrt{x}} - 1 = 0, x = 22{,}500$

$500 \cdot x^{1/2} - (5{,}000 + 200 \cdot x^{1/2} + x)$

$\dfrac{d^2P}{dx^2} = \dfrac{-75}{x^{3/2}} < 0$ for all $x > 0$. Therefore, we have found a maximum.

7.4 Technology Examples for Nonliner Optimization

7.4.1 Excel for Nonlinear Optimization

Excel's Solver

We may use technology, the Excel Solver, to solve these problems numerically. We illustrate by revisiting Example 7.3 (the chemical company example). The set up (Figure 7.2):

In Cell A4, we label the variable x. In cell B4, we initialize the value of x, say $x = 0$. In cell A7, we label P, the value of profit of our objective function. In cell B7, we put the equation,

$P = -100000 + 50x - 0.0025x^2$, in cell notation as
$= -10000 + 5\ 0*b4 - 0.0025*b4^\wedge2$.

Solver

We highlight cell b7, and open the Solver (Figure 7.3).

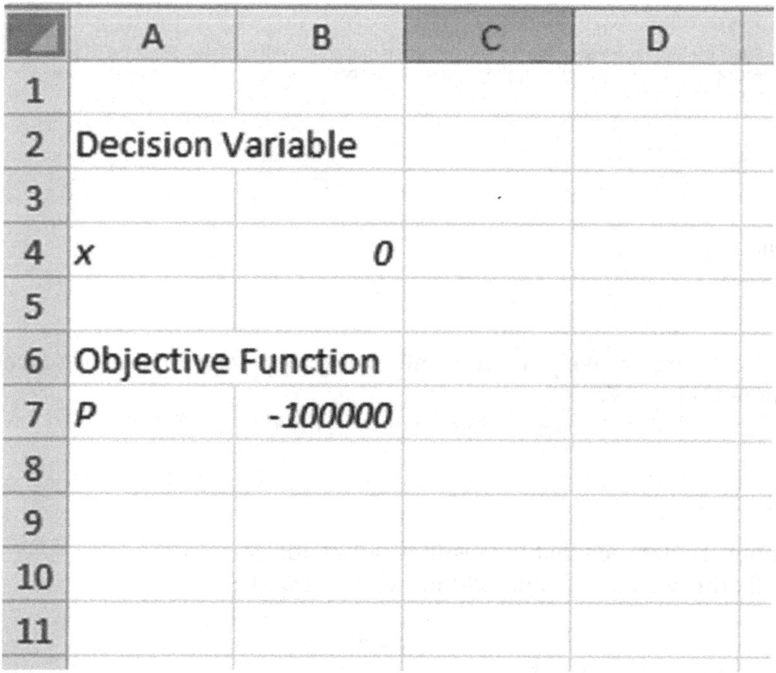

	A	B	C	D
1				
2	Decision Variable			
3			.	
4	x	0		
5				
6	Objective Function			
7	P	-100000		
8				
9				
10				
11				

FIGURE 7.2
Excel screenshot of decision variable and objective function.

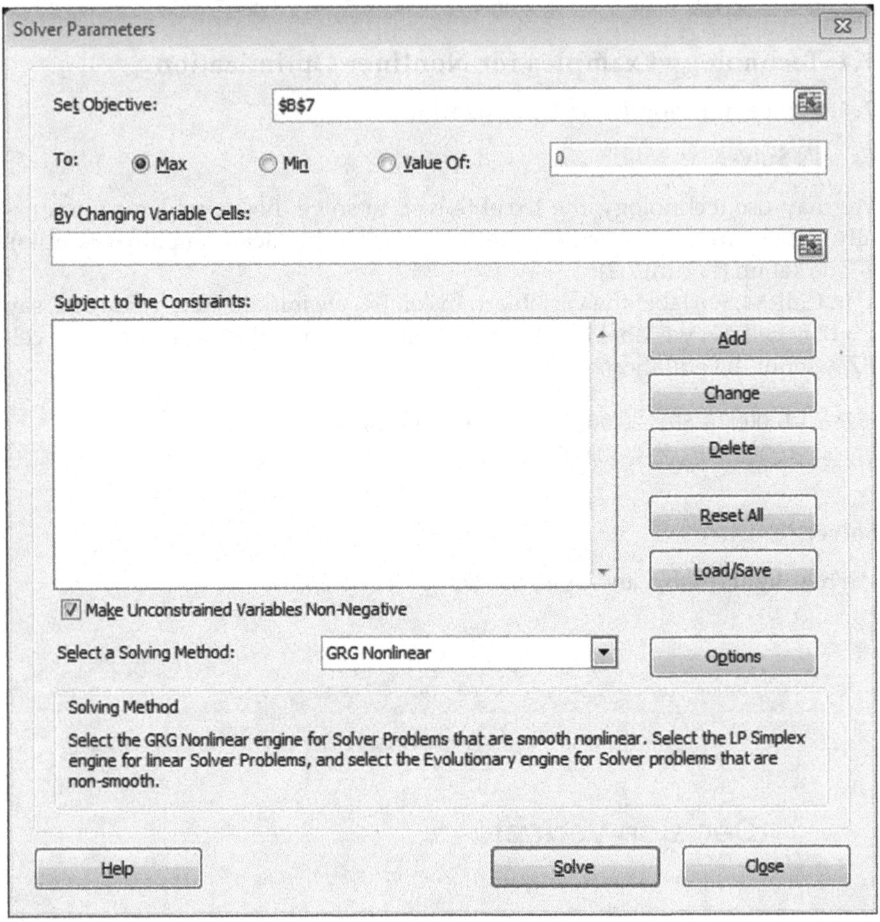

FIGURE 7.3
Excel Solver.

In the dialog box we put cell b4 into the *By Changing Variable Cells* location, as shown (Figure 7.4).

We can solve and check end points manually (Figure 7.5).

Solve

We now check to see that the result is not in our domain.

To fix this we go back and add in two constraints:

$$x \geq a$$
$$x \leq b,$$

where a, b are the domain values.

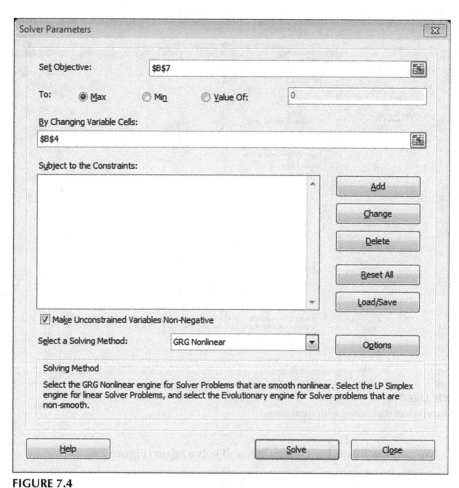

FIGURE 7.4

Excel Solversetting objective function and changing cells.

FIGURE 7.5

Excel screenshot Solver solution.

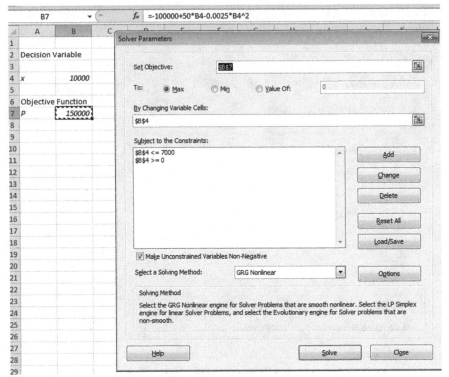

FIGURE 7.6
Excel screenshot Solver with constraints.

We add the following constraints and solve again (Figure 7.6),

$$x \geq 0 \ \& \ x \leq 7{,}000.$$

We now have our solution, $x = 7{,}000$ and $P = \$127{,}500$ (Figure 7.7).

Example 7.6 Oil Rig Location Problem

Consider an oil-drilling rig that is 8.5 miles off shore. The drilling rig is to be connected by underwater pipe to a pumping station. The pumping station is connected by land-based pipe to a refinery, which is 14.7 miles down the shoreline from the drilling rig (Figure 7.1). The underwater pipe costs \$31,575 per mile, and land-based pipe costs \$13,342 per mile. You are to determine where to place the pumping station to minimize cost of the pipe.

Problem Identification: Find a relationship between the location of the pumping station and cost of the installation of the pipe.

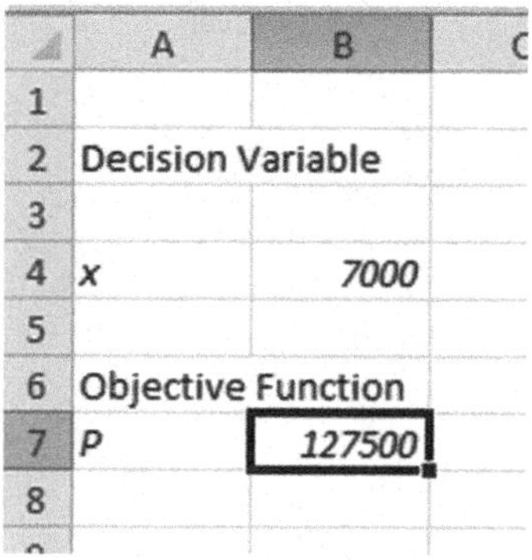

◢	A	B	C
1			
2	Decision Variable		
3			
4	x	7000	
5			
6	Objective Function		
7	P	127500	
8			

FIGURE 7.7
Excel Screenshot updated Solver solution.

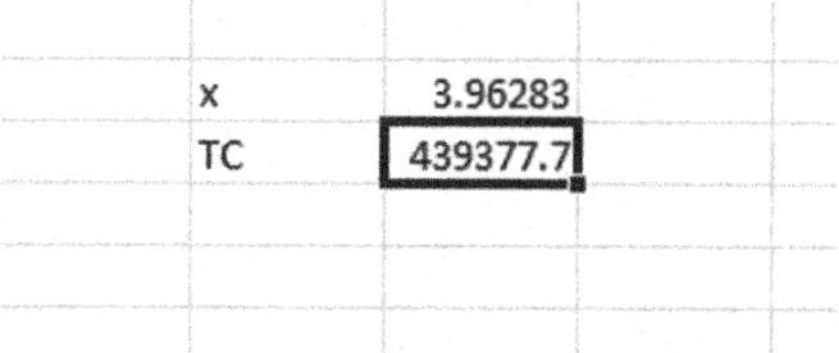

x	3.96283	
TC	439377.7	

FIGURE 7.8
Excel screenshot Solver solution.

Assumptions and Variables: First, we assume no cost saving for the pipe if we purchase in larger lot sizes. We further assume no additional costs are incurred in preparing the terrain to lay the pipe.

Variables:

x = the location of the Pumping station along the horizontal distance from $x = 0$ to $x = 14.7$ miles.
TC = total cost of the pipe for both underwater and on shore piping.

Model Construction

We use Pythagorean's Theorem for the underwater distance of the pipe that is the hypotenuse of the right triangle with height 8.5 miles and base = x. The hypotenuse is $\sqrt{8.5^2 + x^2}$. The length of the pipe on shore is $14.7 - x$.

Total cost = $31{,}575 \sqrt{8.5^2 + x^2} + 13{,}342 \, (14.7 - x)$.

We put his into the solver and minimize Total cost.

Thus, if the pumping station is located at $14.7 - 3.963 = 10.737$ miles from the refinery we will minimize the total cost at a cost of $439,377.69.

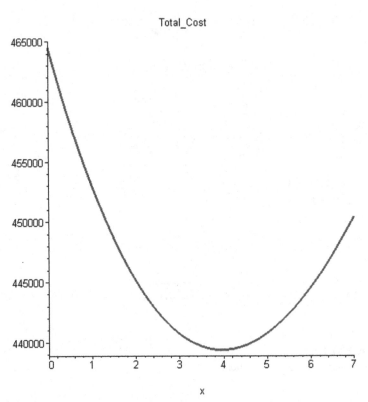

FIGURE 7.9
Plot of total cost.

7.4.2 Maple for Nonlinear Optimization

Having a computer algebra system is a little easier to proceed. Recall Example 7.3 and we show how Maple is used to obtain the solution (Figure 7.10)

>cost:=100*x − (10000 + 50*x + 0.0025*x^2);

$$cost := 50x - 10000 - 0.0025x^2$$

>dc:=diff(cost,x);

$$dc := 50 - 0.0050x$$

>ans:=solve(dc = 0,x);

$$ans := 10000$$

>ddc:= diff(dc,x);

$$ddc := -0.0050$$

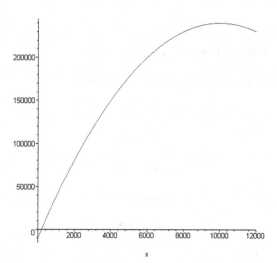

FIGURE 7.10
Maple plot of solution.

>**subs(x = ans,cost);**

240000.0000

>**plot(cost, x = 0..12000);**

Checking the end points:

>**subs(x = 0, cost);**

−10000.

>**subs(x = 7000,cost);**

217500.0000

The end point solution is selected because the critical point is outside the domain. The solution is $x = 7,000$ and $P = 217,500$.

7.4.3 R for Nonlinear Optimization

We can use R to solve the problem to maximize Profit, P. The default in R is a minimization problem, so we can multiple the function to be maximized by −1 and solve. We must remember to change the signs,

$$P = -100000 + 50x - 0.0025x^2$$

Max P = Min −P.

The R Code is:

```
> f <- function (x) 10000-50*x+0.0025*x^2

>xmin<- optimize(f, c(0,15000), tol = 0.0001)

>xmin
```

$minimum

[1] 10000

$objective

[1] –240000

However, the solution of $x = 10,000$ is not in our production interval of [0,7000] so we test the end points.

$f(0) = 10,000$ and we find $f(7,000) = 21,750$.

```
> 10000 – 50*7000 + 0.0025*7000^2
```

[1] – 217500

We accept $x = 7,000$ and $f(7,000) + 21750$ as our solution.

7.5 Single Variable Numerical Search Techniques with Technology

Why the need for numerical methods? There are functions that we cannot solve the derivatives equal to zero in closed form. The need for "good" numerical methods is required.

The basic approach of most numerical methods in optimization is to produce a sequence of improved approximations to the optimal solution according to a specific scheme. We discuss both the elimination (Dichotomous, Golden section, and Fibonacci) and interpolation methods (Newton's and Bisection). We illustrate in Excel only the Golden section and Newton's methods. The others can easily be programmed into Excel.

In numerical methods of optimization, a procedure is used in obtaining values of the objective function at various combinations of the decision variables and conclusions are then drawn regarding the optimal solution. The elimination methods can be used to find an optimal solution for even discontinuous functions. An important relationship (assumption) must be made to use these elimination methods. The function must be unimodal. A unimodal function is one that has only one peak (maximum) or one valley (minimum). This can be stated mathematically as follows:

A function $f(x)$ is unimodal if (1) $x_2 < x^*$ implies that $f(x_2) < f(x_1)$, and (2) $x_1 > x^*$ implies that $f(x_1) < f(x_2)$ where x^* is a minimum and $x_1 < x_2$.

A function $f(x)$ is unimodal if (1) $x_2 > x^*$ implies that $f(x_2) > f(x_1)$, and (2) $x_1 < x^*$ implies that $f(x_1) > f(x_2)$ where x^* is a maximum and $x_1 < x_2$.

Some examples of unimodal functions are shown in Figures 7.11a–7.11c. As seen, unimodal functions may or may not be differentiable.

Thus, a function can be a non-differentiable (corners) or even a discontinuous function. If a function is known to be unimodal in a given interval, then the optimum (maximum or minimum) can be found as a smaller interval.

In this section, we will learn many techniques for numerical searches. For the elimination methods we accept an interval answer. If a single value is required, then we usually evaluate the function at each end point of the final interval and the midpoint of the final interval and take the optimum of those three values to approximate our single value.

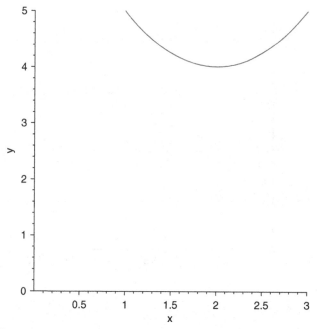

FIGURE 7.11a
unimodal;

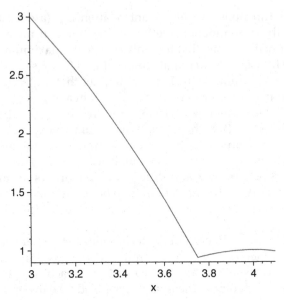

FIGURE 7.11b
non-differentiable (corners);

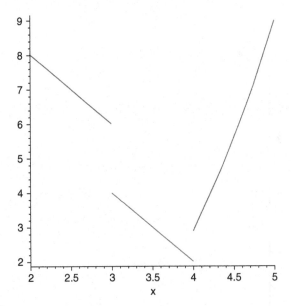

FIGURE 7.11c
discontinuous.

Example 7.7 Unrestricted Search

The method called unrestricted search is used when the optimum needs to be found but we have no known interval of uncertainty. This will involve a search with a fixed step size. This method is not very computationally effective.

Start with a guess, say x_1.

Find $f_1 = f(x_1)$
Assume a step size, S, find $x_2 = x_1 + S$

Find $f_2 = f(x_2)$.

For a minimization problem: if $f_2 < f_1$ then the solution interval cannot possibly lie in $x < x_1$. So we find point's x_3, x_4, x_n this is continued until an increase of the function is found.

The search is terminated at x_i and x_{i-1}.

If $f_2 > f_1$ then the search goes in the reverse direction.
If $f_1 = f_2$, then either x_1 or x_2 is optimum.

If the search proceeds too slowly an acceleration step size, a constant c times S, $c*S$, can be used.

Example 7.8 Find the Minimum of the Function

Using an unrestricted search method with an initial guess as 1.0, and a step size = 0.4.

$$x_1 = 1 \text{ and } f(x_1) = f_1 = -0.5$$

$$x_2 = x_1 + S = 1 + 0.4 = 1.4, f(1.4) = f_2 = -0.7$$

$$f_1 > f_2, \text{ so}$$

$$x_3 = 1.8, f_3 = -0.9, f_2 > f_3$$

$$x4 = 2.2, f4 = -0.8, f_4 > f_3. \text{ Stop}$$

TABLE 7.1

Functions

	$-x/2$	$x \leq 2$
$f(x) =$		
	$x-3$	$x>2$

Thus, the optimum (minimum) must lie between 1.8 and 2.2. If a single value is required we evaluate:

$$f(1.8) = -0.9$$

$$f(2.2) = -0.8$$

$$f((1.8 + 2.2)/2) = f(2) = -1.$$

Since $f(2)$ is the smallest value of $f(x)$, we will use $x = 2$, as our approximation to the minimum yielding a value of $f(2) = -1$.

Example 7.9 Dichotomous Search

This search method is a simultaneous search method in which all the experiments are conducted before any judgment is made concerning the location of the optimum point.

Dichotomous Algorithm:

Initialize

1. A distinguishable constant $2e > 0$ (e is a very small number, like 0.01).

2. Select a length of uncertainty for the final interval, $t > 0$ (t is also small).

3. Calculate the number of iterations required, N, using: $0.5^n = t/(b-a)$.

4. Let $k = 1$.

Main Steps

1. If $(b - a) < t$ then stop because $(b - a)$ is in the final interval

    If $(b - a) > t$ then let

    $$x_1 = \frac{(a+b)}{2} - e, x_2 = \frac{(a+b)}{2} + e$$

2. Perform comparisons of function values at these points.

TABLE 7.2

Minimization problem

If $f(x1) < f(x2)$	If $f(x1) > f(x2)$
$a = a$	$a = x1$
$b = x2$	$b = b$
$k = k + 1$	$k = k + 1$
Return to Main Step 1	Return to Main Step 1

TABLE 7.3

Dichotomous search solution

Iteration	$x1$	$x2$	$f(x1)$	$f(x2)$	Interval
1	1.4900	1.5100	−5.2001	−5.3001	[−3.0000, 6.00]
2	−0.7550	−0.7350	0.9400	0.9300	[−3.0000, 1.51]
3	−1.8775	−1.0000	0.2300	0.2647	[−3.0000, −0.735]
4	−1.3163	−1.2963	0.9000	0.9122	[−1.8775, −0.735]
5	−1.0356	−1.0156	0.9987	0.9998	[−1.3163, −0.735]
6	−0.8953	−0.8753	0.9890	0.9845	[−1.0356, −0.735]
7	−0.9655	−0.9455	0.9988	0.9970	[−1.0356, −0.875]

Example 7.10 Minimize $f(x) = x^2 + 2x$ over the interval [−3,6] using Dichotomous Search.

Let $t = 0.2$ and $e = 0.01$.
Find n using $0.5^n = t/(b - a)$
$0.5^n = 0.2/(6 - (-3)) = 0.2/9$
$0.5^n = (0.2/9)$
$n \ln(0.5) = \ln(0.2/9)$
$n = 5.49$ or 6 (rounding up) iterations.

The interval $[a,b]$ is [−3.00, 6.00] and user specified tolerance level is 0.20000.

The first 2 experimental endpoints are $x_1 = 1.490$ and $x_2 = 1.510$.
We summarize the results in Table 7.3.
The midpoint of the final interval is −0.990547 and $f(midpoint) = 1.000$.
The minimum function is 0.999 and the x value = −1.035625.
The final interval is [−1.03563, −0.87531]. The length of this interval is 0.1603, which is less than our tolerance of $t = 0.2$. The value of n refers to number of x pairs observed.

Example 7.11 Golden Section Search

Golden Section Search is a search procedure that utilizes the *golden ratio*. To better understand the *golden ratio*, consider a line segment over the interval that is divided into two separate regions. These segments are divided into the *golden ratio* if the length of the whole line is to the length of the larger part as the length of the larger part is to the length of the smaller part of the line. Symbolically, this can be written as

$$\frac{1}{r} = \frac{r}{(r-1)}$$

Algebraic manipulation of the golden ratio relationship yields, $r^2 + r - 1 = 0$. Solving this function for its roots (using the quadratic formula) gives us two real solutions:

$$r_1 = \frac{\sqrt{5}-1}{2}, \quad r_2 = \frac{-\sqrt{5}-1}{2}$$

Only the positive root, r_1, satisfies the requirement of residing on the given line segment. The numerical value of r_1 is 0.618. This value is known as the *golden ratio*. This ratio has, among its properties, being the limiting value for the ratio of the consecutive Fibonacci sequences, which we will see in the next method. It is noted here because there is also a Fibonacci search method that could be used in lieu of the golden section method.

Golden Section Method
In order to use the Golden Section search procedure, we must insure that certain assumptions hold.

These key assumptions include:

1)   the function must be unimodal over a specified interval,

2)   the function must have an optimal solution over a known interval of uncertainty, and

3)   we must accept an interval solution since the exact optimal cannot be found by this method.

Only an interval solution, known as the final interval of uncertainty, can be found using this technique. The length of this final interval is controllable by the user and can be made arbitrarily small by the selection of a *tolerance value*. The final interval is guaranteed to be less than this tolerance level.

Line search procedures use an initial interval of uncertainty to iterate to the final interval of uncertainty. The procedure is based, as shown earlier, on solving for the unique positive root of the quadratic equation, $r^2 + r = 1$. The positive root form using the quadratic formula with $a = 1$, $b = 1$, and $c = -1$ is:

$$r = \frac{-b \pm \sqrt{b^2 - 4ac}}{2a} = 0.618.$$

Finding the Maximum of a Function over an Interval with Golden Section
     This search procedure to find a maximum is iterative, requiring evaluations of $f(x)$ at experimental points x_1 and x_2, where $x_1 = b - r(b - a)$ and $x_2 = a + r(b - a)$. These experimental points will lie between the original interval $[a, b]$. These experimental points are used to help determine the new interval of search. If $f(x_1) < f(x_2)$ then the new interval is $[x_1, b]$

and if $f(x_1) > f(x_2)$ then the new interval is $[a,x_2]$. The iterations continue in this manner until the final interval length is less than our imposed tolerance. Our final interval contains the optimum solution. It is the size of this final interval that determines our accuracy in finding the approximate optimum solution. The number of iterations required to achieve this accepted interval length can be found as the smallest integer greater than k where k equals [1,4]:

$$k = \frac{\ln\left(\dfrac{tolerance}{(b-a)}\right)}{\ln(0.618)}$$

Often we are required to provide a point solution instead of the interval solution. When this occurs the method of selecting a point is to evaluate the function, $f(x)$, at the end points of the final interval and at the midpoint of this final interval. For maximization problems, we select the value of x that yields the largest $f(x)$ solution. For minimization problems, we select the value of x that yields the smallest $f(x)$ solution.

To find a maximum solution to a given function, $f(x)$, on the interval $[a, b]$ where the function, $f(x)$, is unimodal.

Input: endpoints a, b; tolerance, t
Output: final interval $[a_1, b_1]$, $f(midpoint)$

Step 1. Initialize the tolerance, $t > 0$.
Step 2. Set $r = 0.618$ and define the test points

$$x_1 = a + (1 - r)(b - a)$$
$$x_2 = a + r(b - a)$$

Step 3. Calculate $f(x_1)$ and $f(x_2)$
Step 4. Compare $f(x_1)$ and $f(x_2)$
    a. If $f(x_1) \le f(x_2)$,then the new interval is $[x_1, b]$:
        a becomes the previous x_1
        b does not change
        x_1 becomes the previous x_2
        Find the new x_2 using the formula in Step 2.
    b. If $f(x_1) > f(x_2)$, then the new interval is $[a, x_2]$:
        a does not change
        b becomes the previous x_2
        x_2 becomes the previous x_1
        Find the new x_1 using the formula in Step 2.

Step 5. If the length of the new interval from Step 4 is less than the tolerance specified, then stop. Otherwise go back to Step 3.

Step 6. Estimate x^* as the midpoint of the final interval and compute, $f(x^*)$, the estimated maximum of the function.

STOP

Example 7.12 Golden Section Search

Although Golden Section can be used with any unimodal function to find the maximum (or minimum) over a specified interval, its main advantage comes when normal calculus procedures fail. Consider the following example:

$Maximize f(x) = -|2 - x| - |5 - 4x| - |8 - 9x|$ over the interval $0 \leq x \leq 3$.

In calculus, absolute values are not differentiable because they have corner points. Thus, taking the first derivative and setting it equal to zero is not an option. We use the Golden Section to solve this problem and other examples later.

Fibonacci Search

Fibonacci Search is a search procedure that utilizes the ratio of Fibonacci numbers to set up experimental points in a sequence. The Fibonacci numbers are a sequence that follows the rules:

$$F_0 = 1$$

$$F_1 = 1$$

$$F_i = F_{i-1} + F_{i-2}$$

This generates the sequence $\{1, 1, 2, 3, 5, 8, 13, 21, 34, 55, 89, \ldots\}$.

The limiting value for the ratio of the consecutive Fibonacci sequences is the golden ratio 0.618. It is noted here because the Golden Section search method could be used in lieu of the Fibonacci method.

However, the Fibonacci Search converges faster than the Golden Section Method.

In order to use the Fibonacci search procedure, we must insure that certain assumptions hold. These key assumptions include:

1) the function must be unimodal over a specified interval,
2) the function must have an optimal solution over a known interval of uncertainty, and

3) we must accept an interval solution since the exact optimal cannot be found by this method.

Only an interval solution, known as the final interval of uncertainty, can be found using this technique. The length of this final interval is controllable by the user and can be made arbitrarily small by the selection of a *tolerance value*. The final interval is guaranteed to be less than this tolerance level.

Line search procedures use an initial interval of uncertainty to iterate to the final interval of uncertainty.

Finding the Maximum of a Function over an Interval with Fibonacci Method

This search procedure to find a maximum is iterative, requiring evaluations of $f(x)$ at experimental points x_1 and x_2, where $x_1 = a + (F_{n-2}/F_n)(b - a)$ and $x_2 = a + (F_{n-1}/F_n)$ $(b - a)$. These experimental points will lie between the original interval $[a, b]$. These experimental points are used to help determine the new interval of search. If $f(x_1) < f(x_2)$ then the new interval is $[x_1, b]$ and if $f(x_1) > f(x_2)$ then the new interval is $[a, x_2]$. The iterations continue in this manner until the final interval length is less than our imposed tolerance. Our final interval contains the optimum solution. It is the size of this final interval that determines our accuracy in finding the approximate optimum solution. The number of iterations required to achieve this accepted interval length can be found as the smallest Fibonacci number from the sequence that satisfies the inequality:

$$F_k > \frac{(b-a)}{tolerance}$$

Often we are required to provide a point solution instead of the interval solution. When this occurs the method of selecting a point is to evaluate the function, $f(x)$, at the end points of the final interval and at the midpoint of this final interval. For maximization problems, we select the value of x that yields the largest $f(x)$ solution. For minimization problems, we select the value of x that yields the smallest $f(x)$ solution.

To find a maximum solution to a given function, $f(x)$, on the interval $[a, b]$ where the function, $f(x)$, is unimodal.

Input: endpoints a, b; tolerance, t, Fibonacci sequence
Output: final interval $[a_i, b_i]$, $f(midpoint)$

Step 1. Initialize the tolerance, $t > 0$.
Step 2. Set $F_n > (b - a)/t$ as the smallest F_n and define the test points

$$x_1 = a + (F_{n-2}/F_n)\,(b - a)$$

$$x_2 = a + (F_{n-1}/F_n)\,(b - a).$$

Step 3. Calculate $f(x_1)$ and $f(x_2)$
Step 4. Compare $f(x_1)$ and $f(x_2)$

    a.   If $f(x_1) \le f(x_2)$, then the new interval is $[x_1, b]$:

        a becomes the previous x_1

        b does not change

        x_1 becomes the previous x_2

        $n = n - 1$.

Find the new x_2 using the formula in Step 2.

    b.   If $f(x_1) > f(x_2)$, then the new interval is $[a, x_2]$:

        a does not change

        b becomes the previous x_2

        x_2 becomes the previous x_1

        $n = n - 1$.

Find the new x_1 using the formula in Step 2.

    Step 5. If the length of the new interval from Step 4 is less than the tolerance specified, the stop. Otherwise go back to Step 3.
    Step 6. Estimate x^* as the midpoint of the final interval and compute, $f(x^*)$, the estimated maximum of the function.

STOP

    Although Fibonacci can be used with any unimodal function to find the maximum (or minimum) over a specified interval, its main advantage comes when normal calculus procedures fail. You will be asked to do this in the exercise set.

Interpolation with Derivatives: Newton's Method

Finding the Critical Points (Roots) of a function

    Newton's Method has been adapted to solve nonlinear optimization problems. For a function of a single variable, the adaptation is straightforward. Newton's Method is applied to the *derivative* of the function we wish to optimize, for the function's critical points occur where the derivative's roots are found. When finding the critical points of the function, Newton's method is based on the derivative of the quadratic approximation of the function $f(x)$ at the point x_k:

$$q(x) = f(x_k) + f'(x_k)(x - x_k) + \tfrac{1}{2} f''(x_k)(x - x_k)^2$$

The result, $q'(x)$, is a linear approximation of $f'(x)$ at the point x_k. Setting $q'(x) = 0$ and solving for x yields the formula

$$x_{k+1} = x_k - \frac{f'(x_k)}{f''(x_k)}$$

where $x_{k+1} \equiv x$.

Newton's method can be terminated when $|x_{k+1} - x_k| < \varepsilon$, where ε is a pre-specified scalar tolerance or when $|f'(x)| < \varepsilon$.

In order to use Newton's Method to find the critical points of a function, the function's first and second derivatives must exist in the neighborhood of interest. Also note that when the second derivative at x_k is zero, the point x_{k+1} cannot be computed.

It is important to first master the computations required in the algorithm. It is also noted that Newton's Method finds only the approximate critical value, it does not know whether it is finding a maximum or a minimum. The sign of the second derivative may be used to determine if we have a maximum or a minimum.

The Basic Application

Consider any simple polynomial, such as $f(x) = 5x - x^2$, whose critical point can easily be found by calculus, taking the first derivative and setting it equal to zero. We find that the critical point $x = 2.5$ yields a maximum of the function. Applying Newton's Method to find critical points requires finding $f'(x)$ and $f''(x)$, then using a computation device to perform the iterations:

$$f'(x) = 5 - 2x \text{ and } f''(x) = -2 \text{ Newton's Method uses}$$

$$x_{k+1} = x_k - \frac{f'(x_k)}{f''(x_k)} \text{ or } x_{k+1} = x_k - \frac{(5 - 2x)}{(-2)}.$$

Starting at $x_0 = 1$ yields:

TABLE 7.4

Evaluations by Newton's Method

k	x_k	$f'(x)$	$f''(x)$	X_{k+1}	$|x_k - x_{k+1}|$
0	1	-3	-2	2.5	1.5
1	2.5	0	-2	2.5	0

Starting at other values also yields $x = 2.5$. Since this simple quadratic function has a derivative that is a linear function, the linear approximation of the derivative will be exact regardless of the starting point, and the answer will be confirmed at the second iteration. Newton's method produces the critical values of $f'(x)$ without regard to the point x_k being a maximum or a minimum. We know we have found a maximum by looking at the entries in the table for $f''(x)$. Since $f''(x)$ at $x = 2.5$ is -2, which is less than or equal to 0.

7.5.1 Excel for Golden Section

Example 7.13 Maximizing a Function that Does Not Have a Derivative Using Golden Section with Excel.

Maximize $f(x) = -|2-x| - |5 - 4x| - |8 - 9x|$ over the interval $0 \le x \le 3$. Assume the tolerance is 0.100 (Figure 7.12A).

The midpoint of the final interval is 0.907364 and f(midpoint) $= -2.629$.

The maximum of the function is -2.629 and the x value $= 0.907364$.

In this example, we want a specific point as our solution. The midpoint yields the maximum value of $f(x)$. Thus we will use 0.907364 as the value of x that maximizes this function.

Example 7.14 Maximizing a Transcendental Function Using Golden Sectionwith Excel

Maximize the function $f(x) = 1 - exp(-x) + 1/(1 + x)$ over the interval $[0,20]$. Assume our tolerance is 0.001.

The first two experimental endpoints are $x_1 = 7.640$ and $x_2 = 12.360$.

The template is provided (Figure 7.12B). You need to enter the end points in cells G6 and H6. Enter the tolerance you want in cell J6. Enter the function in cells H11 and I11 as functions of values in cells F11 and

Golden Section Search										
leftendpt	rightendpt					DV		0	function	16 tolerance	0.1
x1	x2	a	b	f(a)	f(b)	Decision	check	mdpoint	Enter function	Enter Tolerance
0	3	1.164	1.854	-15	-27	0	go	0	0	
0	1.854	0.719352	1.145772	-15	-11.248	1	go	0	0	
0.719352	1.854	1.159595	1.420564	-4.92907	-11.248	0	go	0	0	
0.719352	1.420564	0.991422	1.152701	-4.92907	-6.04677	0	go	0	0	
0.719352	1.152701	0.887492	0.987162	-4.92907	-3.61081	1	go	0	0	
0.887492	1.152701	0.990393	1.051391	-2.57512	-3.61081	0	go	0	0	
0.887492	1.051391	0.951085	0.988782	-2.57512	-3.20556	0	go	0	0	
0.887492	0.988782	0.926792	0.950089	-2.57512	-2.95513	0	go	0	0	
0.887492	0.950089	0.911779	0.926177	-2.57512	-2.80035	0	stop	0.91879	18.13110707	
0.887492	0.926177	0.902501	0.911399	-2.57512	-2.70471	0	stop	0.906834	18.035137	

FIGURE 7.12a
Excel screenshot of Golden Section.

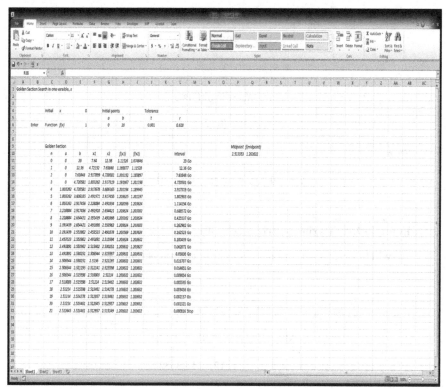

FIGURE 7.12b
Excel screenshot of Golden Section for Example 7.14.

G11 that are calculated automatically. Copy those formulas down and the table fills out. When the word in cell M says "STOP" the algorithm should end manually. Then find both the midpoint of the final interval [a, b] and the value of the f(midpoint).

The midpoint of the final interval is 2.512961 and f(midpoint) = 1.204. The maximum of the function is 1.204 and the x value = 2.512961.

Again, assuming that we desire a specific numerical value as the solution, our solution is x = 2.512705 with f(2.512705) = 1.204.

Example 7.15 Minimize $f(x) = x^2 + 2x$

We will start our guess with x = 4, and use a stopping criteria of t = 0.01.

$$f(x) = x^2 + 2x$$

$$f'(x) = 2x + 2$$

$$f''(x) = 2$$

$$x_2 = x_1 - f'(x_1)/f''(x_1)$$

$$x_2 = 4 - (10/2) = 4 - 5 = -1$$

$$x_3 = x_2 - f'(x_2)/f''(x_2)$$

$$x_2 = -1 - (0/2) = -1$$

Stop

$$|f'(x)| = 0 < 0.01 \text{ or } |x_2 - x_1| = 0 < 0.01$$

Example 7.16 Maximize $f(x) = -2x^3 + 10x - 10$

We will begin at $x = 1$ and use a stopping criteria of t = 0.01

$$f'(x) = -6x^2 + 10$$

$$f''(x) = 12x$$

$$\text{At } x = 1, f'(1) = 4, f''(1) = -12.$$

$$x_2 = 1 - (-4/12) = 1.33333$$

Neither $|f'(x)|$ nor $|x_{k+1} - x_k|$ are less than ε, so we continue. We summarize in Table 7.5.

Since $f(x) = |-2.7E-06| < 0.01$ we stop. Our critical point is $x = 1.290994$. Since $f''(x) < 0$, then we have found a maximum.

Example 7.17 Newton's Method with Excel of the Function $f(x) = -x^2 + 23x + 121$

We will use the solver to perform this action with the GRG nonlinear routine. Simply enter the initial x in a cell at some value, say 0. For

TABLE 7.5

Example 7.16 solution

k	x	f'(x)	f''(x)	x(k+1)
1	1	4	−12	1.333333
2	1.333333	−0.66667	−16	1.291667
3	1.291667	−0.01042	−15.5	1.290995
4	1.290995	−2.7E−06	−15.4919	1.290994

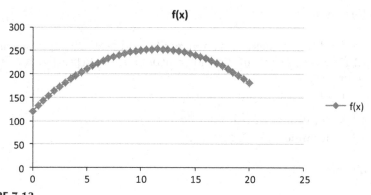

FIGURE 7.13
Solution of Example 7.17 in Excel.

example cell c2, enter the function in cell c3 as a function of cell2. In c3, we would type = -(cell2^2) + 23*c2 + 121 <enter>. Place the cursor in cell c3, open the Solver. Tell the Solver if you desire to maximize or minimize the function in cell c3. The cell changing is cell c2 which holds x. Insure Solver is using the GRG nonlinear routine and that the non-negative of the solution is off.

You read the answer from the spreadsheet as $x = 11.5$ and $f(x) = 253.25$.

To obtain a plot within Excel we have to create x values from some start value to some end-value, perhaps 0 to 20 by some increments in this case. Then we can evaluate the function and use Scatterplot to see the result. We can use the scatterplot to either approximate our solution prior to its evaluation or approximately verify our solution (Figure 7.13).

The Bisection Method with Derivatives

To utilize bisection method with derivatives correctly there are certain properties that must hold for the given function:

Let f' be a function that has opposite sign values at each end of some specified interval. Then, by the *Intermediate-Value Property (IVP)* of continuous functions (mentioned in many basic college algebra texts), we are guaranteed to find a root between the endpoints of the given interval. Specifically, the *IVP* states that given two points (x_1, y_1) and (x_2, y_2) with $y_1 \neq y_2$ on a graph of the continuous function f, the function f takes on every value between y_1 and y_2. Thus, with values having opposite signs, then there must be a value for which $f'(x) = 0$.

The algorithm is:

Step 1. Find two values a and b where $f'(a)$ and $f'(b)$ have opposite signs.
Step 2. Fix a tolerance for the final interval of the solution $[a_f, b_f]$ so that $|b_f - a_f| <$ tolerance.
Step 3. Find the mid-point, $m_i = (b_i + a_i)/2$
Step 4. Compute $f'(a_i)$, $f'(b_i)$, and $f'(m_i)$.
Step 5. Determine if $f'(a_i) * f'(m_i) < 0$.

If true, then

$$a_i = a_i$$

$$b_i = m_i$$

Otherwise,

$$a_i = m_i$$

$$b_i = b_i$$

Step 6. If $f'(m_i) \neq 0$ and the new $|b_i - a_i| >$ tolerance then using the new interval $[a_i, b_i]$ go back to Step 3 and repeat the process. Otherwise, **STOP**

Example 7.18 Minimize $f(x) = x^2 + 2x, -3 \leq x \leq 6$. Let $\varepsilon = 0.2$

The number of required observations is found by solving the formula $(0.5)^n = t/(b-a)$.
We find that $n = 6$ observations.

$$f'(x) = 2x + 2$$

$$k = 1$$

$$x_{mp} = .5(6-3) = 1.5, f'(1.5) = 5, f'(1.5) > 0$$

so

$$a = -3$$

$$b = 1.5$$

$$x_{mp} = -0.75$$

$$f'(-0.75) = 0.5 > 0$$

so

$$a = -3$$

$$b = -0.75$$

$$x_{mp} = -1.875$$

$$f'(-1.875) = -1.75 < 0$$

so

$$a = -1.875$$

$$b = -75$$

$$xmp = -1.3125$$

$$f(-1.3125) = -0.625 < 0$$

so

$$a = -1.3125$$

$$b = -0.75$$

$$xmp = -1.03125$$

$$5.f'(-1.03125) = -0.0625 < 0$$

so

$$a = -1.30125$$

$$b = -0.75$$

$$xmp = -0.89025$$

$$f(-0.89025) = 0.2195 > 0$$

$$a = 1.030125$$

$$b = -0.89025$$

The final interval is less than 0.2 so we stop. Our solution is between −1.030125 and −0.89025.

Just as previously discussed, if we need a single value we can evaluate the derivative of the function, f', at a, at b, and at the midpoint selecting the value that is closest to $f'(x) = 0$ from those points.

$$f'(-1.030125) = -0.0625$$

$$f'(-0.89025) = 0.2195$$

$$f'(-0.96075) = .0705$$

$x = -1.030125$ with $f'(x) = -0.0625$ is best. The exact value, via single variable calculus, for the maximization of f is the solution $x = -1$.

7.5.2 Numerical Method with Maple

We wrote several procedures to perform many of these methods. They are on the companion website. We illustrate Golden Section here using a procedure we wrote to implement for Example 7.12, Golden Section Search.

Example 7.19 Maximizing a Function that Does Not Have a Derivative

Maximize $f(x) = -|2-x| - |5-4x| - |8-9x|$ over the interval $0 \le x \le 3$.

> f := x->-(abs(2-x) + abs(5-4*x) + abs(8-9*x));

$$f := x \rightarrow -\left|2 - x\right| - \left|5 - 4x\right| - \left|8 - 9x\right|$$

> GOLD(f,0,3,.1);

The interval [a,b] is [0.00, 3.00] and user specified tolerance level is 0.10000.

The first 2 experimental endpoints are $x1 = 1.146$ and $x2 = 1.854$.

Iter	x(1)	x(2)	f(x1)	f(x2)	Interval
1	1.1460	1.8540	-3.5840	-11.2480	[0.0000, 1.8540]
2	0.7082	1.1460	-5.0848	-3.5840	[0.7082, 1.8540]
3	1.1460	1.4163	-3.5840	-5.9958	[0.7082, 1.4163]
4	0.9787	1.1460	-2.9149	-3.5840	[0.7082, 1.1460]
5	0.8755	0.9787	-2.7436	-2.9149	[0.7082, 0.9787]
6	0.8116	0.8755	-3.6382	-2.7436	[0.8116, 0.9787]
7	0.8755	0.9149	-2.7436	-2.6594	[0.8755, 0.9787]
8	0.9149	0.9393	-2.6594	-2.7571	[0.8755, 0.9393]

The midpoint of the final interval is 0.907364 and f(midpoint) $= -2.629$.
The maximum of the function is -2.629 and the x value $= 0.907364$.

In this example, we want a specific point as our solution. The midpoint yields the maximum value of $f(x)$. Thus we will use 0.907364 as the value of x that maximizes this function.

7.5.3 R for Numerical Methods

R can be used for numerical methods. We provide several examples here: golden section and Newton's Method.

Newton's Method in R

Enter the derivative of function as $f(x)$.
   newton.method(FUN = function(x) x^2 – 4, init = 10, rg = c(–1, 10), tol = 0.001, interact = FALSE, col.lp = c("blue", "red", "red"), main, xlab, ylab, ...)

Golden Section in R Example

Golden section results using program
   $f(x) = x^2 + 2x, -5 \leq x \leq 5$, tolerance is 0.1
   > golden.section.search(f, –5,5,0.1)

Iteration # 1

   f1 = –0.9674775
   f2 = 3.753882
   f2 > f1
   New Upper Bound = 1.18034
   New Lower Bound = –5
   New Upper Test Point = –1.18034
   New Lower Test Point = –2.63932

Iteration # 2

   f1 = 1.687371
   f2 = –0.9674775
   f2 < f1
   New Upper Bound = 1.18034
   New Lower Bound = –2.63932
   New Lower Test Point = –1.18034
   New Upper Test Point = –0.2786405

Iteration # 3

   f1 = –0.9674775
   f2 = –0.4796404
   f2 > f1

New Upper Bound = –0.2786405
New Lower Bound = –2.63932
New Upper Test Point = –1.18034
New Lower Test Point = –1.737621

Iteration # 4

f1 = –0.4559156
f2 = –0.9674775
f2 < f1
New Upper Bound = –0.2786405
New Lower Bound = –1.737621
New Lower Test Point = –1.18034
New Upper Test Point = –0.8359214

Iteration # 5

f1 = –0.9674775
f2 = –0.9730782
f2 < f1
New Upper Bound = –0.2786405
New Lower Bound = –1.18034
New Lower Test Point = –0.8359214
New Upper Test Point = –0.623059

Iteration # 6

f1 = –0.9730782
f2 = –0.8579155
f2 > f1
New Upper Bound = –0.623059
New Lower Bound = –1.18034
New Upper Test Point = –0.8359214
New Lower Test Point = –0.9674775

Iteration # 7

f1 = –0.9989423
f2 = –0.9730782
f2 > f1

New Upper Bound = –0.8359214
New Lower Bound = –1.18034
New Upper Test Point = –0.9674775
New Lower Test Point = –1.048784

Iteration # 8

f1 = –0.9976201
f2 = –0.9989423
f2 < f1
New Upper Bound = –0.8359214
New Lower Bound = –1.048784
New Lower Test Point = –0.9674775
New Upper Test Point = –0.9172275

Iteration # 9

f1 = –0.9989423
f2 = –0.9931487
f2 > f1
New Upper Bound = –0.9172275
New Lower Bound = –1.048784
New Upper Test Point = –0.9674775
New Lower Test Point = –0.9985337

Iteration # 10

f1 = –0.9999979
f2 = –0.9989423
f2 > f1
New Upper Bound = –0.9674775
New Lower Bound = –1.048784
New Upper Test Point = –0.9985337
New Lower Test Point = –1.017728

Final Lower Bound = –1.048784
Final Upper Bound = –0.9674775
Estimated Minimizer = –1.008131
>

Again, by decreasing the tolerance we more accurately approximate the result.
Newton' Method Modified for finding $f'(x) = 0$
Modified from source (https://rpubs.com/aaronsc32/newton)

```
newton.raphson <- function(f,g, a, b, tol = 1e – 5, n = 1000) {

x0 <- a # Set start value to supplied lower bound
k <- n # Initialize for iteration results

# Check the upper and lower bounds to see if approximations result in 0
fa <- f(a)
if (fa == 0.0) {
   return(a)
}

fb <- f(b)
if (fb == 0.0) {
   return(b)
}

for (i in 1:n) {
   #dx <- genD(func = f, x = x0)$D[1] # First-order derivative f'(x0)
   x1 <- x0 - (f(x0)/ g(x0) # Calculate next value x1
   k[i] <- x1 # Store x1
   # Once the difference between x0 and x1 becomes sufficiently small,
      output the results.
   if (abs(x1 – x0) <tol) {
      root.approx<- tail(k, n = 1)
      res <- list('root approximation' = root.approx, 'iterations' = k)
      return(res)
   }
   # If Newton-Raphson has not yet reached convergence set x1 as x0 and
      continue
   x0 <- x1
}
print('Too many iterations in method')
}
Enter f as f'(x) and f"(x) as g.
>newton.raphson(f,g,-5,5)
```

$ root approximation

[1] -2

$iterations

[1] –3.125 –2.2097794 – 2.034166 – 2.000564 – 2.000000 – 2.00000

7.6 Exercises

1. Use Golden Section, Fibonacci's method, Newton's Method, and Bisection Method to solve the following:

2. Maximize $f(x) = -x^2 - 2x$ on the closed interval $[-2,1]$. Using a tolerance for the final interval of **0.6**. Hint (Start Newton's Method at $x = -0.5$)

3. Maximize $f(x) = -x^2 - 3x$ on the closed interval $[-3,1]$. Using a tolerance for the final interval of **0.6**. (Hint:Start Newton's Method at $x = 1$.)

4. Minimize $f(x) = x^2 + 2x$ on the closed interval $[-3,1]$. Using a tolerance for the final interval of **0.5**. (Start Newton's at x = –3.)

5. Minimize $f(x) = -x + e^x$ over the interval $[-1,3]$ using a tolerance of 0.1. (Newton's start at $x = -1$.)

6. List at least two assumptions required by both golden section and Fibonacci's search methods.

7. Consider minimizing $f(x) = -x + e^x$ over the interval $[-1,3]$. Assume your final interval yielded a solution within the tolerance of $[-.80,.25]$. Report a single best value of x to minimize $f(x)$ over the interval.

8. Each morning during rush hour, 10,000 people travel from New Jersey to New York City. If a person takes the subway, the trip lasts 40 minutes. If x thousand people drive to New York City, it takes $20 + 5x$ minutes to make the trip.

   (a) Formulate the problem with the objective to minimize the average travel time per person. Let x = number of people (in thousands) that drive to New York City from New Jersey.
   (b) Find the optimal number of people that drive so that the average time per person is minimized.

9. Find the optimal solution to

$$MAX\ f(x) = 2x^3 - 1$$

Subject to

$$-1 \le x \le 1$$

10. Given the following plot of the function: $f(x) = 0.5x^3 - 8x^2$

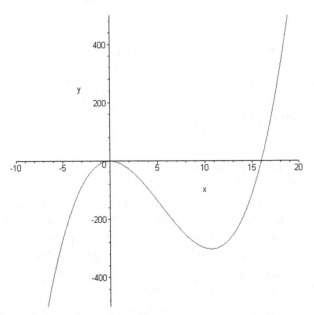

(a) Using analytical techniques (i.e. Calculus), find and classify all *extrema* for $f(x)$.
(b) Apply the definition of convexity to show that the function $f(x)$ is <u>concave</u> over the interval between $x_1 = -6$ and $x_2 = 5$, using $c = 0.5$. (We're fixing the value of c here to keep the algebra from getting ugly.)
(c) Confirm your response to part (b) using the 2nd derivative test to show that $f(x)$ is concave over this entire interval.
(d) Over what interval is the above function convex? concave?
(e) Why is knowing the concavity important in optimization?

11. Find the optimal solution to the bounded single variable optimization problem:

$$\text{MAX } f(x) = 2x^3 - 2$$

$$\text{Subject to } -1 \le x \le 1$$

12.   Dr. E. N. Throat has been taking x-rays of the trachea contracting during coughing. He has found that the trachea appears to contract by 33% (1/3) of its normal size. He has asked the department of mathematics to confirm or deny his claim. You perform some initial research and you find that, under reasonable assumptions about the elasticity of the

tracheal wall and about how air near the wall is slowed by friction, the average flow of velocity v can be modeled by the equation:

$$v = c\,(ro - r)r2 \text{ cm/sec, between } ro/2 \le r \le ro,$$

where c is a positive constant (let $c = 1$), ro is the resting radius of the trachea in centimeters.

Find the value of r that maximizes v and then *support or deny* the claim.

7.7 Projects

1. We are considering buying the new E-phone system, which is computer compatible with the new E-computer system. The company is concerned with how often they will replace the machines and the cost involved. The company's R&D team estimates that when the E-phone is t years old, it will allow the user to earn revenue at a rate of $e-t$ per year. After t years of use, the E-phone can be sold to a third world company for $1/(1+t)$ dollars. Maintenance costs after t years are estimated as .01 t dollars. Build a model for the company and determine how long they should keep the E-phone system before replacing it. Make a recommendation to the CEO as to what you would do based upon your modeling.

2. Previously we discussed the least squares method to fit the parameters to a proposed model. Another choice to fit the model is minimize the sum of the absolute deviations between the proposed model and the data. For example consider the model using $W = KL3$ and the data available.

Data available:

Length	12.5	12.625	12.625	14.125	14.5	14.5	17.27	17.75
Weight	17	16	17	23	26	27	43	49

A model that we could use to find the slope is a search method. Find the value of k that minimizes the function S.

$$\text{Minimize } S = |17 - k{*}12.5\char94 3| + |16 - k{*}12.625\char94 3 + \cdots$$
$$+ |49 - k{*}17.75\char94 3|$$

Use any of our numerical techniques and find the value of k.

3. In the figure below the profit function is given for a sulfuric acid alkyl-
    ation reactor as a function of feed rate and catalyst concentration. Plot
    the profit function as a function of feed rate for a constant catalyst
    concentration of 95%. Place six golden section experiments on the
    interval giving their location, the corresponding value of the profit
    function and the length and locàtion of the final interval of uncer-
    tainty. Determine the optimal solution to a tolerance of 0.25, 0.1, 0.01?

    Profit Function for the Operation of a Sulfuric Acid Alkylation Reactor
    as a Function of Catalyst Concentration and Feed Rate.

4. An economic analysis of a proposed facility is being conducted in order
    to select an operating life such that the maximum uniform annual
    income is achieved. A short life results in high annual amortization
    costs, but the maintenance costs become excessive for a long life. The
    annual income after deducting all operating costs, except mainten-
    ance costs, is $180,000. The installed cost of the facility, C, is $500,000
    borrowed at 10% interest compounded annually. The maintenance
    charges on an annual basis are evaluated using the product of the
    gradient present-worth factor and the capital-recovery factor. In the
    gradient present-worth method, there are no maintenance charges
    the first year, a cost M for the second year, $2M$ for the third year, 3
    M for the fourth year, etc. The second year cost, M, for the problem is
    $10,000. The annual profit is given by the following equation:

    $$P = 180,000 - \{[(i+1)N - 1 - iN]/ i[(1+i)N - 1]\}M$$
    $$- \{i(1+i)N/ [(1+i)N - 1]\}C$$

    where i is the interest rate and N is the number of years. Determine the
    number of years, N, that give the maximum uniform annual income,
    P. For your convenience the following table gives the values of the
    coefficients of M and C for $i = 0.1$ as a function of N.

	Coefficient of			Coefficient of	
Year	M	C	Year	M	C
1	0	1.10	11	4.05	0.154
2	0.476	0.576	12	4.39	0.147
3	0.909	0.403	13	4.69	0.141
4	1.30	0.317	14	5.00	0.136
5	1.80	0.264	15	5.28	0.131
6	2.21	0.230	16	5.54	0.128
7	2.63	0.205	17	5.80	0.125
8	2.98	0.188	18	6.05	0.122
9	3.38	0.174	19	6.29	0.120
10	3.71	0.163	20	6.51	0.117

5. It is proposed to recover the waste heat from exhaust gases leaving a furnace (flow rate, m = 60,000 lb/hr; heat capacity, cp = 0.25 BTU/lb°F) at a temperature of T^{in} = 500°F by installing a heat exchanger (overall heat transfer coefficient, U = 4.0 BTU/hr,ft²,°F) to produce steam at T^s = 220°F from saturated liquid water at 220°F. The value of heat in the form of steam is p = $0.75 per million BTUs, and the installed cost of the heat exchanger is c = $5.00 per ft² of gas side area. The life of the installation is n = 5 years, and the interest rate is i = 8.0%. The following equation gives the net profit P for the 5 year period from the sale of the steam and the cost of the heat exchanger. The exhaust gas temperature T^{out} can be between the upper and lower limits of 500°F and 220°F.

$$P = pqn - cA(1 + i)n$$

where

$$q = mcp(T^{in} - T^{out}) = UADTLM$$

(a) Derive the following equation for this design.

$$P = 91,137 - 492.75T^{out} + 27550ln(T^{out} - 220)$$

(b) Use a Fibonacci search with seven experiments to locate the optimal outlet temperature T^{out} to maximize the profit P on the interval of Tout from 220°F to 500°F. Find the largest value of the profit and the size and location of the final interval of uncertainty for the fractional resolution based on the initial interval of e = 0.01.

(c) Use another numerical method and compare your results.

7.8 References and Suggested Further Reading

Bazarra, M., C. Shetty, and H. D. Scherali (1993). *Nonlinear Programming: Theory and Applications.* New York: Wiley.

Fox, W. P. (1992). Teaching nonlinear programming with Minitab. *COED Journal*, II(1), January–March: 80–84.

Fox, W. P. (1993) Using microcomputers in undergraduate nonlinear optimization. *Collegiate Microcomputer*, XI(3): 214–218.

Fox, W. P. and M. Witherspoon (2001) Single variable optimization when calculus fails: Golden section search methods in nonlinear optimization using Maple. *COED*, XI(2): 50–56.

Giordano, F., W. Fox, and S. Horton (2013). *A First Course in Mathematical Modeling*, 5th ed. Cengage, Boston, MA.

Phillips, D. T., A. Ravindran, and J. Solberg (1976). *Operations Research*. John Wiley & Sons, New York.

Rao, S. S. (1979). *Optimization: Theory and Applications*. Wiley Eastern Limited, New Delhi, India.

Winston, W. (2002). *Introduction to Mathematical Programming: Applications and Algorithms*. 4th ed. Duxbury Press, ITP, Belmont. CA.

8

Multivariable Optimization

Objectives

1. Apply modeling to continuous optimization problems.

2. Understand both unconstrained and constrained optimization in three or more variables.

3. Understand the application of numerical procedures to optimization problems.

8.1 Introduction

Consider a small company that is planning to install a central computer with cable links to five new departments. According to their floor plan, the peripheral computers for the five departments will be situated as shown by the dark circles in the figure below. The company wishes to locate the central computer so that the minimal amount of cable will be used to link to the five peripheral computers. Assuming that cable may be strung over the ceiling panels in a straight line from a point above any peripheral to a point above the central computer, the distance formula may be used to determine the length of cable needed to connect any peripheral to the central computer. Ignore all lengths of cable from the computer itself to a point above the ceiling panel immediately over that computer. That is, work only with lengths of cable strung over the ceiling panels (Figure 8.1).

The coordinates of the locations of the five peripheral computers are listed in Table 8.1.

Assume the central computer will be positioned at coordinates (m, n) where m and n are *integers* in the grid representing the office space. Determine the coordinates (m, n) for placement of the central computer that minimize the

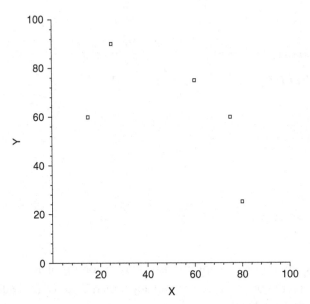

FIGURE 8.1
The grid for the five departments.

TABLE 8.1

Grid coordinates of five departments

X	Y
15	60
25	90
60	75
75	60
80	25

total amount of cable needed. Report the total number of feet of cable needed for this placement along with the coordinates (m, n).

To model and solve problems like this, we need to learn about multivariable unconstrained optimization. We will discuss how to find an optimal solution (if one exists) for the following unconstrained nonlinear optimization problem. In this chapter we will discuss models that require calculus to solve. We will review the calculus concepts for optimization and then apply them to modeling applications. We will discuss how to find an optimal solution (if one exists) for the following unconstrained nonlinear optimization problem:

Maximize (or minimize) $f(x_1, x_2, ..., x_n)$ over R^n.

We assume that the first and second partial derivatives of $f(x_1, x_2, ..., x_n)$ exist and are continuous at all points in the domain of f. Let

$$\frac{\partial f(x_1, x_2, \ldots x_n)}{\partial x_i}$$

be the partial derivative of $f(x_1, x_2, \ldots, x_n)$ with respect to x_i. Candidate critical points (stationary points) are found where

$$\frac{\partial f(x_1, x_2, \ldots x_n)}{\partial x_i} = 0, \text{for } i = 1, 2, \ldots, n.$$

This sets up a system of equations that when solved yields the critical point (if one or more is found) that satisfies all partial derivatives.

Theorem 8.1: If x is a local extremum then x satisfies $\dfrac{\partial f(x_1, x_2, \ldots x_n)}{\partial x_i} = 0$, for

$i = 1, 2, \ldots, n$.

We have previously defined all points that satisfy $\dfrac{\partial f(x_1, x_2, \ldots x_n)}{\partial x_i} = 0$, for $i = 1$,

2, ..., n as critical points (stationary points). Not all critical points (stationary points) are local extremum. If a stationary point is not a local extremum (a maximum or a minimum) then it is called a **saddle point**.

The Hessian Matrix

The Hessian matrix is the matrix of second partial derivatives. The Hessian matrix is used to determine the convexity of the function as well as the type of critical point found from solving the first partial derivatives equal to zero.

How do we determine the convexity of functions of more than one variable?

Definition: A function $f(x)$ is convex if

$$f(x^{(1)}) + \lambda(x^{(2)} - x^{(1)}) \le f(x^{(1)}) + \lambda(f(x^{(2)}) - f(x^{(1)}))$$

for every $x^{(1)}$ and $x^{(2)}$ in its domain and every $\lambda \in [0,1]$. Similarly, $f(x)$ is concave if

$$f(x^{(1)}) + \lambda(x^{(2)} - x^{(1)}) \ge f(x^{(1)}) + \lambda(f(x^{(2)}) - f(x^{(1)}))$$

for every $x^{(1)}$ and $x^{(2)}$ in its domain and every $\lambda \in [0,1]$.

We introduce the Hessian matrix that allows us to determine the convexity of multivariable functions. As we will see the Hessian matrix provides us with additional information about the critical points as well.

Definition: The Hessian Matrix is the $n \times n$ matrix of the second partial derivatives of a multivariable function $f(x_1, x_2, \ldots, x_n)$ where the ij^{th} entry is $\dfrac{\partial^2 f}{\partial x_i x_j}$

$$H = \begin{bmatrix} \dfrac{\partial^2 f}{\partial x_1^2} & \dfrac{\partial^2 f}{\partial x_1 x_2} & \dfrac{\partial^2 f}{\partial x_1 x_3} & \cdots & \dfrac{\partial^2 f}{\partial x_1 x_n} \\ \dfrac{\partial^2 f}{\partial x_2 x_1} & \dfrac{\partial^2 f}{\partial x_2^2} & & & \\ \dfrac{\partial^2 f}{\partial x_3 x_1} & \cdots & & \cdots & \\ \cdots & & & & \\ \dfrac{\partial^2 f}{\partial x_n x_1} & & & & \dfrac{\partial^2 f}{\partial x_n^2} \end{bmatrix}$$

In a 2×2 case,

$$H = \begin{bmatrix} \dfrac{\partial^2 f}{\partial x_1^2} & \dfrac{\partial^2 f}{\partial x_1 x_2} \\ \dfrac{\partial^2 f}{\partial x_2 x_1} & \dfrac{\partial^2 f}{\partial x_2^2} \end{bmatrix}$$

We note that the mixed partials are always equal $\dfrac{\partial^2 f}{\partial x_1 \partial x_2} = \dfrac{\partial^2 f}{\partial x_2 \partial x_1}$ as you will see in our examples.

Example 8.1 If $f(x_1, x_2) = x_1^2 + 3x_2^2$, Find the Hessian

$$\frac{\partial f}{\partial x_1} = 2x_1, \frac{\partial f}{\partial x_2} = 6x_2^2$$

$$\frac{\partial^2 f}{\partial x_1^2} = 2$$

$$\frac{\partial^2 f}{\partial x_2^2} = 6$$

$$\frac{\partial^2 f}{\partial x_1 x_2} = \frac{\partial^2 f}{\partial x_2 x_1} = 0$$

$$H = \begin{bmatrix} 2 & 0 \\ 0 & 6 \end{bmatrix}$$

Example 8.2 If $f(x_1, x_2) = -x_1^2 - 3x_2^2 + 3x_1 \cdot x_2$, **Find the Hessian**

$$\frac{\partial f}{\partial x_1} = -2x_1 + 3x_2, \frac{\partial f}{\partial x_2} = -6x_2 + 3x_1$$

$$\frac{\partial^2 f}{\partial x_1^2} = -2$$

$$\frac{\partial^2 f}{\partial x_2^2} = -6$$

$$\frac{\partial^2 f}{\partial x_1 x_2} = \frac{\partial^2 f}{\partial x_2 x_1} = 3$$

$$H = \begin{bmatrix} -2 & 3 \\ 3 & -6 \end{bmatrix}$$

Definition: The ith leading principal minor of an $n \times n$ matrix is the determinant of any $i \times i$ *matrix* obtained by deleting $n - i$ rows and the corresponding $n - i$ columns of the matrix.

Example 8.3 Given the 3×3 Hessian Matrix,

$$H = \begin{bmatrix} 2 & 0 & 4 \\ 0 & 1 & 5 \\ 4 & 5 & 3 \end{bmatrix}$$

1. There are three 1st leading principal minors ($i = 1$, so eliminate $3 - 1 = 2$ rows and 2 columns);

    a. eliminate rows 2,3 and columns 2,3 yield the matrix [2].

$$\text{Det}[2] = 2.$$

b.   eliminate rows 1,3 and columns 1,3 yield the matrix [1].

Det [1] = 1.

c.   eliminate rows 1,2 and columns 1,2 yield the matrix [3].

Det [3] = 3.

Note: These 1st leading principal minors are the entries of the main diagonal.

2.   There are three 2nd leading principal minors (i = 2 so we eliminate 3 – 2 = 1 row and 1 column).

a.   eliminate row 3 and column 3, yields the matrix

$$\begin{bmatrix} 2 & 0 \\ 0 & 1 \end{bmatrix}, \text{Det.}\begin{bmatrix} 2 & 0 \\ 0 & 1 \end{bmatrix} = 2$$

b.   eliminate row 2 and column 2, yields the matrix

$$\begin{bmatrix} 2 & 4 \\ 4 & 3 \end{bmatrix}, \text{Det}\begin{bmatrix} 2 & 4 \\ 4 & 3 \end{bmatrix} = 6 - 16 = -10.$$

c.   eliminate row 1 and column 1, yields the matrix

$$\begin{bmatrix} 2 & 0 \\ 0 & 1 \end{bmatrix}, \text{Det}\begin{bmatrix} 2 & 0 \\ 0 & 1 \end{bmatrix} = 3 - 25 = -22.$$

2.   There is only one 3rdleading principal minor (i = 0, so eliminate no rows or columns).

$$\text{Det}\begin{bmatrix} 2 & 0 & 4 \\ 0 & 1 & 5 \\ 4 & 5 & 3 \end{bmatrix} = -60$$

Definition: The kth leading principal minor of an $n \times n$ matrix is the determinant of the $k \times k$ matrix obtained by deleting the last $n - k$ rows and $n - k$ columns of the matrix.

Notice that if you examine the matrix, H, the leading principal minors are just the determinant of the square matrices along the main diagonal.

So, how do we use all these determinants that give the principal minor and leading principal minors of the Hessian matrix to determine the convexity of the multivariate function?

Theorem 8.2a: Let $f(x_1, x_2, \ldots x_n)$ be a function with continuous second-order partial derivatives for every point in the domain of f. Then $f(x_1, x_2, \ldots x_n)$ is a convex function if all the leading principal minors of the Hessian matrix are non-negative.

Theorem 8.2b: Let $f(x_1, x_2, \ldots x_n)$ be a function with continuous second-order partial derivatives for every point in the domain of f. $f(x_1, x_2, \ldots x_n)$ is a concave function if all the non-zero leading principal minors of the Hessian matrix follow the sign of $(-1)^k$, where k represents the order of the principle minors ($k = 1, 2, 3, \ldots$).

Theorem 8.2c: If the leading principal minors do not follow either Theorem 8.2a or Theorem 8.2b then $f(x_1, x_2, \ldots x_n)$ is neither a convex function nor a concave function.

Example 8.4 Determine the Convexity of the Function using the Hessian Matrix, $f(x_1, x_2) = x_1^2 + 3x_2^2$.

$$H = \begin{bmatrix} 2 & 0 \\ 0 & 6 \end{bmatrix}$$

The 1st leading principal minor are det [2] = 2 > 0, det[6] = 6 > 0.

The 2nd leading principal minor is $\det\begin{bmatrix} 2 & 0 \\ 0 & 6 \end{bmatrix} = 12 > 0$.

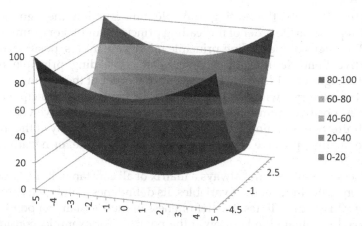

FIGURE 8.2

Graph of $f(x_1, x_2) = x_1^2 + 3x_2^2$.

Since all principal minors are non-negative, we can classify $f(x_1, x_2) = x_1^2 + 3x_2^2$ as a convex function by Theorem 8.2a. The graph is shown in Figure 8.2. We describe obtaining a surface plot in Excel.

$$H(x) = \begin{bmatrix} F_{xx} & F_{xy} & \cdot & \cdot & F_{xz} \\ F_{yx} & F_{yy} & \cdot & \cdot & F_{yz} \\ \cdot & \cdot & \cdot & \cdot & \cdot \\ \cdot & \cdot & \cdot & \cdot & \cdot \\ F_{zx} & F_{zy} & \cdot & \cdot & F_{zz} \end{bmatrix}$$

The Hessian matrix is a symmetric $n \times n$ matrix, where n is the number of independent variables. For example, $z = f(x,y)$ would yield a 2×2 symmetric matrix. The matrix is symmetric because $F_{xy} = F_{yx}$. The function is $x^T H x$. This notation is used to describe the matrix as either:

Positive Definite > 0
Positive Semi-Definite ≥ 0
Negative Definite < 0
Negative Semi-Definite ≤ 0
Indefinite Otherwise.

These also relate to the signs of the leading principal minors (PM), which were previously described.

Positive definite: all leading principle minors are > 0.

Positive Semi-definite: all leading principle minors are non-negative (some are zero).

Negative Definite: the leading principle minors follow the signs of $(-1)^k$, where k represents the order of the leading principle minor. For example, with $k = 1,23$ the signs of the leading principle minors are minus, positive, minus.

Negative Semi-definite: all non-zero valued leading principle minors follow the signs of $(-1)^k$, where k represents the order of the leading principle minor. For example, with $k = 1,23$ the signs of the leading principle minors are minus, positive, minus. Some leading principle minors have value zero.

Indefinite: Some leading principle minors do not follow any of the rules for positive definite, positive semi-definite, negative definite, or negative semi-definite above.

The Hessian matrix is not always a matrix of all constants. If the Hessian is a function of the independent variables, its definiteness might vary from one value of x to another. To test the definiteness of the Hessian at a point x^*, it is necessary to evaluate the Hessian at the point x^*. For example, consider the following Hessian:

TABLE 8.2

Hessian matrix results

Determinants: k, leading Principle Minors (PM)	Results	Conclusions about stationary points
$H_k > 0$	Positive Definite, f convex	Minima
$H_k > 0$	Positive Semi-Definite, f convex	Local minima
H_k follows the signs of $(-1)^k$	Negative Definite, f concave	Maxima
H_k either 0 (not all 0) or follows $(-1)^k$	Negative Definite, f concave	Local maxima
H_k not all 0 and none of the above	Indefinte, f neither	Saddle point
H_k all 0"s	Indefinite	Inconclusive

$$H(X) = \begin{bmatrix} 2x_1 & x_2 \\ x_2 & 4 \end{bmatrix}$$

The values of x_1 and x_2 determine whether the matrix is positive definite, positive semi-definite, negative definite, negative semi-definite, or indefinite.

There exists a relationship between the Hessian matrix definiteness and the classification of stationary points (extrema) as maximum, minimum, saddle points, or inconclusive. Table 8.2 summarizes these results:

where k indicates the order of the leading Principal Minors of the Hessian. The ith PM is found by eliminating the $n - i$ rows and corresponding columns of the matrix. The 1st leading PMs are always the main diagonal of the original Hessian matrix.

8.2 Unconstrained Optimization

We put the theorems and definitions from the previous sections to work in this section.

Example 8.5 Find and classify all the stationary points of

$$f(x,y) = 55 \cdot x - 4 \cdot x^2 + 135 \cdot y - 15 \cdot y^2 - 100$$

$$\frac{\partial f}{\partial x} = 55 - 8x = 0$$

$$\frac{\partial f}{\partial x} = 135 - 30y = 0$$

These solve and we find $x = \dfrac{55}{8}$ and $y = \dfrac{135}{30}$. There is the only stationary point.

$$H\left(\frac{55}{8}, \frac{135}{30}\right) = \begin{bmatrix} -8 & 0 \\ 0 & -30 \end{bmatrix}$$

The 1st PM are –8, –30 which follow $(-1)^1$.
  The 2nd PM is 240 > 0 which follows $(-1)^2$.

The function, f, is concave at $\left(\dfrac{55}{8}, \dfrac{135}{30}\right)$, therefore, $\left(\dfrac{55}{8}, \dfrac{135}{30}\right)$ represents the maximum of f.

$$f\left(\frac{55}{8}, \frac{135}{30}\right) = 329.81$$

We use the Solver in Excel (Figure 8.3).
  The Solver has provided an answer but we need to show it is in fact a maximum. We find the Hessian, evaluated at our critical point and see it is negative definite so that our critical point is a maximum.

$$H = \begin{bmatrix} -8 & 0 \\ 0 & -30 \end{bmatrix}$$

◢	A	B	C	D
1				
2	Decision variables			
3				
4		x	6.874999	
5		y	4.5	
6				
7		z	392.8125	
8				

FIGURE 8.3
Excel screenshot of Solver solution.

Example 8.6 Find and Classify All the Stationary Points of the Function

$$f(x,y) = 2xy + 4x + 6y - 2x^2 - 2y^2$$

$$f(x,y) = 2xy + 4x + 6y - 2x^2 - 2y^2$$

$$\frac{\partial f}{\partial x} = 2y + 4 - 4x$$

$$\frac{\partial f}{\partial x} = 2x + 6 - 4y$$

To find where $\dfrac{\partial f}{\partial x} = \dfrac{\partial f}{\partial y} = 0$, we solve

$$-4x + 2y = -4$$

$$2x - 4y = 6$$

$$\begin{bmatrix} -4 & 2 & -4 \\ 2 & -4 & -6 \end{bmatrix}, \text{ row echelon} \quad \begin{matrix} 1 & 0 & \dfrac{7}{3} \\ 0 & 1 & \dfrac{8}{3} \end{matrix}$$

$$x = \frac{7}{3}, y = \frac{8}{3}$$

$$H = \begin{bmatrix} -4 & 2 \\ 2 & -4 \end{bmatrix}$$

1st leading PM are –4, –4 following $(-1)^k$, $k = 1$.
  2nd leading PM is $16 - 4 = 12 > 0$ following $(-1)^k$, $k = 2$.
  Thus, H is negative definite, f is concave, $(x^*, y^*) = (7/3, 8/3)$ is a global maxima (see Figure 8.4).

Example 8.7 Least Squares Models

In a least squares model, fitting a line, $y = a + bx$, we want to minimize the sum of squared error,

$$f(a,b) = \sum_{i=1}^{i=n} (y_i - a - bx_i)^2$$

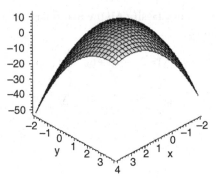

FIGURE 8.4
Graph of $z = 2xy + 4x + 6y - 2x^2 - 2y^2$.

$$\frac{\partial f}{\partial a} = 2\left(\sum_{i=1}^{i=n}(y_i - a - bx_i)(-1)\right) = 0$$

$$\frac{\partial f}{\partial a} = 2\left(\sum_{i=1}^{i=n}(y_i - a - bx_i)(-x_i)\right) = 0$$

$\frac{\partial f}{\partial a} = \frac{\partial f}{\partial b} = 0$ hold when we find (a,b) such that the following hold:

$$na + b\sum x_i = \sum_{i=1}^{i=n} y_i$$

$$a\sum_{i=1}^{i=n} x_i + b\sum_{i=1}^{i=n} x_i^2 = \sum_{i=1}^{i=n} x_i y_i$$

These are called the Normal Equations for Least Squares.

$$H = \begin{matrix} 2n & 2\sum_{i=1}^{i=n} x_i \\ 2\sum_{i=1}^{i=n} x_i & 2\sum_{i=1}^{i=n} x_i^2 \end{matrix}$$

1st leading PM are $2n$, $2\sum_{i=1}^{i=n} x_i^2$ both are > 0.
2nd leading PM is $2n > 0$.
H is positive definite, f is strictly convex so $(a,b)^*$ will be a global minimum.
Now let's look at this with some data (Table 8.3):

$$\text{Minimize } S = (2 - a - b)^2 + (4.8 - 2a - b)^2 + (7 - 3a - b)^2$$

We take the partial derivatives and set them equal to zero. We solve the two simultaneous equations:

TABLE 8.3

Data for Example 8.7 least square method

X	1	2	3
Y	2	4.8	9

$$28a + 12b = 65.2$$

$$12a + 6b = 27.6$$

$$a = 2.5, b = -0.40$$

$$H = \begin{bmatrix} 28 & 12 \\ 12 & 6 \end{bmatrix}$$

The Hessian matrix is positive definite, so we have found the minimum.

8.3 Multivariable Numerical Search Methods for Unconstrained Optimization

In the previous sections we discussed analytical techniques to solve the unconstrained NLP:

$$\text{Max } z = f(x_1, x_2, x_3, \ldots, x_n) \text{ over } R^n. \tag{8.1}$$

In many problems, it is quite difficult to find the stationary points (critical points) and use them to determine the nature of the stationary point. In this chapter, we will discuss several numerical techniques to either maximize or minimize a multivariable function as expressed in Equation (8.1).

Gradient Search Methods

Suppose we want to solve the following unconstrained nonlinear programming problem (NLP):

$$\text{Max } z = f(x_1, x_2, x_3, \ldots, x_n) \tag{8.2}$$

In calculus, if Equation (8.2) is a concave function, then the optimal solution (if there is one) will occur at a stationary point x^* having the following property:

$$\frac{\partial f(x^*)}{\partial x_1} = \frac{\partial f(x^*)}{\partial x_2} = \cdots = \frac{\partial f(x^*)}{\partial x_n} = 0$$

In many problems it is not an easy task to find the stationary point. Thus, the Method of Steepest Ascent (maximization problems) and the Method of Steepest Descent (minimization problems) offers an alternative to finding an approximate stationary point. We will continue to discuss the gradient method for the Steepest Ascent.

Given a function, like the one in Figure 8.5, assume that we want to find the maximum point of the function. If we started at the bottom of the hill then we might proceed by finding the gradient. The gradient is the vector of the partial derivatives that points "up the hill". We define the gradient vector as follows:

$$\nabla f(x) = \left[\frac{\partial f(x^*)}{\partial x_1}, \frac{\partial f(x^*)}{\partial x_2}, ..., \frac{\partial f(x^*)}{\partial x_n} \right]$$

If we were lucky, the gradient would point all the way to the top of the function but the contours of functions do not always cooperate and rarely do. Thus, the gradient "points up hill" but for how far? We need to find the distance along the gradient for which to travel that maximizes the height of the function in that direction. From that new point, we re-compute a new gradient vector to find a new direction that "points up hill." We continue this method until we get to the top of the hill.

From a starting point, we move in the direction of the gradient as long as we continue to increase the value of f. At that point, we move in the direction of a newly calculated gradient as far as we can so long as it continues to improve f. This continues until we achieve our maximum value within some specific tolerance (or margin of acceptable error). Figure 8.6 displays an algorithm for the Method of Steepest Ascent using the gradient.

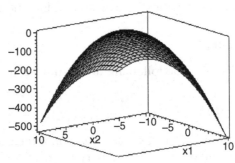

FIGURE 8.5
The function, $Z = f(x1,x2) = 2x_1x_2 + 2x_2 - x_1^2 - 2x_2^2$.

To find a maximum solution to agiven multivariable unconstrained function, $f(x)$

INPUT: starting point x_0; tolerance, t
OUTPUT: Approximate x^*, and $f(x^*)$

Step 1. Initialize the tolerance, $t > 0$.

Step 2. Set $x = x_0$ and define the gradient at that point.
$$\nabla f(x_0)$$

Step 3. Calculate the maximum of the new function $f(x_i + t_i \nabla f(x_i))$, where $t_i \geq 0$, byfinding the value of t_i.

Step 4. Find the new x_i point by substituting t_i into
$$x_{i+1} = x_i + t_i \nabla f(x_i)$$

Step 5. If the length (magnitude) of **x**, defined by
$$\| x \| = (x_1^2 + x_2^2 + \cdots + x_n^2)^{\frac{1}{2}},$$

(difference between two successive points) is less than the tolerance specified or if the **absolute magnitude of gradient is less than our tolerance** (derivative approximately zero), then continue. Otherwise, go back to Step 3.

Step 6. Use x^* as the approximate stationary pointand compute, $f(x^*)$, the estimated maximum of the function.
STOP

FIGURE 8.6
Steepest ascent algorithm.

Examples of Gradient Search

Example 8.8 Maximize $f(x_1, x_2) = 2x_1 x_2 + 2x_2 - x_1^2 - 2x_2^2$

The gradient of $f(x_1, x_2)$, ∇f, is found using the partial derivatives as shown in the last chapter. The gradient is the vector $[2x_2 - 2x_1, 2x_1 + 2 - 4x_2]$.

$\nabla f(0,0) = [0,2]$. From $(0,0)$ we move along (up) the x_2 axis in the direction of $[0,2]$. How far do we go?

We need to maximize the function starting at the point $(0,0)$ using the function

$$f(x_i + t_i \nabla f(x_i)) = f(0 + 0t, x0 + 2t) = 2(2t) - 2(2t)^2 = 4t - 8t^2.$$

This function can be maximized by using any of the one-dimensional search techniques that we discussed previously. This function can also be maximized by simple single variable calculus:

$$\frac{df}{dt} = 0 = 4 - 16t = 0, t = 0.25$$

The new point is found by substitution into $x_{i+1} = x_i + t_i \nabla f(x_i)$.
So, $x_1 = [0 + 0(0.25), 0 + 2(0.25)], [0, 0.5]$.

The magnitude of x_1 is 0.5 which is not less than our tolerance of 0.01 (chosen arbitrarily). Since we are not optimal we continue. We now repeat the calculations from the new point [0, 0.5].

Iteration 2

The gradient vector is $[2\,x_2 - 2x_1,\ 2\,x_1 + 2 - 4x_2]$.

$\nabla f(0,0.5) = [1,1]$. From $(0,0.5)$ we move in the direction of $[1,0]$. How far do we go? We need to maximize the function starting at the new point $(0,0.5)$ using the function

$$f(x_i + t_i \nabla f(x_i)) = f(0 + 1t, 0.5 + 0t) = 2(t)(0.5) + 2(0.5) - t^2 - 2(0.5)^2$$
$$= -t2 + t + 0.5.$$

This function can also be maximized be using any of the one-dimensional search techniques that we discussed in Chapter 9 or be maximized by simple single variable calculus:

$$\frac{df}{dt} = 0 = -2t + 1 = 0, t = 0.50$$

The new point is found by substitution into $x_{i+1} = x_i + t_i \nabla f(x_i)$.

So, $x_1 = [0 + 1(0.5), 0.5 + 0(0.5)]$, [0.5, 0.5].

The magnitude of x_1 is $\sqrt{.5} = 0.707$ which is not less than our tolerance of 0.01 (chosen arbitral lily). The magnitude of $\nabla f = 1$ which is also not less than 0.01. Since we are not optimal we continue. We repeat the calculations from the new point [0.5, 0.5].

Initial Condition: (0.0000, 0.0000)

Iter	Gradient vector G	Magnitude G	x[k]	Step length
1	(0.0000, 2.0000)	2.0000	(0.0000, 0.0000)	.25
2	(1.0000, 0.0000)	1.0000	(0.0000,.5000)	.50
3	(0.0000, 1.0000)	1.0000	(.5000,.5000)	.25
4	(.5000, 0.0000)	.5000	(.5000,.7500)	.50
5	(0.0000,.5000)	.5000	(.7500,.7500)	.25
6	(.2500, 0.0000)	.2500	(.7500,.8750)	.50
7	(0.0000,.2500)	.2500	(.8750,.8750)	.25
8	(.1250, 0.0000)	.1250	(.8750,.9375)	.50
9	(0.0000,.1250)	.1250	(.9375,.9375)	.25
10	(.0625, 0.0000)	.0625	(.9375,.9688)	.50
11	(0.0000,.0625)	.0625	(.9688,.9688)	.25
12	(.0313, 0.0000)	.0313	(.9688,.9844)	.50
13	(0.0000,.0313)	.0313	(.9844,.9844)	.25
14	(.0156, 0.0000)	.0156	(.9844,.9922)	.50
15	(0.0000,.0156)	.0156	(.9922,.9922)	.25
16	(.0078,.0000)	.0078		

Approximate Solution: (0.9922, 0.9961)
Maximum Functional Value: 1.0000
Number gradient evaluations: 17
Number function evaluations: 16

The solution, via calculus, is as follows:
  Maximize $f(x_1,x_2) = 2x_1x_2 + 2x_2 - x_1^2 - 2x_2^2$

$$\frac{\partial f}{\partial x_1} = 0 = 2x_2 - 2x_1$$

$$\frac{\partial f}{\partial x_2} = 0 = 2x_1 + 2 - 4x_2$$

Solving these two equations simultaneously yields:

$$x_1 = 1$$

$$x_2 = 1$$

$$f(x_1,x_2) = 1.$$

The Hessian matrix, $\begin{bmatrix} -2 & 2 \\ 2 & -4 \end{bmatrix}$, is negative definite so the point x^* is a

maximum. Note that our approximate solutions, x, (0.9922, 0.9961), and $f(x) = 1.000$, are close to the exact value of x^*, (1,1) and $f(x^*) = 1$. To get a closer approximation, we should make our tolerance smaller. A look at the contour plot confirms a hill at approximately (1,1) (Figure 8.7).

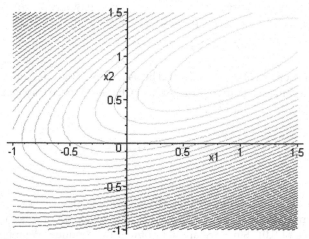

FIGURE 8.7
Contour plot of $2x_1x_2 + 2x_2 - x_1^2 - 2x_2^2$.

◢	A	B	C	D	E
1					
2					
3		x1	1		
4		x2	0.999999		
5					
6		Z	1		

FIGURE 8.8
Contour plot of $2x_1x_2 + 2x_2 - x_1^2 - 2x_2^2$.

In Excel using the GRG routine, which is a good approximation (Figure 8.8).

Example 8.9 Maximize $f(x1, x2)$ $= 55 \times 1 - 4 \times 1^2 + 135 \times 2 - 15 \times 2^2 - 100$ **from the point (1,1).**

We will maximize $f(x1,x2)$ starting from the point (1,1) using a tolerance of 0.01. Figure 8.9 provides a visual reference.

Iteration 2
The gradient vector is $[55 - 8x_1, 135 - 30x_2]$.
    $\nabla f(1,1) = [47,105]$. From (1,1) we move in the direction of [47,105]. How far do we go? We need to maximize the function starting at the point (1,1) using the function

$$f(x_i+t_i\nabla f(x_i)) = f(1 + 47t, 1 + 1055t) = 55(1 + 47t) - 4(1 + 47t)^2$$
$$+ 135(1 + 1055t) - 15(1 + 105t)^2 - 100$$

$$= 71 + 13234t - 174211t^2$$

This function can also be maximized be using any of the one-dimensional search techniques that we discussed in Chapter 3 or be maximized by simple single variable calculus:

$$\frac{df}{dt} = 0 = 13234 - 348422t, t = 0.0379$$

The new point is found by substitution into $x_{i+1} = x_i + t_i\nabla f(x_i)$.
    So, $x_1 = [1 + 47(..0379), 1 + 105(.0379)], [2.785, 4.98]$.
    The magnitude of the gradient is 115.73, which is not less than our tolerance of 0.01. Since we are not optimal we continue. We repeat the calculations from the new point [2.785, 4.98].

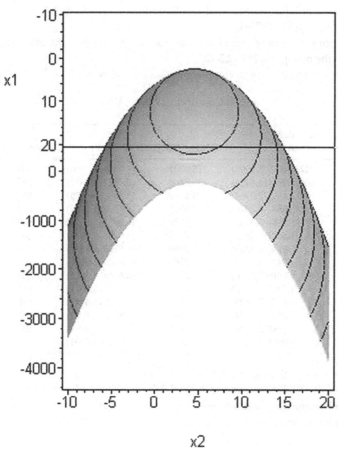

FIGURE 8.9
Plot of $55 \times 1 - 4 \times 1^2 + 135 \times 2 - 15 \times 2^2 - 100$.

Iteration 2

The gradient vector is $[55 - 8x_1, 135 - 30x_2]$.

$\nabla f(2.785, 4.98) = [32.72 - 14.4]$. From (2.785, 4.98) we move in the direction of $[32.72 - 14.4]$. How far do we go? We need to maximize the function starting at the new point (2.785, 4.98) using the function

$$f(x_i + t_i \nabla f(x_i)) = f(2.785 + 32.72t, 4.98 - 14.4t) =$$
$$= 322.444100 + 1277.95840t - 7392.7936t^2$$

This function can also be maximized by simple single variable calculus:

$$\frac{df}{dt} = 0, t = .0864$$

The new point is found by substitution into $x_{i+1} = x_i + t_i \nabla f(x_i)$.

So, $x_1 = [5.612, 3.736]$.

Since we are not optimal we continue. We now repeat the calculations from the new point [5.612,3.736].

We use technology to complete the solution process.

Initial Condition: (1.0000, 1.0000)

Iter	Gradient vector G	Magnitude G	x[k]	Step length
1	(47.0000,105.0000)	115.0391	(1.0000, 1.0000)	.0380
2	(32.7185,-14.6454)	35.8468	(2.7852, 4.9882)	.0857
3	(10.2936, 22.9964)	25.1951	(5.5883, 3.7335)	.0380
4	(7.1658, -3.2075)	7.8509	(5.9793, 4.6069)	.0857
5	(2.2544, 5.0365)	5.5180	(6.5932, 4.3321)	.0380
6	(1.5694, -.7025)	1.7195	(6.6788, 4.5234)	.0857
7	(.4938, 1.1031)	1.2085	(6.8133, 4.4632)	.0380
8	(.3437, -.1539)	.3766	(6.8320, 4.5051)	.0857
9	(.1081,.2416)	.2647	(6.8615, 4.4919)	.0380
10	(.0753, -.0337)	.0825	(6.8656, 4.5011)	.0857
11	(.0237,.0529)	.0580	(6.8720, 4.4982)	.0380
12	(.0165, -.0074)	.0181	(6.8729, 4.5002)	.0857
13	(.0052,.0116)	.0127	(6.8744, 4.4996)	.0380
14	(.0036, -.0016)	.0040	(6.8745, 4.5001)	

Approximate Solution: (6.8745, 4.5001)

Maximum Functional Value: 392.8125

Number gradient evaluations: 15

Number function evaluations: 14.

We point out that the solution process might zig-zag and converge more slowly to an optimal solution as shown in Figure 8.10.

Example 8.10 Maximizing a Transcendental Multivariable Function when Calculus Fails

$$f(x_1, x_2) = 2x_1 x_2 + 2x_2 - e^{x_1} - e^{x_2} + 10$$

These partial derivative equations,

$$2x_2 - e^{x_1} = 0$$

$$2x_1 + 2 - e^{x_2} = 0$$

are not solvable in closed form for (x_1, x_2) without numerical methods. So, we will use the gradient method to approximate the solution. Here we have no analytical solution for comparison.

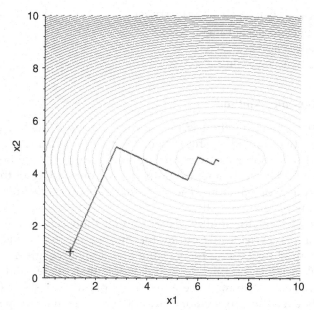

FIGURE 8.10
Zig-zagging to a solution.

We will start at the point (0.5,0.5).
The gradient is $\nabla f = [2x_2 - e^{x1}, 2x_1 + 2 - e^{x2}]$.

$$\nabla f(0,0) = [-.6487, 1.3513]\text{ with magnitude} = 1.4989.$$

New point is found by maximizing

$$f[0.5 - .6487t, 0.5 + 1.3513t] = 2(0.5 - .6487)(0.5 + 1.3513t)$$
$$+ 2(0.5 + 1.3513t) - e^{0.5 - .6487t} - e^{0.5 + 1.3513t}$$

$$df/dt = 0 = 3.40520 - 3.50635324t + 0.6487e^{(0.5 - 0.6487t)} - 1.3513e^{(0.5 + 1.3513t)}$$

$t = 0.2873853295 = 0.2874$ (rounded to 4 decimals)
This moves us to the next point (.3136,.8883). We continue the search using technology.
Initial Condition: (0.5000, 0.5000)

Iter	Gradient vector G	Magnitude G	x[k]	Step length
1	(-.6487, 1.3513)	1.4989	(.5000,.5000)	.2874
2	(.4084,.1961)	.4530	(.3136,.8883)	1.7041
3	(-.2993,.6235)	.6917	(1.0095, 1.2224)	.2001
4	(.1097,.0526)	.1216	(.9496, 1.3472)	.7345
5	(-.0298,.0621)	.0689	(1.0302, 1.3859)	.1868
6	(.0090,.0043)	.0099	(1.0246, 1.3975)	

Approximate Solution: (1.0246, 1.3975)
Maximum Functional Value: 8.8277
Number gradient evaluations: 7
Number function evaluations: 6.

Modified Newton's Method

An alternative search method is the Newton–Raphson numerical method illustrated in two variables. This numerical method appears to do a more efficient and faster job in converging to the near optimal solution. It is an iterative root finding technique using the partial derivatives of the function as the new system of equations. The algorithm uses Cramer's Rule to find the solution of the system of equations.

Newton's Method for multivariable optimization searches is based on Newton's single variable algorithm for finding the roots and Newton–Raphson Method for finding roots of the first derivative, given a x_0, iterate $x_{n+1} = x_n - f'(x_n)/f''(x_n)$ until $|x_{n+1} - x_n|$ is less than some small tolerance. In several variables, we may use a vector $\mathbf{x_0}$, or two variables, (x_0,y_0). The algorithm is expanded to include partial derivatives with respect to each variable's dimension. In two variables (x,y), this would yield a system of equations where F is the derivative of $f(x,y)$ with respect to x and G is the derivative of $f(x,y)$ with respect to y. Thus, we need to find both $F = 0$ and $G = 0$ simultaneously.

This yields a matrix equation $\sum_{j=1}^{N} \alpha_{ij}\delta x_j = \beta_i$, where

$$\alpha_{ij} = \frac{\partial f_i}{\partial x_i}, \beta_i = -f_i.$$

The matrix equation can be solved by LU decomposition or in the case of a 2 x 2 by Cramer's Rule. The corrections are then added to the solution vector

$$x_i^{new} = x_i^{old} + \delta x_i, i = 1,...N$$

and iterated until it converges within a tolerance. The algorithm is shown in Figure 8.11.

Modified Newton with Technology

Let's repeat our examples with technology.

INPUT: x(0), y(0), N, Tolerance

OUTPUT: x(n), y(n)

Step 1. For n = 1 to N do

Step 2. Calculate the new estimate for x(n) and y(n) as follows:

$$\frac{\partial F}{\partial x}(x(n-1),y(n-1)) \to q$$

$$\frac{\partial F}{\partial y}(x(n-1),y(n-1)) \to r$$

$$\frac{\partial G}{\partial x}(x(n-1),y(n-1)) \to s$$

$$\frac{\partial G}{\partial x}(x(n-1),y(n-1)) \to t$$

$$-F(x(n-1),y(n-1)) \to u$$

$$-G(x(n-1),y(n-1)) \to v$$

$$qt-rs \to D$$

$$x(n-1)+(ut-vr)/D \to x(n)$$

$$y(n-1)+(qv-su)/D \to y(n)$$

Step 3. If $((x(n)-x(n-1))^2 + (y(n)-y(n-1))^2)^{1/2}$ < tolerance,

   Then **Stop**

   Else, Go back to Step 2.

STOP

FIGURE 8.11
Pseudocode for Newton's Method in two variables.

Example 8.11 Maximize $f(x_1,x_2) = 2x_1x_2 + 2x_2 - x_1^2 - 2x_2^2$ **starting at the point (0,0) with a tolerance of 0.01.**

The routine converges to the point (1,1) after 2 iterations shown in Figure 8.12.

Example 8.12 $2x_1x_2 + 2x_2 - e^{x1} - e^{x2} + 10$ **start at (0,0)**

$$F = 2x2 - e^{x1}$$

$$G = 2x_1 + 2 - e^{x2}$$

$$F_{x1} = -e^{x1}$$

$$F_{x2} = 2$$

$$G_{x1} = 2$$

$$G_{x2} = -e^{x2}$$

The solution converges quickly to $x_1 = -0.2667$, $x_2 = 0.382953$, $Z = -1.67087$ shown in Figure 8.13.

Comparisons of Methods

We compared these two routines and found that the Newton's Method converges faster than the gradient method. This is displayed in Table 8.4.

8.4 Constrained Optimization

A company manufactures new E-phones that are supposed to capture the market by storm. The two main inputs components of the new E-phone are the circuit board and the relay switches that make the phone faster, smarter, and have more memory. The number of E-phones to be produced is estimated to equal $E = 200a^{\frac{1}{2}}b^{\frac{1}{4}}$ where E is the number of phones produced while a and b are the number of circuit broad hours and the number of relay hours worked respectively. Such a function is known to Economists as a *Cobb–Douglas function*. Our laborers are paid by the type work they do: the circuit boards and the relays for \$5 and \$10 an hour. We want to maximize the number of E-phones to be made if we have \$150,000 to spend on these components in the short run.

Problems such as this can be modeled using constrained optimization. We begin our discussion with equality constrained optimization and then we discuss inequality constrained optimization. We address solving of these using technology.

	A	B	C	D	E	F	G	H	I	J	K	L
1	t		1									
2												
3	n	x	y	q	r	s	t	u	v	D	New-X	New-y
4	0	0	0	-2	2	2	-4	0	-2	4	1	1
5	1	1	1	-2	2	2	-4	0	0	4	1	1
6	1	1	1	-2	2	2	-4	0	0	4	1	1

FIGURE 8.12
Excel screenshot of modified Newton.

35	n	x	y	q	r	s	t	u	v	D	New-X	New-y
36	0	0	0	-1	2	2	-1	1	-1	-3	-0.33333	0.333333
37	1	-0.33333	0.333333	-0.71653	2	2	-1.39561	0.049865	0.062279	-3	-0.26862	0.381451
38	1	-0.26862	0.381451	-0.76444	2	2	-1.46441	0.001533	0.001642	-2.88055	-0.2667	0.382952
39	1	-0.2667	0.382952	-0.7659	2	2	-1.46661	1.41E-06	1.65E-06	-2.87672	-0.2667	0.382953
40	1	-0.2667	0.382953	-0.76591	2	2	-1.46661	1.33E-12	1.48E-12	-2.87671	-0.2667	0.382953
41	1	-0.2667	0.382953	-0.76591	2	2	-1.46661	-1.1E-16	0	-2.87671	-0.2667	0.382953
42	1	-0.2667	0.382953	-0.76591	2	2	-1.46661	0	0	-2.87671	-0.2667	0.382953
43	1	-0.2667	0.382953	-0.76591	2	2	-1.46661	0	0	-2.87671	-0.2667	0.382953

FIGURE 8.13

Excel screenshot of modified Newton for $2x_1x_2 + 2x_2 - e^{x1} - e^{x2} + 10$.

TABLE 8.4

Comparison of Newton and gradient methods

Function 1	Initial Condition	Iterations	Feval	gevals	Soln	Max F
Steepest Ascent	(0,0)	16	16	17	$x = .9922$ $y = .9961$	1.0
Newton's Method	(0,0)	2	2		$x = 1$ $y = 1$	1.00000

Function 2	Initial condition	Iterations	Feval	gevals		
Steepest Ascent	(0,0)	4	5	4	$x = -.26638$, $y = .3853$	8.3291
Newton's Method	(0,0)	2			$x = -.269$, $y = .381$	8.329

In general we are solving

$$Max\ (Min)\ z = f(x_1, x_2, x_3, ..., x_n).$$

Subject to:

$$g_1(x_1, x_2, ..., x_n) \begin{Bmatrix} \le \\ = \\ \ge \end{Bmatrix} ..b_1$$

$$g_2(x_1, x_2, ..., x_n) \begin{Bmatrix} \le \\ = \\ \ge \end{Bmatrix} ..b_2$$

$$g_m(x_1, x_2, ..., x_n) \begin{Bmatrix} \le \\ = \\ \ge \end{Bmatrix} ..b_m$$

We begin with equality constraints and then transition to inequality constraints.

Equality Constraints – Method of Lagrange Multipliers

Lagrange multipliers can be used to solve nonlinear optimization problems (called NLPs) in which all the constraints are equality constrained. We consider the following type NLPs:

$$\text{Max (Min) } z = f(x_1, x_2, x_3, ..., x_n)$$

Subject to:

$$g_1(x_1, x_2, ..., x_n) = b_1$$

$$g_2(x_1, x_2, ..., x_n) = b_2$$

$$...$$

$$g_m(x_1, x_2, ..., x_n) = b_m$$

In our E-phones example, we find we can build an equality constrained model. We want to maximize

$$E = 200a^{\frac{1}{2}}b^{\frac{1}{4}}$$

Subject to the constraint

$$5a + 10b = 150,000.$$

Basic Theory

In order to solve NLPs in the form of (1), we associate a Lagrangian multiplier, λ_i, with the ith constraint and form the Lagrangian equation:

$$L(X, \lambda) = f(X) + \sum_{i=1}^{m} \lambda_i (b_i - g_i(X))$$

The computational procedure for Lagrange Multipliers requires that all the partials of this Lagrangian function must equal zero. These partials are the *necessary conditions* of the NLP problem. These are the conditions required for $\mathbf{x} = \{x_1, x_2, ..., x_n\}$ to be a solution.

The Necessary Conditions (8.3)

(3a) $\partial L/\partial Xj = 0$ ($j = 1, 2, ..., n$ variables)
(3b) $\partial L/\partial \lambda_i = 0$ ($i = 1, 2, ..., m$ constraints).

Definition: x is a regular point if and only if $\nabla g_i(x)$, $i = 1, 2, ..., m$ are linearly independent.

Theorem 8.3.
(a) Let (1) be a maximization problem. If f is a concave function and each $g_i(x)$ is a linear function, then any point satisfying (8.3) will yield an optimal solution.
(b) Let (1) be a minimization problem. If f is a convex function and each $g_i(x)$ is a linear function, then any point satisfying (8.3) will yield an optimal solution.

Recall from Section 8.2 that we used the Hessian matrix to determine if a function was convex, concave, or neither. We also note that the above theorem limits our constraints to linear functions. What if we have nonlinear constraints?

We can use the Bordered Hessian in the sufficient conditions. Given the bivariate Lagrangian function as in

$$L(x_1, x_2, \lambda) = f(x_1, x_2) + \sum_{i=1}^{m} \lambda_i (b_i - g(x_1, x_2))$$

The bordered Hessian is

$$BdH = \begin{bmatrix} 0 & g_1 & g_2 \\ g_1 & f_{11} - \lambda g_{11} & f_{12} - \lambda g_{12} \\ g_2 & f_{21} - \lambda g_{21} & f_{22} - \lambda g_{22} \end{bmatrix}$$

We find the determinant of this Bordered Hessian as Equation (8.4)

$$|BdH| = \det \begin{bmatrix} 0 & g_1 & g_2 \\ g_1 & f_{11} - \lambda g_{11} & f_{12} - \lambda g_{12} \\ g_2 & f_{21} - \lambda g_{21} & f_{22} - \lambda g_{22} \end{bmatrix} = g_1 g_2 (f_{21} - \lambda g_{21})$$
$$+ g_2 g_1 (f_{12} - \lambda g_{12}) - g_2^2 (f_{11} - \lambda g_{11}) - g_1^2 (f_{22} - \lambda g_{22})$$

(8.4)

The sufficient condition for a *maximum*, in the bivariate case with one constraint, is the determinant of its bordered Hessian is positive when evaluated at the critical point.

The sufficient condition for a *minimum*, in the bivariate case with one constraint, is the determinant of its bordered Hessian is negative when evaluated at the critical point.

If x is a regular point and $g_i(x) = 0$ (constraints are satisfied) then $M = \{y \mid \nabla g_i(x) \bullet y = 0\}$ defines a plane tangent to the feasible region at x.

Lemma: If x is regular and $g_i(x) = 0$, and $\nabla g_i(x) \bullet y = 0$, then $\nabla f(x) \bullet y = 0$.

Note that the Lagrange Multiplier conditions are exactly the same for a minimization problem as a maximization problem. This is the reason that these conditions alone are not sufficient conditions. Thus, a given solution can either be a maximum or a minimum. In order to determine whether the point found is a maximum, minimum, or saddle point we will use the Hessian.

The Lagrange Multiplier, λ, has an important modeling interpretation. It is the "shadow price" for scarce resources. Thus, λ_i is the shadow price of the ith constraint. Thus, if the right hand side is increased by a small amount Δ, in a maximization or a minimization problem, then the optimal solution will change by $\lambda_i \Delta$. We will illustrate the shadow price both graphically and computationally.

Graphical Interpretation of Lagrange Multipliers

The method of Lagrange multipliers is based on its geometric interpretation. This geometric interpretation involves the gradients of both the function and the constraints. Initially, let's consider only one constraint,

$$g(x_1, x_2, ..., x_n) = b$$

so that the Lagrangian equation simplifies to

$$\nabla f = \lambda \nabla g.$$

The solution is the point in x where the gradient vector, $\nabla g(x)$, is perpendicular to the surface. The gradient vector, ∇f, always points in the direction in which f increases fastest. At both maximums and minimums, this direction must also be perpendicular to S. Thus, since both ∇f and ∇g both point along the same perpendicular line, then $\nabla f = \lambda \nabla g$.

In the case of multiple constraints, the geometrical arguments are similar, see Figure 8.14.

Let's preview a graphical solution to our first computational example.

$$\text{Maximize } z = -2\,x^2 + -2\,y^2 + xy + 8x + 3y$$

$$\text{s.t.} \quad 3x + y = 6.$$

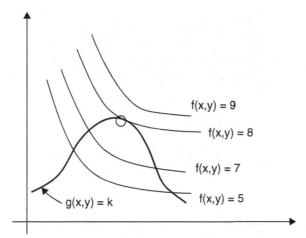

FIGURE 8.14
One equality constraint.

We obtained a contour plot of z and overlaid the single constraint onto the contour plot, see Figure 8.15. What information can we obtain from this graphical representation? First, we note that the unconstrained optimal does not lie on the constraint. We can estimate the unconstrained optimal $(x^*, y^*) = (2.3, 1.3)$. The optimal constrained solution lies at the point where the constraint is tangent to a contour of the function, f. This point is labeled, X^*, and is estimated as $(1.8, 1.0)$. We see clearly that the resource does not pass through the unconstrained maximum and thus, it can be modified (if feasible) until the line passes through the unconstrained solution. At that point, we would no longer add (or subtract) any more resources (see Figure 8.16). We can gain valuable insights about the problem if we are able to plot the information.

Computational Method of Lagrange Multipliers

Consider the set of equations in (8.3). This gives $m + n$ equations in the $m + n$ unknowns (x_j, λ_i). Generally speaking this is a difficult problem to solve without a computer except for simple problems. Also, since the Lagrange Multipliers are necessary conditions only (not sufficient) we may find solutions (x_j, λ_i) that are not optimal for our NLP. We need to be able to determine the classification of points found in the solution to the necessary conditions. Commonly used methods as justification include:

a. Hessian matrix
b. Boarded Hessian, Det [BH] = 0.

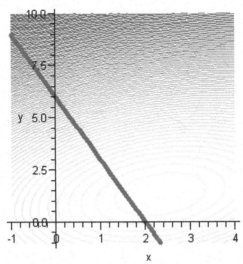

FIGURE 8.15
Contour plot of function, *f*, and the equality constraint example $g(x) = 3x + y = 6$.

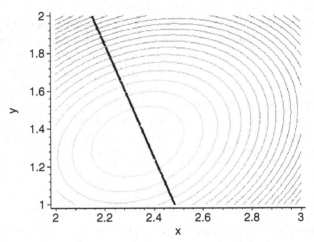

FIGURE 8.16
Added to resource, $g(x) = 8.45 = 3x + y$.

Example 8.13 Maximize z = $-2x^2 - 2y^2 + xy + 8x + 3y$

> s.t. $3x + y = 6$.

The Lagrangian function is:

$$L(x, y, \lambda) = -2x^2 - y^2 + xy + 8x + 3y + \lambda[6 - 3x - y].$$

The necessary conditions are:

$$L_x = -4x - +y + 8 - 3\lambda = 0$$

$$L_y = -2y - +x + 3 - \lambda = 0$$

$$L_\lambda = 6 - 3x - y = 0$$

We have three equations and three unknowns to solve. We solve to find

$$X = 1.46428, y = 1.60714, l = 1.25, Z = 12.0178.$$

Thus, our solution is

$$X = 1.46428, y = 1.60714, l = 1.25, Z = 12.0178.$$

We have a solution but we need to know whether this solution represents the maximum or the minimum of the Lagrangian. We use the Hessian matrix in our justification. We could use either the Hessian or the Bordered Hessian described below to justify that we have found the correct solution to our problem to maximize L.

(1) Hessian is negative definite so we have found the maximum.

$$H = \begin{bmatrix} -4 & 1 \\ 1 & -4 \end{bmatrix}$$

(2) Boarded Hessian.

$$BH := \begin{bmatrix} -4 & 1 & -3 \\ 1 & -4 & -1 \\ -3 & -1 & 0 \end{bmatrix}$$

We find the determinate is equal to 46. Since the determinant is positive we have found the maximum at the critical point.

Either method works in this example to determine that we have found the maximum of L.

Now, let's interpret the shadow price, $\lambda = 1.25$. If the right hand side of the constraint is increased by a small amount Δ, then the function will increase by $1.25\ \Delta$. Since this is a maximization problem we would add to the resource if possible, because it improves the value of the objective function. From the graph, it can be seen that the incremental change must be small or the objective function will begin to increase. If we increase the RHS by one unit so that $g(x) = 3x + y = 7$ the solution at the new point

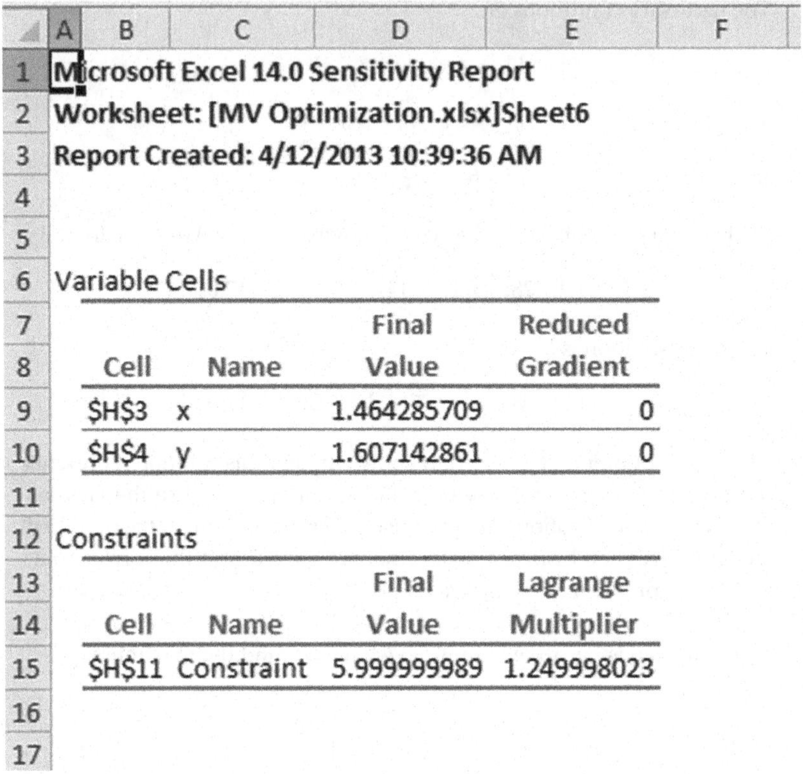

◢	A	B	C	D	E	F
1	Microsoft Excel 14.0 Sensitivity Report					
2	Worksheet: [MV Optimization.xlsx]Sheet6					
3	Report Created: 4/12/2013 10:39:36 AM					
4						
5						
6	Variable Cells					
7				Final	Reduced	
8		Cell	Name	Value	Gradient	
9		$H$3	x	1.464285709	0	
10		$H$4	y	1.607142861	0	
11						
12	Constraints					
13				Final	Lagrange	
14		Cell	Name	Value	Multiplier	
15		$H$11	Constraint	5.999999989	1.249998023	
16						
17						

FIGURE 8.17
Excel screenshot of Solver sensitivity report.

(x^{**}, y^{**}) should yield a functional value, $f(x^{**}, y^{**}) \approx$ old $f + \lambda = 12.0176 + 1.25 = 13.2676$.

We may use the Solver to obtain the solution, shown in Figure 8.17.

Example 8.14 Multiple Constraints in Excel

$$\text{Minimize } w = x^2 + y^2 + 3^* z$$

$$\text{s.t.} \quad x + y = 3$$

$$x + 3y + 2z = 7$$

We will illustrate only in Excel (Figure 8.18–8.19).

The solution is $x = 0.75$, $y = 2.25$, $z = -0.25$, $\lambda_1 = 0$, $\lambda_2 = 1.5$, $w = 4.875$.

Justification with the Hessian:

	A	B	C	D	E	F
1						
2		x	0.75			
3		y	2.25			
4		z	-0.25			
5						
6						
7		W	4.875			
8						
9						
10			CONST	Used	Limit	
11				3	3	
12				7	7	
13						

FIGURE 8.18
Excel screenshot of optimization problem.

$$h2 := Hessian(L,[x,y,z]);$$

$$h2 := \begin{bmatrix} 2 & 0 & 0 \\ 0 & 2 & 0 \\ 0 & 0 & 0 \end{bmatrix}$$

The Hessian is always positive semi-definite. The function is convex and our critical point is a minimum.

Let us interpret the shadow prices, λ_1 and λ_2. If we could only spend an extra dollar on one of the two resources, which one would we spend it on? The values of the shadow prices are 0 and 1.5 respectively. Since the shadow price is $\partial w/\partial b$, we would not spend an extra dollar on resource number 2 because it will cause the objective function to increase by approximately $1.50.

Modeling and Applications with Lagrange Multipliers

Example 8.15 Cobb–Douglas Function

Recall the problem suggested in the introduction of the chapter. A company manufactures new E-phones that are supposed to capture the

	A	B	C	D	E
1	Microsoft Excel 14.0 Sensitivity Report				
2	Worksheet: [MV Optimization.xlsx]Sheet9				
3	Report Created: 4/12/2013 10:45:59 AM				
4					
5					
6	Variable Cells				
7				Final	Reduced
8		Cell	Name	Value	Gradient
9		$C$2	x	0.750000008	0
10		$C$3	y	2.249999992	0
11		$C$4	z	-0.249999992	0
12					
13	Constraints				
14				Final	Lagrange
15		Cell	Name	Value	Multiplier
16		$D$11	Used	3	1.01328E-06
17		$D$12	Used	7	1.500000775
18					

FIGURE 8.19
Excel screenshot of Solver SA report.

market by storm. The two main inputs components to the E-phone are the circuit board and the relay switches. The number of E-phones produced is estimated to equal $E = 200a^{\frac{1}{2}}b^{\frac{1}{4}}$ where E is the number of phones produced while a and b are the number of circuit broad hours and the number of relay hours worked respectively. Such a function is known to Economists as a *Cobb–Douglas function*. Our laborers are paid by the type work they do: the circuit boards and the relays for $5 and $10 an hour. We want to maximize the number of E-phones to be made if we have $150,000 to spend on these components in the short run (Figure 8.20–8.21).

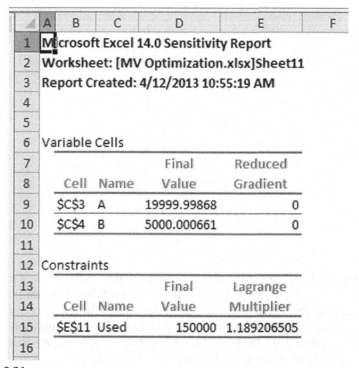

	A	B	C	D	E	F
1						
2						
3		A	20000			
4		B	5000.001			
5						
6		COBB	237841.4			
7						
8						
9						
10			Constraint		Used	
11					150000	150000
12						

FIGURE 8.20
Excel screenshot of optimization problem.

	A	B	C	D	E	F
1	Microsoft Excel 14.0 Sensitivity Report					
2	Worksheet: [MV Optimization.xlsx]Sheet11					
3	Report Created: 4/12/2013 10:55:19 AM					
4						
5						
6	Variable Cells					
7				Final	Reduced	
8		Cell	Name	Value	Gradient	
9		$C$3	A	19999.99868	0	
10		$C$4	B	5000.000661	0	
11						
12	Constraints					
13				Final	Lagrange	
14		Cell	Name	Value	Multiplier	
15		$E$11	Used	150000	1.189206505	
16						

FIGURE 8.21
Excel screenshot of Solver SA report.

$$L(A, B, \lambda) = 200A^{0.5} B^{.25} + \lambda(150000 - 5A - 10B).$$

We find we can make 237841.42 E-phones using 20,000 relays and 5,000 circuit boards hours of labor.

8.5 Inequality Constraints – Kuhn-Tucker (KTC) Necessary/Sufficient Conditions

Introduction to KTC

In the previous sections we investigated procedures to solve problems with equality constraints. However, in most realistic problems many of the constraints are inequalities. These constraints form the boundaries for the solution. The generic form of the NLP we will study in this section is:

$$Max\ (Min)\ z = f(x_1, x_2, x_3, ..., x_n) \tag{8.5}$$

Subject to:

$$g_1(x_1, x_2, ..., x_n) \begin{Bmatrix} \leq \\ = \\ \geq \end{Bmatrix} b_1$$

$$g_2(x_1, x_2, ..., x_n) \begin{Bmatrix} \leq \\ = \\ \geq \end{Bmatrix} b_2$$

$$\cdots \tag{8.6}$$

$$g_m(x_1, x_2, ..., x_n) \begin{Bmatrix} \leq \\ = \\ \geq \end{Bmatrix} b_m$$

One method to solve NLPs of this type, Equation (8.5), is the *Kuhn-Tucker Conditions* (KTC). In this section we describe the KTC first graphically and then analytically. We discuss the necessary and sufficient conditions for $X = \{x_1, x_2, ..., x_n\}$ to be an optimal solution to the NLP of Equation (8.6). We illustrate how to use Maple to solve these KTC problems. We then present some applications using the KTC solution methodology.

Basic Theory of Constrained Optimization

During these sections on KTC, we are concerned with problems of the form:

$$\text{MAX (MIN) } f(X)$$

Subject to:

$$g_i(X) \left\{ \geq \right\} b_i \qquad i = 1, 2, \ldots, m \qquad (8.7)$$

$$\left\{ \leq \right\}$$

$$\left\{ = \right\}$$

We allow the constraints to be either "less than or equal to" or "greater than or equal to". We have previously completed a block on Lagrange multipliers to solve problems with equality constraints. You recall that during the Lagrange block that the optimal solution actually fell on one constraint or at an intersection of several constraints. With the inequality constraints, the solution no longer must lie on a constraint or at an intersection point of constraints. This concept poses new problems for you. You need a method for accounting for the position of the optimal solution relative to each constraint. This KTC Procedure involves setting up a Lagrangian function of the decision variables X, the Lagrange multipliers λ, and the slack or surplus variables U_i^2. The X are the decision variables $(x_1, x_2, \ldots, x_n)$, the $-\lambda_i$ are the shadow prices for the ith constraint, and the U_i^2 are either added (slack variables from $\leq$ constraints) or subtracted (surplus variables from $\geq$ constraints). Thus, with the sign of U_i^2 we are able to accommodate both $\leq$ and $\geq$ constraints.

We set up this generic Lagrangian function,

$$L(X, \lambda, U^2) = f(X) + \sum_{i=1}^{m} \lambda_i (g(X_i) + (\pm U^2) - b_i) \qquad (8.8)$$

Note: $\pm U_i^2$ depends on the type of inequality constraint as we will explain more in detail.

The computational procedure for the KTC requires that all the partials of this Lagrangian function equal zero. These partials are the *necessary conditions* of the NLP problem. These are the conditions required for $x = \{x_1, x_2, \ldots, x_n\}$ to be a solution to (8.7).

The Necessary Conditions (8.9)

$$\partial L / \partial Xj = 0 \ (j = 1, 2, \ldots, n) \qquad (8.9a)$$

$$\partial L / \partial \lambda_i = 0 \ (i = 1, 2, \ldots, m) \qquad (8.9b)$$

$$\partial L / \partial U_i = 0 \text{ or } 2 U_i \lambda_i = 0 \ (i = 1, 2, \ldots, m). \qquad (8.9c)$$

The following two theorems give the sufficient conditions for $x^* = \{x_1, x_2, ...,$ $x_n)$ to be an optimal solution to the NLP given in Equation (8.9).

The Sufficient Conditions

Minimum: If $f(x)$ is a convex function and each of the $g_i(x)$ are convex functions, then any point that satisfies the necessary conditions is an optimal solution. An optimal point is a point that minimizes the function subject to the constraints. λ_i is greater than or equal to zero for all i.

Maximum: If $f(x)$ is a concave function and each of the $g_i(x)$ are convex functions, then any point that satisfies the necessary conditions is an optimal solution. An optimal point is a point that maximizes the function subject to the constraints. λ_i is less than or equal to zero for all i.

If these above conditions are not completely satisfied, then we may use another method to check the nature of a potential stationary or regular point, such as the bordered Hessian.

Boarded Hessian: The boarded Hessian is a symmetric matrix of second partials of the Lagrangian.

$$H_B = \partial^2 L / \partial (X_j^2, \lambda_k^2) \ \{j = 1, 2, ..., n \ k = 1, 2, ..., m\}.$$

We can determine, if possible, the nature of the stationary point by classifying the boarded Hessian. This method is only valid leading to max or min points. If the boarded Hessian is indefinite then another method should be used.

Complementary Slackness

The KTC computational solution process uses these necessary conditions and solving 2^m possible cases, where m equals the number of constraints. The value 2 comes from the possible conditions placed on λ_i, either it equals zero or it does not equal zero. There is actually more to this process since it really involves the complementary slackness condition imbedded in the necessary condition, $2 U_i \lambda_i = 0$. Thus, either U_i equals zero and λ_i does not equal zero or U_i is greater than or equal to zero and λ_i is equal to zero. This insures that the complementary slackness conditions are satisfied while solving the other necessary conditions from Equations (8.9a) and (8.9b).

It is these complimentary slackness necessary conditions, Equation (8.9c) that leads to the solution process that we focus our computational and geometric interpretation. We have defined U_i^2 as a slack or surplus variable. Therefore, if U_i^2 equals zero then our point is on the ith constraint and if U_i^2 is greater than zero then the point does not lie on the ith constraint. Furthermore, if the value of U_i^2 is undefined because it equals a negative number, then the point of concern is infeasible. Figures 8.22–8.24 illustrate these conditions.

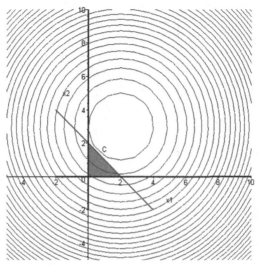

FIGURE 8.22

$U^2 = 0$, point C is on the constraint.

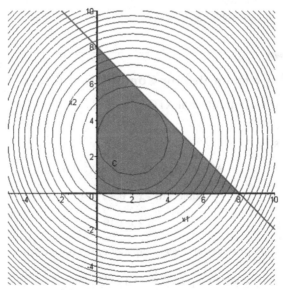

FIGURE 8.23

The point C is inside the feasible region, $U^2 > 0$.

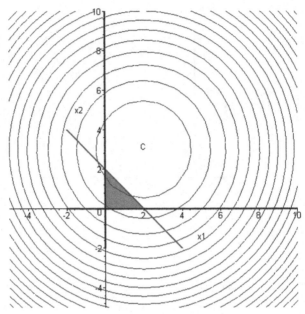

FIGURE 8.24
The point C is not in the feasible region, $U^2 < 0$.

8.6 Technology Examples for Computational KTC

8.6.1 Excel for Computational KTC

Example 8.16

$$\text{Minimize } z = (x - 14)^2 + (y - 11)^2$$

$$\text{subject to } 2x + 0.5y \leq 16$$

$$x + y \leq 19$$

Figure 8.25 provides an Excel snapshot of the optimization model.

The sensitivity analysis report is shown as Figure 8.26
Our solution is $z = 72.05882$ when $x = 5.764706$, $y = 8.941176$, $\lambda_1 = -8.235294342$, $\lambda_2 = 0$.

NLPSolve(obj1, {constr});

$[72.0588235294117538, [x = 5.76470588235294, y = 8.94117647058824]]$

	x	5.764706	
	y	8.941176	
	obj	72.05882	
	constraints		
		16	16
		14.70588	19

FIGURE 8.25
Excel screenshot of optimization problem.

Microsoft Excel 16.0 Sensitivity Report
Worksheet: [KTC optimization.xlsx]Sheet1
Report Created: 6/24/2020 3:13:02 PM

Variable Cells

Cell	Name	Final Value	Reduced Gradient
$C$5	x	5.764705882	0
$C$6	y	8.941176471	0

Constraints

Cell	Name	Final Value	Lagrange Multiplier
$C$11		16	-8.235294342
$C$12		14.70588235	0

FIGURE 8.26
Excel screenshot of Solver SA report.

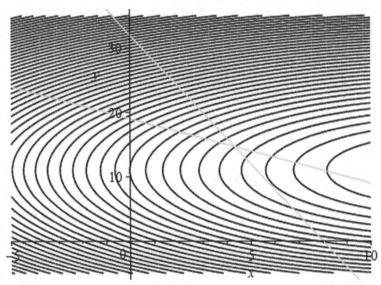

FIGURE 8.27
Contour plot of Example 8.16.

8.6.2 Maple for Computational KTC

We provide a contour plot over-laid with the constraints in Figure 8.27.
We may use NLPSolve to obtain the solution.

$NLPSolve(obj1, \{constr\});$
$$[72.0588235294117538, [x = 5.76470588235294,$$
$$y = 8.94117647058824]]$$

It does not provide the Lagrange multipliers that we might need for sensitivity analysis.

Therefore, we try the longer way to compute these in Maple.

$$fl := obj + lam1(2x + 0.5y - 16 + U1^2) + lam2(x + y - 19 + U2^2)$$
$$fl := (x - 14)^2 + (y = 11)^2$$
$$+ lam1(2x + 0.5y - 16 + U1^2) + (U2^2 + x + y - 19)lam2$$

$fd1 := diff\,(fl, x);$ $fd1 := 2x - 28 + 2lam1 + lam2$
$fd2 := diff\,(fl, y);$ $fd2 := 2y - 22 + 0.5\,lam1 + lam2$
$fd3 := diff\,(fl, lam1);$ $fd3 := 2x + 0.5y - 16 + U1^2$
$fd4 := diff\,(fl, lam2);$ $fd4 := U2^2 + x + y - 19$

$fsolve(\{fd1 = 0, fd2 = 0, fd3 = 0, U1\ lam1 = 0, lam2 = 0\},$
$\qquad \{x, y, lam2, U1, U2\}); \{U1 = 0, U2 = 0,$
$\qquad lam1 = 17.77777778, lam2 = -16.22222222,$
$\qquad x = 4.333333333, y = 14.6666667\}$

$solve(\{fd1 = 0, fd2 = 0, fd3 = 0, U1\ lam1 = 0, lam2 = 0\}, \{x, y, lam2, U1, U2\})$
$\{U1 = 0, U2 = 0, lam1 = 17.77777778, lam2 = -16.22222222, x = 4.333333333,$
$\quad y = 14.66666667\}, \{U1 = 0, U2 = 2.072225289, lam1 = 8.235294118,$
$\quad lam2 = 0,\ x = 5.764705882, y = 8.941176471\},\ \{U1 = 0., U2 = -2.0722252,$
$\quad lam1 = 8.235294118, lam2 = 0, x = 5.764705882, y = 8.941176471\},$
$\quad \{U1 = 3.162277660I,\ U2 = 0.,\ lam1 = 0., lam2 = 6,\ x = 11, y = 8.\},$
$\quad \{U1 = -3.162277660I, U2 = 0,\ lam1 = 0, lam2 = 6,\ x = 11, y = 8\},$
$\quad \{U1 = 4.183300133I, U2 = 2.449489743I, lam1 = 0., lam2 = 0, x = 14, y = 11\},$
$\quad \{U1 = -4.183300133I, U2 = 2.449489743I,\ lam1 = 0, lam2 = 0, x = 14, y = 11.\},$
$\quad \{U1 = 4.183300133I, U2 = -2.449489743I,\ lam1 = 0, lam2 = 0, x = 14, y = 11.\},$
$\quad \{U1 = -4.183300133I, U2 = -2.449489743I,\ lam1 = 0, lam2 = 0, x = 14, y = 11.\}$

The **fsolve** did not provide the answer we seek, so we use the **solve** command. We then canvass the output for feasible solutions. The solution we seek is:

$\{U1 = 0, U2 = 2.072225289, lam1 = 8.235294118, lam2 = 0.,$
$\qquad x = 5.764705882, y = 8.941176471\}.$

8.6.3 R for Computational KTC

R Code to find Optimized Solution (default is a minimum so we multiplied by –1)

```
> f <-function(x) -1*(-100-15*x[2]^2+135*x[2]+55*x[1]-4*x[1]^2)

>

> r <-optim(c(1, 1), f)

>

>r$convergence

[1] 0

>r$par

[1] 6.875081 4.500145

>r$value

[1] -392.8125
```

Recall, that R's default is a minimum. We modified the maximization function to a

Minimization function by multiplying by –1. Therefore, we must multiply our answer of $f(x,y)$.

By (–1).

The function $f(6.875081,4.500145) = 393.8125$.

Contour Plots in R

```
# Contains rotate.persp()# Evaluate z on a grid given by x and y
x <- y <- seq(-1, 10, len = 25)
z <- outer(x, y, FUN = function(x,y) 55*x-4*x^2 +135*y-15*y^2-100)
# Contour plots
contour(x,y,z, main = "Contour Plot")
```

Inequality Constrained Optimization in R

You must load the package Rsolnp first. It does not load on all versions of R at the moment,

This example is from Neto, at source: www.di.fc.ul.pt/~jpn/r/optimization

$$\text{Minimize } f(x, y) = 4x^2 + 10y^2$$

$$s.t. \qquad x^2 + y^2 \le 4$$

The R Code for this problem is:

```
fn<- function(x) {# f(x,y) = 4x^2 + 10y^2
4*x[1]^2 + 10*x[2]^2
}
# constraint z1: x^2 + y^2 < = 4
ineq<- function(x) {
z1 = x[1]^2 + x[2]^2
return(c(z1))
}

lh<- c(0)
uh <- c(4)

x0 = c(1, 1) # setup init values
sol1 <- solnp(x0, fun = fn, ineqfun = ineq, ineqLB = lh, ineqUB = uh)

##
## Iter: 1 fn: 2.8697 Pars: 0.68437 0.31563
```

Iter: 2 fn: 0.6456 Pars: 0.39701 0.03895

Iter: 3 fn: 0.1604 Pars: 0.200217 0.002001

Iter: 4 fn: 0.04009 Pars: 0.10011818 0.00005323

Iter: 5 fn: 0.01002 Pars: 0.0500591336 0.0000006785

Iter: 6 fn: 0.002506 Pars: 0.02502959475 -0.00000004495

Iter: 7 fn: 0.0006265 Pars: 0.01251488097 -0.00000004998

Iter: 8 fn: 0.0001566 Pars: 0.00625751 -0.00000005

Iter: 9 fn: 0.00003916 Pars: 0.00312878 -0.00000005

Iter: 10 fn: 0.000009791 Pars: 0.00156452 -0.00000005

Iter: 11 fn: 0.000002448 Pars: 0.00078235 -0.00000005

Iter: 12 fn: 0.0000006137 Pars: 0.00039171 -0.00000005

Iter: 13 fn: 0.0000001564 Pars: 0.00019772 -0.00000005

Iter: 14 fn: 0.00000004006 Pars: 0.00010007 -0.00000005

Iter: 15 fn: 0.00000001307 Pars: 0.00005716 -0.00000005

Iter: 16 fn: 0.000000006833 Pars: 0.00004133 -0.00000005

solnp--> Completed in 16 iterations

sol1$pars

[1] 4.133035e-05 -5.000058e-08

The result is quite close to (0,0)

Necessary and Sufficient Conditions for Computational KTC

We have shown how the use of visual interpretation can reduce the amount of work required to solve the problem. By seeing the plot you can interpret the conditions involved at the optimal point and then solve directly for that point. However, sometimes we cannot obtain the graphical interpretation so we must rely on the computational method alone. When this occurs we must solve all the cases and interpret the results. Perhaps, we also need to check the Hessian and/or Boarded Hessian,

8.6.4 Modeling and Applications of KTC

Example 8.17 Maximizing Profit from Perfume Manufacturing

A company manufactures perfumes and can purchase up to 1925 oz of the main chemical ingredient for $10 per oz. At a cost of $3 per oz the chemical can be manufactured into an ounce of perfume #11, and at a cost of

$5 per oz the chemical can be manufactured into an ounce of the higher priced perfume #2. An advertising firm estimates that if x ounces of perfume #1 are manufactured, it will sell for $30–.01x$ per ounce. If y ounces of perfume #2 are produced, it can sell for $50–.02y$ per ounce. The company wants to maximize profits for the company.

Formulation:

x = ounces of perfume 1 produced
y = ounces of perfume 2 produced
z = ounces of main chemical purchased

$$\text{Max } f(x,y,z) = x(30 - .01x) + y(50 - .02y) - 3\,x - 5\,y - 10\,z$$

Subject to: $x + y \le z$

$$z \le 1925.$$

Solution (Figure 8.28–8.29):

Consider the significance of the shadow price for λ_1. How do we interpret the shadow price in terms of this scenario? If we could obtain an extra ounce($\Delta = 1$) of chemical at no cost, it would improve the profit to about $22,537.50 + \$10\ (1)$.

	A	B	C	D	E
1					
2					
3					
4			x	849.9994	
5			y	874.9997	
6			z	1724.999	
7					
8			F	22537.5	
9					
10					
11				Used	Limit
12				0	0
13				1724.999	1925
14					

FIGURE 8.28
Excel screenshot of optimization problem.

	A	B	C	D	E	F
1	Microsoft Excel 14.0 Sensitivity Report					
2	Worksheet: [MV Optimization.xlsx]Sheet17					
3	Report Created: 4/12/2013 11:19:19 AM					
4						
5						
6	Variable Cells					
7				Final	Reduced	
8		Cell	Name	Value	Gradient	
9		$D$4	x	849.9994415	0	
10		$D$5	y	874.9996983	0	
11		$D$6	z	1724.99914	0	
12						
13	Constraints					
14				Final	Lagrange	
15		Cell	Name	Value	Multiplier	
16		$D$12	Used	0	10.00000286	
17		$D$13	Used	1724.99914	0	
18						

FIGURE 8.29
Excel screenshot of Solver SA report.

8.7 Exercises

1. Using the Hessian Matrix, H, determine the convexity and then find the critical points and classify them for the following:

   (a) $f(x,y) = x^2 + 3xy - y^2$
   (b) $f(x,y) = x^2 + y^2$
   (c) $f(x,y) = -x^2 - xy - 2y^2$
   (d) $f(x,y) = 3x + 5y - 4x^2 + y^2 - 5xy$
   (e) $f(x,y,z) = 2x + 3y + 3z - xy + xz - yz - x^2 - 3y^2 - z^2$
   (f) Determine the values of a, b, &c such that $ax^2 + bxy + cy^2$ is convex? Is concave?

2. Find and classify all the extreme points for the following:

   (a) $f(x,y) = x^2 + 3xy - y^2$
   (b) $f(x,y) = x^2 + y^2$
   (c) $f(x,y) = -x^2 - xy - 2y^2$
   (d) $f(x,y) = 3x + 5y - 4x^2 + y^2 - 5xy$
   (e) $f(x,y,z) = 2x + 3y + 3z - xy + xz - yz - x^2 - 3y^2 - z^2$

3. Find and classify all critical points of $f(x,y) = e(x-y) + x2 + y2$

4. Find and classify all critical points of $f(x,y) = (x2 + y2)\ 1.5 - 4(x2 + y2)$

5. Consider a small company that is planning to install a central computer with cable links to five departments. According to their floor plan, the peripheral computers for the five departments will be situated as shown by the dark circles in Figure 8.1. The company wishes to locate the central computer so that the minimal amount of cable will be used to link to the five peripheral computers. Assuming that cable may be strung over the ceiling panels in a straight line from a point above any peripheral to a point above the central computer, the distance formula may be used to determine the length of cable needed to connect any peripheral to the central computer. Ignore all lengths of cable from the computer itself to a point above the ceiling panel immediately over that computer. That is, work only with lengths of cable strung over the ceiling panels.

The Grid for the Five Departments
   The coordinates of the locations of the five peripheral computers are listed in Table 8.5.
   Assume the central computer will be positioned at coordinates (m, n) where m and n are integers in the grid representing the office space. Determine the coordinates (m, n) for placement of the central computer that minimize the total amount of cable needed. Report the total number of feet of cable needed for this placement along with the coordinates (m, n).

6. Find all the extrema and then classify the extrema for the following functions:

   (a) $f(x, y) = x3 - 3xy2 + 4y4$.
   (b) $w(x, y, z) = x2 + 2xy - 4z + yz2$.

TABLE 8.5

Grid coordinates of five departments

X	Y
15	60
25	90
60	75
75	60
80	25

7. Use the KKT Conditions to find the optimal solution to the following nonlinear problem:

$$\text{Maximize } f(x,y) = -x^2 - y^2 + xy + 7x + 4y$$
$$\text{subject to:} \quad 2x + 3y \leq 24$$
$$-5x + 12y \geq 20$$

8. Use the KKT Conditions to find the optimal solution to the following nonlinear problem:

$$\text{Maximize } f(x,y) = -x^2 - y^2 + xy + 7x + 4y$$
$$\text{subject to:} \quad 2x + 3y \geq 16$$
$$-5x + 12y \leq 20$$

9. Use the KKT Conditions to find the optimal solution to the following nonlinear problem:

$$\text{Minimize } f(x,y) = 2x + xy + 3y$$
$$\text{subject to:} \quad x^2 + y \geq 3$$
$$2.5 - 0.5x - y \leq 0$$

10. Use the KTC to solve

$$\text{Minimize } f(x,y) = 2x + xy + 3y$$
$$\text{subject to:} \quad x^2 + y \geq 3$$
$$x + 0.5 \geq 0$$
$$y \geq 0$$

11. Solve the following:

$$max -(x - 0.5)^2 - (y - 5)^2$$
$$\text{subject to:} -x + 2y \leq 4$$
$$x^2 + y^2 \leq 14$$
$$0 \leq x, 0 \leq y$$

12. Apply the modified Newton's method (Multivariable) to find the following:

(a) MAX $f(x,y) = -x3 + 3x + 84y - 6y2$
start at (1,1).
Why can't we start at (0,0)?

(b) MIN $f(x,y) = -4x + 4x2 - 3y - + y2$
start at (0,0).

(c) Perform 3 iterations to
MIN $f(x,y) = (x-2)4 + (x - 2y)2$, start at (0,0).
Why is this problem not converging as quickly as problem (b)?

13. Use the gradient search to find the approximate minimum to

$f(x,y) = (x-2)2 + x + y2$. Start at (2.5,1.5).

14. Solve the following constrained problems

(a) Minimize $x2 + y2$
subject to $x + 2y = 4$.

(b) Maximize $(x - 3)2 + (y - 2)2$
subject to $x + 2y = 4$.

(c) Maximize $x2 + 4xy + y2$
subject to $x2 + y2 = 1$.

(d) Maximize $x2 + 4xy + y2$
subject to $x2 + y2 = 4$.

15. Find and classify the extrema for

$f(x,y,z) = x2 + y2 + z2$
s.t. $x2 + 2y2 - z2 = 1$.

16. MAX $Z = -2x2 - y2 + xy + 8x + 3y$

s.t. $3x + y = 10$
$x2 + y2 = 16$.

17. Use the Method of LaGrange multipliers to find the maxima for

MAX $f(x,y,w) = xyw$
s.t. $2x + 3y + 4w = 36$.

Determine how much $f(x, y, w)$ would change if one more unit was added to the constraint.

8.8 Projects

1. Suppose a newspaper publisher must purchase three kinds of paper stock. The publisher must meet their demand but desire to minimize their costs in the process. They decide to use an Economic Lot Size model to assist them in their decisions. Given an Economic Order Quantity Model (EOQ) with constraints where the total cost is the sum of the individual quantity costs:

$$C(Q1,Q2,Q3) = C(Q1) + C(Q2) + C(Q3)$$

$$C(Qi) = ai\ di/\ Qi + hi\ Qi/2 \text{ where}$$

d is the order rate

h is the cost per unit time (storage)

$Q/2$ is the average amount on hand

a is the order cost.

The constraint is the amount of storage area available to the publisher so that he can have the three kinds of paper on hand for use. The items cannot be stacked, but can be laid side by side. They are constrained by the available storage area, S.
The following data is collected:

	TYPE I	TYPE II	TYPE III
d	32 rolls/week	24	20
a	$25	$18	$20
h	$1/roll/week	$1.5	$2.0
s	4 sqft/roll	3	2

You have 200 sq. ft of storage space available.
Required:

(a) Find the levels of quantity that are the unconstrained minimum total cost and show that these values would not satisfy the constraint. What purpose do these values serve?

(b) Find the constrained optimal solution by using the Lagrange Multipliers assuming we will use all 200 sq. ft.

(c) Find and interpret the shadow prices.

2. Resolve the tank storage problem to determine whether it is better to have a cylindrical storage space or rectangular storage space of 50 cubic units.

3. Suppose, you want to use the Cobb–Douglas function $P(L, K) = A$ La Kb to predict output in thousands, based upon amount of capital and labor used. Suppose you know the price of capital and labor per year is $10,000 and $7,000 respectively. Your company estimates the values of A as 1.2, $a = 0.3$ and $b = 0.6$. Your total cost is assumed to be $T = PL*L + Pk*k$, where PL and Pk are the price of capital and labor. There are three possible funding levels: $63,940, $55,060, or $71,510. Determine which budget yields the best solution for your company. Interpret the Lagrange Multiplier.

8.9 References and Suggested Further Reading

Bazarra, M., C. Shetty, and H. D. Scherali (1993). *Nonlinear Programming: Theory and Applications*. John Wiley & Sons, New York.

Ecker, J. and M. Kupperschmid (1988). *Introduction to Operations Research*. John Wiley & Sons, New York.

Fox, W. P. (1992). Teaching nonlinear programming with Minitab. *COED Journal*, II(1), January–March: 80–84.

Fox, W. P. (1993). Using microcomputers in undergraduate nonlinear optimization. *Collegiate Microcomputer*, XI(3): 214–218.

Fox, W. P. and J. Appleget (2000). Some fun with Newton's Method. *COED Journal*, X(4), October–December: 38–43.

Fox, W. P. and W. Richardson (2000). Mathematical modeling with least squares using Maple. *Maple Application Center, Nonlinear Mathematics*, October.

Fox, W. P., F. Giordano, S. Maddox, and M. Weir (1987). *Mathematical Modeling with Minitab*. Brooks/Cole, Monterey, CA.

Fox, W. P., F. Giordano, and M. Weir (1997). *A First Course in Mathematical Modeling*, 2nd ed. Brooks/Cole, Monterey, CA.

Hiller, F. S. and G. J. Liberman (1990). *Introduction to Mathematical Programming*, McGraw Hill Publishing Company, New York.

Meerschaert, M. (1993). *Mathematical Modeling*, Academic Press, San Diego, CA.

Phillips, D.T., A. Ravindran, and J. Solberg (1976). *Operations Research*. John Wiley & Sons, New York.

Press, W. H., B. Flannery, S. Teukolsky, and W. Vetterling (1987). *Numerical Recipes*. Cambridge University Press, New York, pp. 269–271.

Rao, S. S. (1979). *Optimization: Theory and Applications*. Wiley Eastern Limited. New Delhi, India.

Winston, W. L. (2002). *Introduction to Mathematical Programming Applications and Algorithms*, 4th ed. Duxbury Press, Belmont, CA.

9

Simulation Models

Objectives

1. Understand the power and limitation to simulations.
2. Understand random numbers.
3. Understand the concept of algorithms.
4. Build simple deterministic and stochastic simulations in Excel.
5. Understand the law of large numbers in simulations.

9.1 Introduction

Consider an engineering company that conducts vehicle inspections for a specific state. We have data for times of vehicle arrivals and departures, service times for inspectors under various conditions, numbers of inspection stations, and penalties levied for failure to meet state inspection standards in terms of waiting time for customers. The company wants to know how it can improve its inspection process throughout the state in order to both maximize its profit and minimize the penalties it receives. This type of analysis for a complex system has many variables, and we could use a computer simulation to model this operation.

A modeler may encounter situations where the construction of an analytic model is infeasible because of the complexity of the situation. In instances where the behavior cannot be modeled analytically or where data are collected directly, the modeler might simulate the behavior indirectly and then test various alternatives to estimate how each affects the behavior. Data can then be collected to determine which alternative is best. Monte Carlo simulation is a common simulation method that a modeler can use, usually with the aid of a computer. The proliferation of today's computers in the academic and business worlds makes Monte Carlo simulation very attractive. It is imperative that students have at least a basic understanding of how to use and interpret Monte Carlo simulations as a modeling tool.

There are many forms of simulation ranging from building scale models such as those used by scientists or designers in experimentation to various types of computer simulations. One preferred type of simulation is the Monte Carlo simulation. Monte Carlo simulation deals with the use of random numbers. There are many serious mathematical concerns associated with the construction and interpretation of Monte Carlo simulations. Here we are concerned only with reinforcing the techniques of simulations with these random variates.

A principal advantage of Monte Carlo simulation is the ease with which it can be used to approximate the behavior of very complex systems. Often, simplifying assumptions must be made to reduce this complex system into a manageable model. In the environment forced on the system, the modeler attempts to represent the real system as closely as possible. This system is probably a stochastic system; however, simulation can allow either a deterministic or stochastic approach. We will concentrate on the stochastic modeling approach to deterministic behavior.

Many undergraduate mathematical science, engineering, and operations research programs currently require or offer a course involving simulation. Typically, such a course will use a high-level simulation language such as C++, Java, FORTRAN, SLAM, Prolog, STELLA, SIMAN, or GPSS as the tool to teach simulation. Here we will use Excel to simulate some simple modeling scenarios.

Our emphasis is twofold. First, we want you (the student) to think in terms of an algorithm, not a specific language. Second, we want you to understand that *more* is better in Monte Carlo simulations. The "More Is Better" rule is based on the law of large numbers where probabilities are assigned to events in accordance with their limiting relative frequencies.

In this chapter, our focus is on Monte Carlo Simulation. The concept of Monte Carlo Simulation stems from the study of games of chance. These type of simulations can be accomplished using three distinct steps: generate a random number, define how the random numbers relates to an event, execute the event as shown in Figure 9.1. These three steps are repeated lots of times as we will illustrate in our examples.

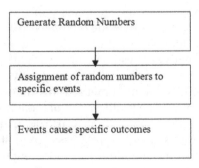

FIGURE 9.1
Three steps in Monte Carlo simulations.

One advantage of dealing with simulations is there ease to examine "what if" analysis to the systems with actually altering the real system. For example, if we want to design a sensor to detect an illness it is easier to test on a computer simulation than to affect many people and actually experiment.

9.2 Random Number and Monte Carlo Simulation

A Monte Carlo simulation model is a model that uses random numbers to simulate behavior of a situation. Using a known probability distribution (such as uniform, exponential, or normal) or an empirical probability distribution, a modeler assigns a behavior to a specific range of random numbers. The behavior returned from the random number generated is then used in analyzing the problem. For example, if a modeler is simulating the tossing of a fair coin using a uniform random number generator that gives numbers in the range $0 \leq x < 1$, then he or she may assign all numbers less than 0.5 to be a head while numbers from 0.5 to 1 are tails.

A Monte Carlo simulation can be used to model either stochastic or deterministic behavior. It is possible to use a Monte Carlo simulation to determine the area under a curve (a deterministic problem) or stochastic behavior like the probability of winning in craps (a stochastic problem). In this chapter, we will introduce both a deterministic problem and a stochastic problem. We discuss how to create algorithms to solve both. We will start with the deterministic simulation modeling.

First, Monte Carlo simulation deals with the use of generated random numbers to cause specific events to occur within the simulation according to a specific scheme. Basically, the flow from

Random number $\rightarrow$ Assignment $\rightarrow$ Event

is observed within the simulation. The most important aspect of the simulation process is the algorithm. The algorithm is the step by step process to go from INPUTS to OUTPUTS. We will illustrate with a few examples.

Steps of a Monte Carlo simulation include:

(1) Establish a probability distribution for each variable that is subject to chance. Obtain the CDF of the distribution.
(2) Generate a random number from this distribution for each variable in step 1.
(3) Make assignments from random numbers to the appropriate events.
(4) Repeat the process for a series of replications (trials).

Random-Number Generators

Using random numbers is of paramount importance in running Monte Carlo simulations, so a good random-number generator is critical. In particular, a modeler must have a method of generating uniform, U(0,1), random numbers– that is, numbers that are uniformly distributed between 0 and 1. All other distributions, known and empirical, can be derived from the U(0,1) distribution. At the graduate level, a lot of class time is spent on the theory behind good and bad random-number generators, and the tests that can be made on them. More and more is being learned about what does and does not make up a true random-number generator. At the undergraduate level, this is not necessary, provided the students have access to either random numbers or a good algorithm for generating pseudo-random numbers.

In addition, most computer languages now use good pseudo-random-number generators (although this has not always been the case – the old RANDU generator distributed by IBM was statistically unsound). These good generators use the recursive sequence $X_i = (aX_{i-1} + c) \bmod m$ where a, c, and m determine the statistical quality of the generator. Because we do not discuss the testing of random-number generators in our course, we trust the generators provided by our software packages. Serious study of simulation must, of course, include a study of random-number generators because a bad generator will provide output from which a modeler may make poor conclusions.

9.2.1 Excel and Random Number Generation

In Excel, here are commands to obtain random numbers

To Simulate	Excel Formula to use
Random number, uniform [0,1]	= rand()
Random number between [a, b]	= a+ (b-a)*rand()
Discrete integer random number between [a, b]	= randbetween(a,b)
Normal random number	= NORMINV(rand(),μ,σ)
Exponential random number with mean rateλ	= (-1/μ)* ln(rand())
Discrete general distribution with only two outcomes (like a flip of a coin): A and B Probability of outcome is p.	= if(rand()<p, A, B)
Discrete general distribution for more than two outcomes.	= lookup(RAND(), Range1,Range2)
Range1 = cell range for lower limits of the random number intervals	
Range2 = cell range containing the variable values	

Note: in the command RAND() there is not space between the two parentheses.

Examples in Excel

We need uniform random number between [0,1]. To get these we type = rand() in cell D6. We obtained a random number, 0.317638748. We can copy this down for as many random numbers as we need. In cell E6 we create a random number between [1,10] using $1 + (10 - 1)*$rand(). We obtained 1.157956 and we copy down for as many random number as we need in our simulation. Figure 9.2 provides a screenshot of the Excel formulas and then the values for obtaining ten random numbers.

The following algorithms might be helpful to obtain other types of random numbers.

(i) Uniform [a,b]
   a. Generate a random uniform number U from [0,1]
   b. Return $X = a + (b - a)*U$
   c. $X = a + (b - a)*$rand()

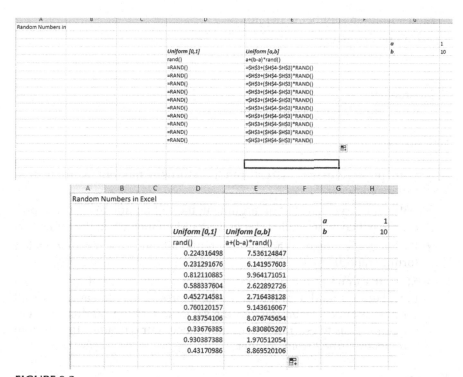

FIGURE 9.2
Screenshots to obtain random numbers.

(ii) Exponential with mean β

    a.   Generate a random uniform number U from $[0,1]$

    b.   Return $X = -\beta\ln(U)$

    c. $X = -\beta ln(rand())$

(iii) Normal$(0,1)$

    a.   Generate U_1 and U_2 from uniform $[0,1]$.

    b.   Let $V_i = 2U_i - 1$ for $i = 1,2$.

    c.   Let $W = V_1^2 + V_2^2$

    d.   If $W > 1$, go back to step a. Otherwise, let
$$Y = \sqrt{(-2\ln(W)/W)}, X_1 = V_1 Y, X_2 = V_2 Y.$$

    e. X_1 and X_2 are normal $(0,1)$.

9.2.2 Maple and Random Number Generation

Maple has several choices to generate random numbers. We will present a few of Maple's commands. We point out that the command rand(r) returns a 12 digit non-negative number.

    rand – random number generator

        Calling Sequence

        rand(r)

    Parameters

        r – (optional) integer range or integer

Description

- With no arguments, the call rand() returns a random 12 digit non-negative integer.

- With an integer range as an argument, the call rand(a..b) returns a procedure which, when called, generates random integers in the range a..b.

- With a single integer as an argument, the call rand(n) is the abbreviated form of rand(0..n–1).

- More than one random number generator may be used at the same time, because rand(a..b) returns a Maple procedure, However, since all random number generators use the same underlying random number sequence, calls to one random number generator will affect the random numbers returned from another.

Examples

    >rand();

                         395718860534

```
>rand();
```

$$193139816415$$

```
>roll:=rand(1..6):
roll();
```

$$6$$

```
>roll();
```

$$2s$$

```
>RandomTools[Generate](integer(range=1..6));
```

$$3$$

Random
Description

- The **stats[random]**subpackage provides random numbers with a given distribution.
- The **random** subpackage provides filters that transform a uniformly distributed stream of random numbers to a stream that has the required distribution. The underlying uniform generator can be specified by the parameter **uniform**.
- The seed of the default generator can be modified by using the function **randomize()**.
- Random number generators are obtained by specifying the keyword **'generator'** as the parameter **quantity**. The number of digits (if it is not the default value given in **Digits**) in the generated numbers is indicated as an index to **'generator'**. If **randomize()** is used, it must be done prior to creating the generators.
- The uniform generator specified using the parameter **uniform** must return one number between 0 and 1. The parameter **uniform** allows the generation of independent streams of random numbers, provided independent uniform generators are used. It must be noted that the design of good uniform generators, independent of each other, is not an easy task. A randomly chosen random number generator will most likely be of poor quality.

Examples

```
>with(stats):
```

Generate 20 numbers with the standard normal distribution

```
>stats[random, normald](20);
```

1.275676034,0.4761860865,-2.620054215,0.8547207946,
-1.089001803, 0.7165427609,-0.02996436194,1.330855643,
-1.296345120,-0.4703304005, -0.7122630822,-0.08717900722,
1.282253011,-1.683720684,03265473243, 0.08179818206,
0.2127564698,-0.7279468705,0.6855773713,-0.3760280547 -0.+

This time, use the 'inverse' method

>stats[random, normald](20,'default','inverse');

0.2942905914,1.023730745,−0.07256069586,−0.2726670238,
0.9797884030, -0.06456731317,−0.7584692928,−1.4145009920,
0.5848741507,0.1502221444, 0.7052134613,-0.1760249535,
1.114921264,1.111477077,1.837138817, -0.007014805706,
0.1412349839,-0.3487795280,2.269571409,0.6290512530

Generate 20 numbers with the Poisson[3] distribution, using a user specified generator.

>seed:=1;

$$seed := 1$$

>uniform_generator:=proc() # not a very good generator

   global seed;

   seed: = irem(seed*11,101);

   return(seed/101)

end proc:

random[gamma[3]](20,uniform_generator);

1.171878004,1.514461967,3.368200854,1.997566185,4.818133461,
6.550753566,0.6309471660,3.556933956,3.761058678,2.869958271,
3.026686005,1.117441781,3.623114178,1.001527868,3.492458052,
4.230152601,4.063203183,1.9111760304,2.259044006,3.691114038

Random number from [0,1] in Maple. In the next example we generate 20 random number from [0,1].

$$> random := \mathbf{proc}(n)\,\mathbf{option}\,remember\;\mathbf{for}\,i\,\mathrm{from}\,1\;\mathbf{to}\,n\;\mathrm{do}\,x(i):=$$
$$evalf\left(\frac{rand(\ )}{99999999999}\right)\mathbf{end\,do\,end};$$

Warning, `i` is implicitly declared local to procedure `random`

```
random: = proc(n)
    option remember;
    local i;
    for i to n do
        x(i) := evalf (1/999999999999* rand( ))
    end do
end proc
```

> $random(20);$

$$0.8950382012$$

> $seq(x(i), i = 1 \ 20);$

0.2240171515, 0.2008401063, 0.8685719066, 0.5704134665,
0.9920881460, 0.04437752746, 0.4780291372,
0.01294305199, 0.6408831565, 0.2487102377,
0.9700117830, 0.04705144220, 0.5745759906,
0.1717448313, 0.6488303816, 0.4114303688,
0.09704265386, 0.7719464925, 0.2477991384, 0.8950382012

We can generate many distributions that we might need in modeling from uniform [0,1] random numbers.
Uniform [a,b]

a. Generate a random uniform number U from [0,1]
b. Return $X = a + (b - a)*U$

Exponential with mean β

a. Generate a random uniform number U from [0,1]
b. Return $X = -\beta \ln(U)$

Normal(0,1)

a. Generate U_1 and U_2 from uniform [0,1].
b. Let $V_i = 2U_i - 1$ for $i = 1,2$.
c. Let $W = V_1^2 + V_2^2$
d. If $W > 1$, go back to step a. Otherwise, let
   $Y = \sqrt{(-2\ln(W)/W)}, X_1 = V_1Y, X_2 = V_2Y.$
e. X_1 and X_2 are normal (0,1).

9.2.3 R and Random Number Generation

Random number in R
Uniform random numbers

$$\text{runif(n, min} = 0, \text{max} = 1)$$

Normal random number

$$\text{rnorm(n,mean} = 0,\text{sd} = 1)$$

Binomial random number (Figure 9.3)

```
rbinom(#,n,p)
> n=5;p=.25    # change as appropriate
> x=rbinom(100,n,p)   # 100 random numbers
> hist(x,probability=TRUE,)
## use points, not curve as dbinom wants integers only for x
>xvals=0:n;points(xvals,dbinom(xvals,n,p),type="h",lwd=3)
> points(xvals,dbinom(xvals,n,p),type="p",lwd=3)
    ... repeat with n=15, n=50
```

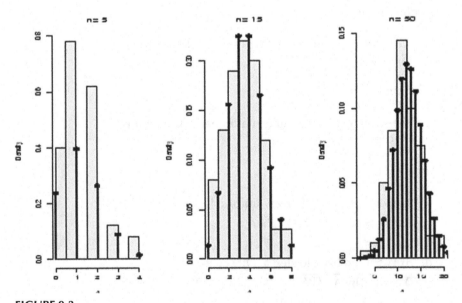

FIGURE 9.3
Random binomial data with the theoretical distribution.

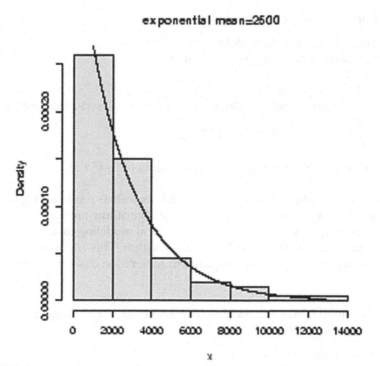

FIGURE 9.4
Exponential random number.

Exponential Radom Numbers
rexp(n,rate = 1). Here is an example with the rate being 1/2500 (Figure 9.4).

> x = rexp(100,1/2500)

> hist(x,probability = TRUE,col[[[I_02601]]]

9.3 Probability and Monte Carlo Simulation Using Deterministic Behavior

One key to good Monte Carlo simulation is an understanding of the axioms of probability. *Probability* is a long-term average. For example, if the probability of an event occurring is 1/5, this means that "in the long term, the chance of the event happening is 1/5 = 0.2." not that it will occur exactly once out of every five trials.

Deterministic Simulation Examples

Let's consider the following deterministic examples.
  Compute the area under a non-negative curve.

1. The curve $y = x^3$ from $0 \le x \le 2$.
2. The curve (which does not have a closed-form solution to $\int_{x=0}^{1.4} \cos(x^2) \cdot \sqrt{x} \cdot e^{x^2} dx$ from [0,1.4]).
3. Compute the volume in the first octant of $x^2 + y^2 + z^2 \ge 1$.

We will present algorithms for their models as well as produce output of the Monte Carlo simulation to analyze. These algorithms are important to the understanding of simulation as a mathematical modeling tool.
  Here is a generic framework for an algorithm. This framework includes inputs, outputs, and the steps required to achieve the desired output.

Example 9.1 Monte Carlo Algorithm Area under the Non-negative Curve (for Excel)

Input: Total number of points
  Output: AREA = approximate area under a specified curve $y = f(x)$ over the given interval $a \le x \le b$, where $0 \le f(x) \le M$.

Step 1. In Column 1, list n = 1,2,...N from cell $a1$ to aN. Create columns 2–5.

Step 2. In Column 2, generate a random x_i between a and b using, $a + (b - a)*rand()$. These are listed in cells $b1$ to bN.

Step 3. In Column 3, generate a random y_i between 0 and M using, $0 + (M - 0)*rand()$. These are listed in cells $c1$ to cN.

Step 4. In Column 4, compute $f(x_i)$. These are listed in cells $d1$ to dN.

Step 5. In Column 5, check to see if each random coordinate (x_i, y_i) point is below curve. Compute $f(x_i)$ and see if $y_i \le f(x_i)$. Use a logical IF statement, If $y_i \le f(x_i)$ then let the cell value = 1, otherwise let the cell value equal 0. In cells $d1$ to dn, put, **IF(cell c1< = d1, 1,0)**. These are listed in cells $e1$ to eN.

Step 6. Count the cell values that equal 1, use Sum(e1:eN).

Step 7. Calculate area in g4. Area = M(b-a) Sum/N.

Repeat the process and increase N to get better approximations. You can plot the (x_i, y_i) coordinate and $f(x_i)$ for a visual representation.

Example 9.2 $y = x^3$ from $0 \le x \le 2$ Using 100 Random Numbers

We applied our area under the curve algorithm to $y = x^3$ from [0,2]. We see a visual representation of this in Figure 9.5. In this example, with only 100 random points, we find our simulated area is about 4.64 in Figure 9.6. The real area is found by integration, $\int_{0}^{2} x^3 dx = 4$, when our function can be integrated.

We present a method for repeating the process in Excel in order to obtain more iterations.

For example, go to cell M1 and enter 1 and iterate to cell M1000 for 1,000 trials. They are number from 1 to 1,000. In cell N1, reference your cell g4 (see Figure 9.6). Highlight cells M1 to N1000. Go to Data→What if Analysis→Data Table and enter. In the dialog box that comes up, put nothing in Rows and put a used cell reference in the column (like P1). Press OK. The table fills in running the area simulation previous written 1,000 times. For this example we now have 100 runs 1,000 times or 10,000 results. Copy M1 to N1000 and paste as *values* into another location such as AA1. You do this so the values do not keep changing. Then highlight the column of simulated area values and obtain their description statistics.

The descriptive statistics table, Figure 9.7, would look like this:

Now we are ready to *approximate* the area by using Monte Carlo simulation. The simulation only approximates the solution. We increase the number of trials attempting to get closer to the value. We present the results in Table 9.1. Recall that we introduced randomness into the procedure with the Monte Carlo simulation area algorithm. In our output, we provide graphical output as well so that the algorithm may be seen as

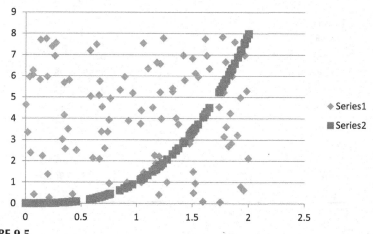

FIGURE 9.5

Area under curve graphical representation for $y = x^3$ from [0,2].

Function y=x^3 from [0,2]

n	rand()	Rand_X x	y	Rand_Y y	count	Area
1	0.245527	0.491055	0.118410534	4.487261	0	4.64
2	0.980981	1.961963	7.552178557	7.117628	1	
3	0.830105	1.66021	4.576028846	4.358678	1	
4	0.921794	1.843587	6.266011943	6.465696	0	
5	0.164428	0.328856	0.035564438	5.399423	0	
6	0.065181	0.130363	0.002215439	5.388175	0	
7	0.292613	0.585226	0.200433862	5.11966	0	
8	0.453498	0.906995	0.746131043	5.937811	0	
9	0.224417	0.448834	0.090418732	4.260345	0	
10	0.497616	0.995232	0.985763098	5.59896	0	
11	0.081892	0.163784	0.004393533	2.951267	0	
12	0.708622	1.417245	2.846652288	2.954609	0	
13	0.824472	1.648944	4.483507308	2.02574	1	
14	0.459856	0.919711	0.777954851	6.481868	0	
15	0.981421	1.962841	7.562327224	1.76143	1	

FIGURE 9.6
Screenshot of simulation of area showing only 1–15 random trials.

A	B	C
Column1		
Mean	3.98896	
Standard Error	0.022034274	
Median	4	
Mode	3.68	
Standard Deviation	0.696784922	
Sample Variance	0.485509228	
Kurtosis	-0.228529708	
Skewness	0.197260113	
Range	4.32	
Minimum	1.76	
Maximum	6.08	
Sum	3988.96	
Count	1000	

FIGURE 9.7
Screenshot of descriptive statistics for our simulated areas.

TABLE 9.1

Summary of output for the area under x^3 from 0 to 2

Number of trials	Approximate area	Absolute percent error (%)
100	3.36	16
500	3.872	3.2
1000	4.32	8
5000	4.1056	2.64
10,000	3.98896	0.275

a process. In our graphical output, each generated coordinate (x_i, y_i) is a point on the graph. Points are randomly generated in our intervals $[a,b]$ for x and $[0,M]$ for y. The curve for the function $f(x)$ is overlaid with the points. The output also includes the approximate area.

We need to stress that in modeling deterministic behavior with stochastic features, we (not nature) have introduced the randomness into the problem. Although more runs is generally better, it is not true that the deterministic solution becomes closer to reality as we increase the number of trials, $N, \to \infty$. It is generally true that more runs is better than a small number of runs (16 percent was the worst by almost an order of magnitude, and that occurred at $N = 100$). In general, more trials are better.

Example 9.3 The curve (which does not have a closed-form solution)

$$\int_{x=0}^{1.4} \cos(x^2) \cdot \sqrt{x} \cdot e^{x^2} dx \text{ from [0,1.4].}$$

We can tell from the integral that x varies from 0 to 1.4. But what about y? Take the function for y and obtain the plot as x varies from 0 to 1.4. In Figure 9.8, we can estimate the maximum value as about 9. Thus we generate random values for y from 0 to 9.

We ran 1,000 iterations and obtain a numerical approximation for the area of 2.9736. Since we cannot find the integral solution directly in this case, we use the trapezoidal method to approximate the solution to see how well our simulation faired. The numerical method provides an approximate solution of 3.0414. Our simulation's error compared to the trapezoidal method was 2.29%.

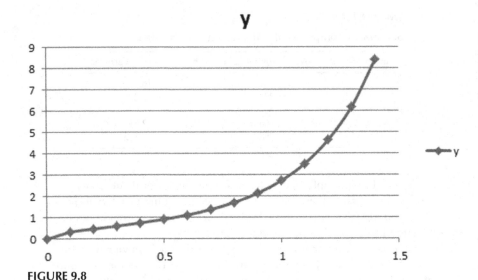

FIGURE 9.8
Plot of $\cos(x^2)\sqrt{x}\,e^{x^2}$ from 0 to 1.4.

TABLE 9.2

Percent errors in finding the volume in first octant

Number of points	Approximate volume	Percent error ($|\%|$)
100	0.47	10.24
200	0.595	13.64
300	0.5030	3.93
500	0.514	1.833
1,000	0.518	1.069
2,000	0.512	2.21
5,000	0.518	1.069
10,000	0.5234	0.13368
20,000	0.5242	0.11459

Example 9.4 Finding the Volume in the First Octant

We can also extend this concept to multiple dimensions. We develop an algorithm for the volume under a surface in the first octant.

Table 9.2 provides the numerical output. The actual volume in the first octant is $\pi/6$ (with radius as 1). We take $\pi/6$ to four decimals as 0.5236 cubic units. Figure 9.9 graphically displays the algorithm.

Generally, though not uniformly, the percentage errors become smaller as the number of points, N, is increased.

Monte Carlo Volume Algorithm

INPUT   The total number of random points, N. The nonnegative function, f(x),

the interval for x [a,b], interval for y [c,d] and an interval for z [0,M] where M > Max

f(x,y),a<x<b, c<y<d

OUTPUT   The approximate volume enclosed for the function f(x,y) in the first octant,

x>0, y>0, and z >0.

Step 1.   Set all counters at 0

Step 2.   For i from 1 to N do step 3- 5

Step 3.   Calculate random coordinates in the rectangular region:

$$a < x_i < b, \quad c < y_i < d, \quad 0 < z_i < M$$

Step 4.   Calculate f(x_i, y_i)

Step 5.   Compare f(x_i, y_i) and z_i . If z_i < f(x_i, y_i) then increment counter by 1.

Otherwise, do not increment counter.

Step 6.   Estimate the Volume by $$V = (M - 0) \cdot (c - d) \cdot (b - a) \cdot \frac{counter}{N}$$

Stop

FIGURE 9.9
Algorithm for volume in first octant.

9.4 Deterministic Simulations in R and Maple

9.4.1 R for Deterministic Simulation

We can use R to provide the output of the area under a curve for a function (Figure 9.10 and Table 9.3).

```
runs<-1000000
> #runif samples from a uniform distribution
>xs<-runif(runs,min=0,max=2)
>ys<-runif(runs,min=0,max=8)
>in.curve<-(xs^3) <=ys
```

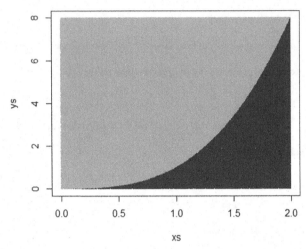

FIGURE 9.10
Plot of x^3.

TABLE 9.3

Summary of output for the area under x^3 from 0 to 2

Number of trials	Approximate area	Percent error
100	3.36	16%
500	3.872	3.2%
1000	4.32	8%
5000	4.1056	2.64%
10,000	4.136	3.4%
1,000,000	3.99452	0.137%

```
>mc.area<- 2*8*(1-(sum(in.curve)/runs))
> plot(xs,ys,pch='.',col=ifelse(in.curve,"grey","blue"))
>
>mc.area
[1] 3.9952
```

We stress that in modeling deterministic behavior with stochastic features, we have introduced the randomness into the problem (not nature). Although more runs is better, it is not true that as we increase the number of trials, N, →∞ the solution becomes closer to reality. It is generally true that more runs is better than a small number of runs (16% was the worst by almost an order of magnitude and that occurred at N = 100).

Example 9.5 Area under a More Complicated Curve in Figure 9.11

We modify the algorithm to compute the area under the curve for the function $e^{x^2} \bullet \cos(x) \bullet \sqrt{x}$ over the interval [0,1.4]. We provide the complete step-by-step algorithm, your formulas, and the final output for runs of 50, 100,500, and 1,000 trials.

We provide one graphical output from our program in Figure 9.12 with N = 1,000,000. Since this function has no closed from solution, we used Simpson's Method to obtain a numerical solution to:

$$\int_0^{1.4} e^{(x^2)} \cos(x)\sqrt{x}\ dx$$

The numerical integration solution is 1.448293361.

```
runs<-1000000
> #runif samples from a uniform distribution
>xs<-runif(runs,min=0,max=1.4)
>ys<-runif(runs,min=0,max=1.6)
>in.curve<-(exp(xs^2)*cos(xs)*sqrt(xs))<=ys
>mc.area<-1.4*1.6*(1-(sum(in.curve)/runs))
> plot(xs,ys,pch='.',col = ifelse(in.curve,"grey","blue"))
>mc.area
[1] 1.437453
```

The percent error for this run is only 0.8535%, less than 1 % error.

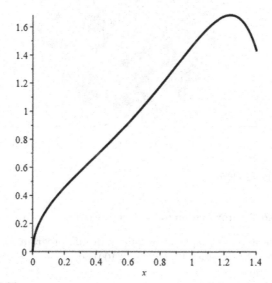

FIGURE 9.11
Plot of f(x) from [0,1.40].

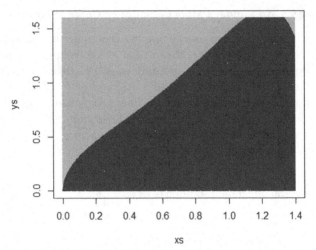

FIGURE 9.12
Graphical output of simulation, N = 1,000,000 with approximate area as 1.437433.

MC Approximation of Pi = 3.14076

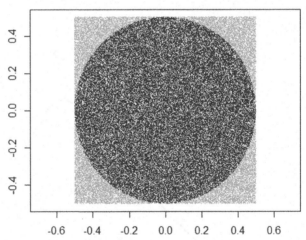

FIGURE 9.13
Approximation of *Pi* = 3.14076.

Example 9.6 Approximation to Pie from a Circle

We provide another graphical output from our program for approximating Pie (Figure 9.13) with N = 100,000.

runs <- 100000

#runif samples from a uniform distribution

```
xs<- runif(runs,min = -0.5,max = 0.5)
ys<- runif(runs,min = -0.5,max = 0.5)
in.circle<- xs^2 + ys^2 < = 0.5^2
mc.pi<- (sum(in.circle)/runs)*4
plot(xs,ys,pch = '.',col = ifelse(in.circle,"blue","grey")
    ,xlab = ",ylab = ",asp = 1,
    main = paste("MC Approximation of Pi  = ",mc.pi))
```

9.4.2 Maple for Deterministic Simulation

Our program for Area (Figure 9.14).

area: = proc(expr::algebraic,an::numeric, bn::numeric, ymin::numeric, ymax::numeric,N::posint)option remember;

local   A,B,y1,y2,areaR,count,i,area,xpt,ypt,xrpt,yrpt,gen_x,gen_y,c1,c2,c3, c4,c5;

y1: = ymin;

y2: = ymax;

A: = an;

B: = bn;

```
>
> f1:= x->cos(x);
```
$$x \rightarrow \cos(x)$$
```
> f:=cos(x);
```
$$\cos(x)$$
```
>
> Area(f,-1.57,1.57,0,2,2000);
```

6.28, 597, 2000, 1.874580000

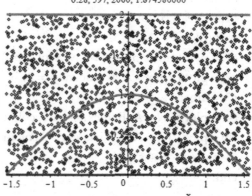

FIGURE 9.14
Screenshot of Maple area under cos(x).

```
areaR: = (B-A)*(y2-y1);
count: = 0;
i: = 0;
gen_x: = evalf(an+(bn-an)*rand(0..10^5)/10^5):
gen_y: = evalf(ymin+(ymax-ymin)*rand(0..10^5)/10^5):
xrpt: = []:
yrpt: = []:
while i<N do
```

Step 1

```
xpt[i]: = A+(B-A)*(rand()/10^12);
ypt[i]: = y1+(y2-y1)*(rand()/10^12);
```

Step 2

```
xrpt: = [op(xrpt),gen_x()]:
yrpt: = [op(yrpt),gen_y()]:
if ypt[i]<evalf(f1(xpt[i])) then
count: = count + 1;
i: = I + 1;
else
i: = i + 1;
fi;
od;
area: = areaR*count/N;
print(areaR,count,N,area);
with(plots):
c1: = pointplot(zip((x,y)->[x,y],xrpt,yrpt)):
c2: = plot(expr,x = A..B, thickness = 3):
c3: = area:
c4: = plot(y2,x = A..B,thickness = 3, colour = red):
c5: = plot(y1,x = A..B,thickness = 3,colour = blue):
display(c1,c2,c4,c5);
end:
```

Example> Area under cos(x) from

f1:=x->cos(x);

x->cos(x)

f:=cos(x);

$$\cos(x)$$

Area(f,-1.57,1.57,0,2,2000);

6.28, 597, 2000, 1.874580000

9.5 Probability and Monte Carlo Simulation Using Probabilistic Behavior

Let's consider the following simple probabilistic examples.

1. Compute the probability of getting a head or a tail if you flip a fair coin.
2. Compute the probability of rolling a number from 1 to 6 using a fair die.

Example 9.7 Flip a Fair Coin

Algorithm
Input: The number of trials, N
Output: The probability of a head or a tail

Step 1. Initialize counters to 0.

Step 2. For $i = 1, 2, \ldots, N$ do.

Step 3. Generate a random number, x, $U(0,1)$.

Step 4. If $0 \leq x < 0.5$ increment heads, $H = H + 1$, otherwise $T = T + 1$.

Step 5. Output H/N and T/N, the probabilities for heads and tails.

Example 9.8 Roll of a Fair Die

Rolling a fair die adds the additional process of multiple assignments (six for a six-sided die). The probability will be the number of occurrences of each number divided by the total number of trials.

INPUT: Number of rolls

Output: Probability of getting a {1,2,3,4,5,6}

Step 1. Initialize all counters (counter1 through counter 6) to 0.

Step 2. For $i = 1, 2, \ldots, n$, do steps 3 and 4.

Step 3. Obtain a random number j from integers (1,6).

Step 4. Increment the counter for the value of j so that Counter j = counter $j + 1$

Step 5. Calculate the probability of each roll {1,2,3,4,5,6} by Counter j/n

Step 6. Output probabilities.

Step 7. Stop.

Roll-a-Fair-Die Program

The expected probability is $1/6$ or 0.1667. We note that as the number of trials increases, the closer our probabilities are to the expected long-run values. We offer the following concluding remarks. When you have to run simulations, run them for a very large number of trials.

Example 9.9 Discrete Probability Distribution

Assume we have a distribution as follows:

X	0	1	2	3	4	5	6
P(X = x)	0.33	0.25	0.19	.09	0.05	0.05	0.04

Our algorithm to produce random numbers for a simulation is (Figure 9.15):

	A	B	C
1	Trial Number	rand()	result
2	1	=RAND()	=IF(B2<=0.33,0,IF(B2<=0.58,1,IF(B2<=0.77,2,IF(B2<=0.86,3,IF(B2<=0.91,4,IF(B2<=0.96,5,6))))))
3	=A2+1	=RAND()	=IF(B3<=0.33,0,IF(B3<=0.58,1,IF(B3<=0.77,2,IF(B3<=0.86,3,IF(B3<=0.91,4,IF(B3<=0.96,5,6))))))
4	=A3+1	=RAND()	=IF(B4<=0.33,0,IF(B4<=0.58,1,IF(B4<=0.77,2,IF(B4<=0.86,3,IF(B4<=0.91,4,IF(B4<=0.96,5,6))))))
5	=A4+1	=RAND()	=IF(B5<=0.33,0,IF(B5<=0.58,1,IF(B5<=0.77,2,IF(B5<=0.86,3,IF(B5<=0.91,4,IF(B5<=0.96,5,6))))))
6	=A5+1	=RAND()	=IF(B6<=0.33,0,IF(B6<=0.58,1,IF(B6<=0.77,2,IF(B6<=0.86,3,IF(B6<=0.91,4,IF(B6<=0.96,5,6))))))
7	=A6+1	=RAND()	=IF(B7<=0.33,0,IF(B7<=0.58,1,IF(B7<=0.77,2,IF(B7<=0.86,3,IF(B7<=0.91,4,IF(B7<=0.96,5,6))))))
8	=A7+1	=RAND()	=IF(B8<=0.33,0,IF(B8<=0.58,1,IF(B8<=0.77,2,IF(B8<=0.86,3,IF(B8<=0.91,4,IF(B8<=0.96,5,6))))))
9	=A8+1	=RAND()	=IF(B9<=0.33,0,IF(B9<=0.58,1,IF(B9<=0.77,2,IF(B9<=0.86,3,IF(B9<=0.91,4,IF(B9<=0.96,5,6))))))
10	=A9+1	=RAND()	=IF(B10<=0.33,0,IF(B10<=0.58,1,IF(B10<=0.77,2,IF(B10<=0.86,3,IF(B10<=0.91,4,IF(B10<=0.96,5,6))))))
11	=A10+1	=RAND()	=IF(B11<=0.33,0,IF(B11<=0.58,1,IF(B11<=0.77,2,IF(B11<=0.86,3,IF(B11<=0.91,4,IF(B11<=0.96,5,6))))))
12	=A11+1	=RAND()	=IF(B12<=0.33,0,IF(B12<=0.58,1,IF(B12<=0.77,2,IF(B12<=0.86,3,IF(B12<=0.91,4,IF(B12<=0.96,5,6))))))
13	=A12+1	=RAND()	=IF(B13<=0.33,0,IF(B13<=0.58,1,IF(B13<=0.77,2,IF(B13<=0.86,3,IF(B13<=0.91,4,IF(B13<=0.96,5,6))))))
14	=A13+1	=RAND()	=IF(B14<=0.33,0,IF(B14<=0.58,1,IF(B14<=0.77,2,IF(B14<=0.86,3,IF(B14<=0.91,4,IF(B14<=0.96,5,6))))))
15	=A14+1	=RAND()	=IF(B15<=0.33,0,IF(B15<=0.58,1,IF(B15<=0.77,2,IF(B15<=0.86,3,IF(B15<=0.91,4,IF(B15<=0.96,5,6))))))
16	=A15+1	=RAND()	=IF(B16<=0.33,0,IF(B16<=0.58,1,IF(B16<=0.77,2,IF(B16<=0.86,3,IF(B16<=0.91,4,IF(B16<=0.96,5,6))))))
17			

FIGURE 9.15

Screenshot from Excel of formuals.

A	B	C
Trial Number	rand()	result
1	0.207393982	0
2	0.127847961	0
3	0.560264474	1
4	0.567940969	1
5	0.16113914	0
6	0.417507081	1
7	0.538948905	1
8	0.387619174	1
9	0.747548053	2
10	0.476099542	1
11	0.932428906	5
12	0.577012606	1
13	0.598187466	2
14	0.986144446	6
15	0.408557927	1

FIGURE 9.16
Screenshot of Excel for Example 9.9.

We note that as the number of trials increases, the closer our probabilities are to the expected long-run values in the probability table (Figure 9.16). We offer the following concluding remarks. When you have to run simulations, run them for a very large number of trials.

9.5.1 Maple and Monte Carlo Simulation

Example 9.10 Flip a Fair Coin

Algorithm
    Input: The number of trials, N
    Output: The probability of a head or a tail

Step 1. Initialize counters to 0.
Step 2. For i = 1, 2, ..., N do
    Step 3. Generate a random number, x, U(0,1)
    Step 4. If $0 \leq x < .5$ Increment Heads, H = H + 1 otherwise T = T + 1
Step 5. Output H/N and T/N the probabilities for heads and tails.

```
>restart;
>flip:=proc(n)option remember;
>head:=0:tail:=0;
>i:=0;
>while i<n do
>seed:=n:
>x[i]:=(rand()/10^12);
```

```
>if x[i]<=0.500 then
>head:=head+1;
>i:=i+1;
>else;
>tail:=tail+1:i:=i+1;
>end if;
>end do;
>prob_head:=evalf(head/n);
>prob_tail:=evalf(tail/n);
>print (prob_head,prob_tail);
>end;
```

$$flip := \mathbf{proc}(n)$$
$$\mathbf{local}\ head, tail, i, seed, x, prob\_head, prob\_talk;$$
$$\mathbf{option}\ remember;$$
$$head := 0;$$
$$tail := 0;$$
$$i := 0;$$
$$\mathbf{while}\ i < n\ \mathbf{do}$$
$$seed := n;$$
$$x[i] := 1 / 1000000000000 < \mathrm{rand}(\ );$$
$$\mathbf{if}\ x[i] \le 0.500\ \mathbf{then}\ head := head + 1; i = i + 1$$
$$\mathbf{else}\ tail := tail + 1; i = i + 1\ \ print(prob\_head, prob\_tail)$$
$$\mathbf{end\ if} \qquad\qquad\qquad \mathbf{end\ proc}$$
$$\mathbf{end\ do};$$
$$prob\_head := \mathrm{evalf}(head/n);$$
$$prob\_tail := \mathrm{evalf}(tail/n);$$

```
>flip(10);
```
                           0.5000000000,0.5000000000
```
>flip(1000);
```
                           0.5310000000,0.4690000000
```
>flip(2000);
```
                           0.5175000000,0.4825000000
```
>flip(5000);
```
                           0.5058000000,0.4942000000
```
>flip(10000);
>flip(20000);
```
                           0.5016000000,0.4984000000

Roll of a Fair Die

Rolling a fair die adds the additional process of multiple assignments (six for a six-sided die). The probability will be the number of occurrences of each number divided by the total number of trials.

INPUT: Number of rolls

Output: Probability of getting a {1,2,3,4,5,6}

Step 1. Initialize all counters (counter1 through counter 6) to 0.

Step 2. For $i = 1, 2, ..., n$, do steps 3 and 4.

    Step 3. Obtain a random number j from integers (1,6).

    Step 4. Increment the counter for the value of j so that Counter j = counter $j + 1$

    Step 5. Calculate the probability of each roll {1,2,3,4,5,6} by Counter j / n

Step 6. Output probabilities.

Step 7. Stop.

Roll-a-Fair-Die Program

The expected probability is 1/6 or 0.1667. We note that as the number of trials increases, the closer our probabilities are to the expected long-run values. We offer the following concluding remarks. When you have to run simulations, run them for a very large number of trials.

INPUT: Number of rolls
Output: Probability of getting a {1,2,3,4,5,6}

Step 1. Initialize all counters (Counter1 through Counter 6) to 0

Step 2. For i = 1,2...n do steps 3 and 4

    Step 3. Obtain a random number j from Integers (1,6)

    Step 4. Increment the counter for the value of j so Counter j = Counter j + 1

    Step 5. Calculate the probability of each roll {1,2,3,4,5,6} by Counter j / n

Step 6. Output probabilities

Step 7. Stop

Roll a Fair Die Program

```
>rolldie:=proc(n)
>c1:=0:c2:=0:c3:=0:c4:=0:c5:=0:c6:=0;
># Loop
>for i from 1 to n do
>roll:=rand(1..6);
>x:=roll();
>if x=1 then
>c1:=c1+1;
>else
```

```
>if x=2 then
>c2:=c2+1;
>else
>if x=3 then
>c3:=c3+1;
>else
>if x=4 then
>c4:=c4+1;
>else
>if x=5 then
>c5:=c5 + 1;
>else
>c6:=c6+1;
>end if;
>end if;
>end if;
>end if;
>end if;
>end do;
>r1: = evalf(c1/n):r2: = evalf(c2/n):r3: = evalf(c3/n):r4: = evalf(c4/n):r5: =
    evalf(c5/n):r6: = evalf(c6/n);
>print(r1,r2,r3,r4,r5,r6);
>end;
```

$rolldie := \mathbf{proc}(n)$
$\mathbf{local}\ c1,c2,c3,c4,c5,c6,i,roll,x,r1,r2,r3,r4,r5,r6;$
$\quad c1 := 0;$
$\quad c2 := 0;$
$\quad c3 := 0;$
$\quad c4 := 0;$
$\quad c5 := 0;$
$\quad c6 := 0;$
$\mathbf{for}\ i\ \mathbf{to}\ n\ \mathbf{do}$
$\quad roll := rand(1...6)$
$\quad x := 1\ roll(\ );$
$\quad \mathbf{if}\ x = 1\ \mathbf{then}\ c1 := c1 + 1$
$\quad \mathbf{else}$
$\quad\quad \mathbf{if}\ x = 2\ \mathbf{then}\ c2 := c2 + 1$
$\quad\quad \mathbf{else}$
$\quad\quad\quad \mathbf{if}\ x = 3\ \mathbf{then}\ c3 := c3 + 1$
$\quad\quad\quad \mathbf{else}$
$\quad\quad\quad\quad \mathbf{if}\ x = 4\ \mathbf{then}\ c4 := c4 + 1$
$\quad\quad\quad\quad \mathbf{else}$
$\quad\quad\quad \mathbf{end\ if}$
$\quad\quad \mathbf{end\ if}$

```
      end if
    end if
  end do
  r1 := evalf (c1 / n);
  r2 := evalf (c2 / n);
  r3 := evalf (c3 / n);
  r4 := evalf (c4 / n);
  r5 := evalf (c5 / n);
  r6 := evalf (c6 / n);
  print(r1, r2, r3, r4, r5, r6)
end proc
```

>rolldie(10);
    0.3000000000,0.1000000000,0.4000000000,0.1000000000,0.1000000000,0.
>rolldie(100);
    0.1500000000,0.1900000000,0.1800000000,0.1200000000,0.1400000000,
    0.2200000000
>rolldie(1000);
    0.1800000000,0.1770000000,0.1470000000,0.1550000000,0.1690000000,
    0.1720000000
>rolldie(10000);
    0.1649000000,0.1709000000,0.1615000000,0.1679000000,0.1643000000,
    0.1705000000

The expected probability is $1/6$ or 0.1667. We note that as the number of trials increases the closer our probabilities are to the expected long run values. We offer the following concluding remarks. When you have to run simulations, run them for a very large number of trials.

9.5.2   R and Monte Carlo Simulation

Example 9.11 Flip a Fair Coin

Algorithm
    Input: The number of trials, N
    Output: The probability of a head or a tail

Step 1. Initialize counters to 0.

Step 2. For i = 1, 2, …, N do

   Step 3. Generate a random number, x, U (0, 1)

   Step 4. If $0 \leq x < 0.5$ Increment Heads, H = H + 1 otherwise T = T + 1

Step 5. Output H/N and T/N the probabilities for heads and tails.

Example 9.12 Roll of a Fair Die

Rolling a fair die adds the additional process of multiple assignments (6 for a six sided die). The probability will be the number of occurrences of each number/ total number of trails
 INPUT: Number of rolls
 Output: Probability of getting a {1,2,3,4,5,6}

Step 1. Initialize all counters (Counter 1 through Counter 6) to 0

Step 2. For i = 1,2...n do steps 3 and 4

 Step 3. Obtain a random number j from Integers (1,6)

 Step 4. Increment the counter for the value of j so Counter j = Counter j + 1

Step 5. Calculate the probability of each roll {1,2,3,4,5,6} by Counter j/ n

Step 6. Output probabilities

Step 7. Stop

We might use a simple, sample function in R to do this.
```
> sample(1:6,10,replace = T)
[1] 6 4 4 3 5 2 3 3 5 4
```
or with a function
```
>RollDie  = function(n) sample(1:6,n,replace = T)
>RollDie(5)
[1] 3 6 1 2 2
```
In fact, R can create lots of different types of random numbers ranging from familiar families of distributions to specialized ones.

Example 9.13 Games of Chance with a Spinning Machine

```
runs < -100,000
> #simulates on game of 10 spins, returns whether the sum of all the
      spins is < 1
>play.game<- function(){
+ results <- sample(c(1,1,-1,2),10,replace = T)
+ return(sum(results) < 0)
+}
>mc.prob<- sum(replicate(runs,play.game()))/run
Error: object 'run' not found
>mc.prob<-sum(replicate(runs,play.game()))/runs
>mc.prob
[1] 0.01327
```

The expected probability is 1/6 or 0.1667. We note that as the number of trials increases the closer our probabilities are to the expected long run values.

We show how to compute the probability of simple events using simulation. Suppose we rolled two fair dice. What is the probability that their sum is at least 7? We will approach this by simulating many throws of two fair dice, and then computing the fraction of those trials whose sum is at least 7. It will be convenient to write a function that simulates the trials and returns TRUE if the sum is at least 7 (we call this an event), and FALSE otherwise.

```
isEvent = function(numDice, numSides, targetValue, numTrials){
apply(matrix(sample(1:numSides, numDice*numTrials, replace = TRUE),
    nrow = numDice), 2, sum) > = targetValue
}
```

Now that we have our function, we are ready to do the Monte Carlo. It is good practice to set the random seed for reproducibility and debugging.

```
set.seed(0)
#try 5 trials
outcomes = isEvent(2, 6, 7, 5)
mean(outcomes)
## [1] 1
```

This is far from the theoretical answer of $\frac{21}{36} = 0.58333$. Now try with 10,000 trials:

```
set.seed(0)
outcomes = isEvent(2, 6, 7, 10000)
mean(outcomes)
## [1] 0.5843
```

9.6 Applied Simulations and Queuing Models

In this section, we present algorithms and Excel output for the following simulations.

1. an aircraft missile attack
2. the amount of gas that a series of gas stations will need
3. a simple single barber in a barbershop queue

Example 9.14 Missile Attack

An analyst plans a missile strike using F-15 aircraft. The F-15 must fly through air-defense sites that hold a maximum of eight missiles. It is vital to ensure success early in the attack. Each aircraft has a probability of 0.5 of destroying the target, assuming it can get to the target through the air-defense systems and then acquire and attack its target. The probability that a single F-15 will acquire a target is approximately 0.9. The target is protected by air-defense equipment with a 0.30 probability of stopping the F-15 from either arriving at or acquiring the target. How many F-15 are needed to have a successful mission assuming we need a 99percent success rate?

Algorithm: Missiles

Inputs:	N = number of F-15s
	M = number of missiles fired
	P = probability that one F-15 can destroy the target
	Q = probability that air defense can disable an F-15
Output:	S = probability of mission success

Step 1. Initialize $S = 0$

Step 2. For $I = 0$ to M do

  Step 3. $P(i) = [1 - (1 - P)^{N-I}]$

  Step 4. $B(i)$ = binomial distribution for (m, i, q)

  Step 5. Compute $S = S + P(i) * B(i)$

Step 6. Output S.

Step 7. Stop

We run the simulation letting the number of F-15s vary and calculate the probability of success. We guess $N = 15$ and find that we a probably of success greater than 0.99 when we send 9 planes. Thus, any number greater than 9 works (Figure 9.17).

We find that nine F-15s gives us $P(s) = 0.99313$.

Actually, any number of F-15 greater than nine provides a result with the probability of success we desire. Fifteen F-15s yielding a $P(s) = 0.996569$. Any more would be overkill.

Example 9.15 Gasoline-Inventory Simulation

You are a consultant to an owner of a chain of gasoline stations along a freeway. The owner wants to maximize profits and meet consumer demand for gasoline. You decide to look at the following problem.

					p	0.5	T		0.9	P*T		0
18												
19		Initial S		Bombers	N	q	0.3					
20	S	0			15	Quess				S > 99	good	
21									S_Final	0.99313666		
22	i	B	P	P*B	New S							
23	0	0.004747562	0.9999	0.004747	0.004747							
24	1	0.030520038	0.9998	0.030513	0.03526							
25	2	0.091560115	0.9996	0.091522	0.126781							
26	3	0.170040213	0.9992	0.16991	0.296691							
27	4	0.218623131	0.9986	0.218319	0.51501							
28	5	0.206130381	0.9975	0.205608	0.720618							
29	6	0.147235986	0.9954	0.146558	0.867176							
30	7	0.081130033	0.9916	0.080451	0.947627							
31	8	0.034770014	0.9848	0.034241	0.981867							
32	9	0.011590005	0.9723	0.011269	0.993137							
33	10	0.002980287	0.9497	0.00283	0.995967							
34	11	0.000580575	0.9085	0.000527	0.996494							
35	12	8.29393E-05	0.8336	6.91E-05	0.996564							
36	13	8.20279E-06	0.6975	5.72E-06	0.996569							
37	14	5.02212E-07	0.45	2.26E-07	0.996569							
38	15	1.43489E-08	0	0	0.996569							

FIGURE 9.17
Excel screenshot of missile attack.

Problem Identification Statement

Minimize the average daily cost of delivering and storing sufficient gasoline at each station to meet consumer demand.

Assumptions

For an initial model, consider that, in the short run, the average daily cost is a function of demand rate, storage costs, and delivery costs. You also assume that you need a model for the demand rate. You decide that historical date will assist you. This is displayed in Table 9.4–9.5.

Model Formulation

We convert the number of days into probabilities by dividing by the total and we use the midpoint of the interval of demand for simplification.

Because cumulative probabilities will be more useful we convert to a cumulative distribution function (CDF) (Table 9.6).

We might use cubic splines to model the function for demand (see Chapter 4 for a discussion of cubic splines).

Inventory Algorithm

Inputs:	Q = delivery quantity in gallons
	T time between deliveries in days
	D = delivery cost in dollars per delivery
	S = storage costs in dollars per gallons
	N = number of days in the simulation
Output:	C = average daily cost

TABLE 9.4

Gasoline demand

Demand: Number of gallons	Number of occurrences (days)
1,000–1,099	10
1,100–1,199	20
1,200–1,299	50
1,300–1,399	120
1,400–1,499	200
1,500–1,599	270
1,600–1,699	180
1,700–1,799	80
1,800–1,899	40
1,900–1,999	30
Total number of days =	1000

TABLE 9.5

Gasoline demand probabilities

Demand: Number of gallons	Probabilities
1,000	0.010
1,150	0.020
1,250	0.050
1,350	0.120
1,450	0.200
1,550	0.270
1,650	0.180
1,750	0.080
1,850	0.040
2,000	0.030
Total number of days =	1.000

TABLE 9.6

CDF of demand

Demand: Number of gallons	Probabilities
1,000	0.010
1,150	0.030
1,250	0.080
1,350	0.20
1,450	0.4
1,550	0.670
1,650	0.850
1,750	0.93
1,850	0.97
2,000	1.0

Step 1. Initialize: Inventory→$I = 0$ and $C = 0$.

Step 2. Begin the next cycle with a delivery:

$$I = I + Q$$

$$C = C + D$$

Step 3. Simulate each day of the cycle.

For $i = 1, 2,..., T$, do steps 4–6.

Step 4. Generate a demand, q_i. Use cubic splines to generate a demand based on a random CDF value, x_i.

Step 5. Update the inventory: $I = I - q^i$.

Step 6. Calculate the updated cost: $C = C + s * I$ if the inventory is positive. If the inventory is ≤ 0, then set $I = 0$ and go to step 7.

Step 7. Return to step 2 until the simulation cycle is completed.

Step 8. Compute the average daily cost: $C = C/n$.

Step 9. Output C.
Stop.

We run the simulation and find that the average cost is about \$5,753.04, and the inventory on hand is about 199,862.4518 gallons.

Example 9.16 Queuing Model

A queue is a waiting line. An example would be people in line to purchase a movie ticket or in a drive through line to order fast food. There are two important entities in a queue: customers and servers. There are some important parameters to describe a queue:

1. The number of servers available
2. Customer arrival rate: average number of customers arriving to be serviced in a time unit.
3. Server rate: average number of customers processed in a time unit.
4. Time

In many simple queuing simulations, as well as theoretical approaches, assume that arrivals and service times are exponentially distributed with a mean arrival rate of λ_1 and a mean service time of λ_2.

Theorem 9.3 If the arrival rate is exponential and the service rate is given by any distribution, then the expected number of customers waiting in line, L_q, and the expected waiting time, W_q, are given by

$$L_q = \frac{\lambda^2 \sigma^2 + \rho^2}{2(1-\rho)} \text{ and } W_q = \frac{L_q}{\lambda}$$

where λ is the mean number of arrival per time period; μ is the mean number of customers serviced per time unit, $\rho = \lambda/\mu$ and $\sigma\sigma$ is the standard deviation of the service time.

Here we have a barber shop where we have two customers arrive every 30 minutes. The service rate of the barber is three customers every 60 minutes. This implies the time between arrivals is 15 minutes and the mean service time is 1 customer every 20 minutes. How many customers will be in the queue and what is their average waiting time?

Possible Solution with Simulation.

We provide an algorithm for use.

Algorithm:

For each customer 1...N

Step 1. Generate an inter-arrival time, an arrival time, start time based on finish time of the previous customer, service time, completion time, amount of time waiting in a line, cumulative wait time, average wait time, number in queue, average queue length.

Step 2. Repeat N times.

Step 3. Output average wait time and queue length
Stop

You will be asked to calculate the theoretical solution in the exercise set. We illustrate the simulation.

We will use the following to generate exponential random numbers, $x = -1/\lambda \ln(1 - \text{rand}())$

We generate a sample of 5,000 runs and plot customers versus average weight time (Figure 9.18–9.19).

We note that the plot appears to be converging at values slightly higher than .66. Thus, we will run 100 more trials of 5,000 and compute the average.

We obtain the descriptive statistics from Excel. We note the mean is 0.6601 that is very close to our theoretical mean.

	D	E	F	G	H
	Customer number	Time between arrivals	Arrival time	Start time	Service time
1		=-(1/$B$1)*LN(1-RAND())	=E2	=F2	=-(1/$B$2)*LN(1-RAND())
=D2+1		=-(1/$B$1)*LN(1-RAND())	=F2+E3	=MAX(I2,F3)	=-(1/$B$2)*LN(1-RAND())
=D3+1		=-(1/$B$1)*LN(1-RAND())	=F3+E4	=MAX(I3,F4)	=-(1/$B$2)*LN(1-RAND())
=D4+1		=-(1/$B$1)*LN(1-RAND())	=F4+E5	=MAX(I4,F5)	=-(1/$B$2)*LN(1-RAND())
=D5+1		=-(1/$B$1)*LN(1-RAND())	=F5+E6	=MAX(I5,F6)	=-(1/$B$2)*LN(1-RAND())
=D6+1		=-(1/$B$1)*LN(1-RAND())	=F6+E7	=MAX(I6,F7)	=-(1/$B$2)*LN(1-RAND())
=D7+1		=-(1/$B$1)*LN(1-RAND())	=F7+E8	=MAX(I7,F8)	=-(1/$B$2)*LN(1-RAND())
=D8+1		=-(1/$B$1)*LN(1-RAND())	=F8+E9	=MAX(I8,F9)	=-(1/$B$2)*LN(1-RAND())
=D9+1		=-(1/$B$1)*LN(1-RAND())	=F9+E10	=MAX(I9,F10)	=-(1/$B$2)*LN(1-RAND())

B	C	D	E	F	G	H	I	J	K	L
2		Customer number	Time between arrivals	Arrival time	Start time	Service time	Completion time	wait time	Cumulative wait time	average wait
3		1	1.934408754	1.934408754	1.9344088	0.071524668	2.005933422	0	0	0
		2	0.116601281	2.051010035	2.05101	0.714947959	2.765957994	0	0	0
		3	0.055768834	2.106778869	2.765958	0.36811946	3.134077454	0.659179	0.659179125	0.219726375
		4	0.879801355	2.986580224	3.1340775	0.206478939	3.340556393	0.147497	0.806676355	0.201669089
		5	0.095844504	3.082424728	3.3405564	1.055590069	4.396146462	0.258132	1.06480802	0.212961604
		6	1.043442803	4.125857531	4.3961465	0.63308224	5.029228702	0.270289	1.335096951	0.222516159
		7	0.223185659	4.34904319	5.0292287	0.818579146	5.847807847	0.680186	2.015282463	0.287897495
		8	2.251848324	6.600891514	6.6008915	0.393204228	6.994095741	0	2.015282463	0.251910308
		9	0.384299775	6.985191288	6.9940957	0.320344496	7.314440237	0.008904	2.024186916	0.224909657
		10	0.163595249	7.148786537	7.3144402	0.066657268	7.381097506	0.165654	2.189840616	0.218984062
		11	0.000502847	7.149289384	7.3810975	0.792646337	8.173743842	0.231808	2.421648738	0.220149885
		12	0.102456472	7.251745856	8.1737438	1.062486891	9.236230734	0.921998	3.343646724	0.278637227
		13	0.384817067	7.636562923	9.2362307	0.2167925	9.453023234	1.599668	4.943314534	0.380254964
		14	0.625581112	8.262144036	9.4530232	0.342525761	9.795548994	1.190879	6.134193732	0.438156695
		15	0.5489886	8.811132636	9.795549	0.392217518	10.18766651	0.984416	7.118610091	0.474574006
		16	0.540099845	9.351232481	10.187667	0.500628779	10.68829529	0.836434	7.955044122	0.497190258
		17	0.025796165	9.377028647	10.688295	0.051349505	10.7396448	1.311267	9.266310766	0.545077104
		18	0.199860228	9.576888875	10.739645	0.497924558	11.23756935	1.162756	10.42906669	0.579392594
		19	0.422003799	9.998892674	11.237569	0.610221593	11.84779095	1.238677	11.66774337	0.614091756
		20	1.086641979	11.08553465	11.847791	0.139853034	11.98764398	0.762256	12.42999966	0.621499983
		21	0.085067941	11.17080259	11.987644	0.304856673	12.29250065	0.817041	13.24704105	0.630811478
		22	0.688452558	11.85905515	12.292501	0.13728232	12.42978297	0.433446	13.68048655	0.621840298

I	J	K	
Completion time	wait time	Cumulative wait time	
=G2+H2	=G2-F2	=J2	=K2/D2
=G3+H3	=G3-F3	=K2+J3	=K3/D3
=G4+H4	=G4-F4	=K3+J4	=K4/D4
=G5+H5	=G5-F5	=K4+J5	=K5/D5
=G6+H6	=G6-F6	=K5+J6	=K6/D6
=G7+H7	=G7-F7	=K6+J7	=K7/D7

FIGURE 9.18
Excel screenshot of customers average wait time.

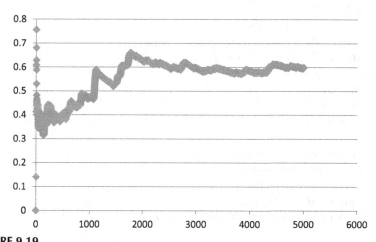

FIGURE 9.19
Customers average wait time.

	Column1	
1	2	3
4	Mean	0.660147135
1.1	Standard Error	0.006315375
1.2	Median	0.658168429
1.3	Mode	#N/A
1.4	Standard Deviation	0.063153753
1.5	Sample Variance	0.003988397
1.6	Kurtosis	-0.319393469
1.7	Skewness	0.155656707
1.8	Range	0.318586462
1.9	Minimum	0.500642393
1.10	Maximum	0.819228855
1.11	Sum	66.01471348
1.12	Count	100

9.6.1 Maple Applied Simulation

Algorithm:

Example 9.17 Missiles

INPUTS:	N = number of F-15s
	M = number of missiles fired
	P = probability the one F-15 can destroy the target
	Q = probability the air-defense can disable the F-15
OUTPUT	S = probability of mission success

Step 1. Initialize S = 0

Step 2. For I = 0 to M do

　　Step 3. $P(i) = (1-(1-P)^{N-I}$

　　Step 4. B(i) = Binomial Distribution for (m,i,q)

　　Step 5. Compute S = S + P(i) * B(i)

Step 6. Output S

Step. 7 Stop

```
>restart;
>with(Statistics):

>bombsaway: = proc(n,m,p,q)
>s: = 0:qn: = q:nn: = n:pn: = p;
>for i from 0 to m do
>pn: = 1-(1-pn)^(nn-i);
>x: = RandomVariable(Binomial(m,qn));
```

```
>b: = ProbabilityFunction(x, i);
>s: = s + pn*b;
>end do;
>print(s);
>end;
```

$bombsaway := \textbf{proc}\left(n,m,p,q\right)$ $pn := p;$
$\quad \textbf{local } s, qn, nn, pn, i, x, b;$ $\quad \textbf{for } i \textbf{ from } 0 \textbf{ to } m \textbf{ do}$
$\quad\quad s := 0;$ $\quad\quad pn := 1 - \left(1 - pm\right)\left(nn - i\right);$
$\quad\quad qn := q;$ $\quad\quad x := Statistics :- Random Variable$
$\quad\quad nn := n;$ $\quad\quad\quad \left(\text{Binomial}\left(m, qn\right)\right);$
$\quad\quad\quad s := s + pn \times b$ $\quad\quad b := Statistics : -Probability$
$\quad\quad \textbf{end do};$ $\quad\quad\quad Function\left(x, i\right);$
$\quad\quad print\left(s\right)$
$\textbf{end proc}$

```
>
```

We run the simulation letting the number of F-15 vary and calculate the probability of success.

```
>for
  i from 1 to 10 do
>bombsaway(i,8,.45,.4);
>end do;
```

$$
\begin{aligned}
&0.0075582720 \\
&0.07419703680 \\
&0.3043347039 \\
&0.5924805227 \\
&0.8254837008 \\
&0.9497277101 \\
&0.9912246093 \\
&0.9992039991 \\
&0.9999226475 \\
&0.9999574561
\end{aligned}
$$

We find that the number of F-15s equaling 7 gives us $P(s) = .99122$.

Actually, any number of F-15 greater than 7 works to provide a result with the probability of success we desire. We would think the 10 F-15s yielding a $P(s) = 0.999957$ would suffice. Any more would be overkill.

Example 9.18 Gasoline Inventory Simulation

Background: This has been provided earlier, so we just show the program.
Inventory Algorithm

INPUTS: Q = delivery quantity in gallons

   T = time between deliveries in days

   D = delivery cost in dollars per delivery

   S = storage costs in dollars per gallons

   N = number of days in the simulation

OUTPUT: C = average daily cost

Step 1. Initialize: Inventory $\rightarrow$ I = 0, and C = 0
Step 2. Begin the next cycle with a delivery

$$I = I + Q$$

$$C = C + D$$

Step 3. Simulate each day of the cycle

$$\text{For } i = 1,2,,....T \text{ do steps 4-6}$$

Step 4. Generate a demand, q_i. Use cubic splines to generate a demand based on a random CDF value x_i.
Step 5. Update the inventory: $I = I - q^i$
Step 6. Calculate the updated cost, $C = C + s*I$ if the Inventory is positive. If the inventory is ≤ 0 then set I = 0 go to Step 7.
Step 7. Return to Step 2 until the simulation cycle is completed.
Step 8. Compute the average daily cost: $C = C/n$
Step 9. Output C

Stop
Maple program

```
>inventorygas: = proc(q,t,d,s,n,xdat,qdat)
>#print(xdat,qdat);
>k: = n:i: = 0:c: = 0:Flag: = 0:nq: = q: nt: = t:nd: = d:ns: = s;
>label_2:i: = i+nq;#print(i,nt);
>c: = c+nd:#print(c);
>if(nt> = k) then
>nt: = k: Flag: = 1;
>readlib(spline):
>spline(xdat,qdat,x,cubic);
>nfunc: = unapply(%,x):
```

```
>end if;
>for j from 1 to nt do
>readlib(spline):
>spline(xdat,qdat,x,cubic);
>nfunc: = unapply(%,x):
>nx: = evalf(rand()/(1.0*10^12));
>newq: = evalf(nfunc(nx));
>i: = i-newq:#print(i);
>if (i< = 0) then
>i: = 0;
>goto(label_9);
>else
>c: = c+i*ns:#print(c);
>end if;
>label_9;
>k: = k-1: #print(k);
>if k>0 then
>goto(label_2);
>else
>newc: = c/n:print(newc);
>goto(label_3);
>
>
>end if;
>print(newc);
>end do;
>label_3;
>print(i,newc);
>end;
>
```

$$>xdat: = [0,.01,.03,.08,.2,.4,.67,.85,.93,.97,1];$$
$$xdat := [0, 0.01, 0.03, 0.08, 0.2, 0.4, 0.67, 0.85, 0.93, 0.97, 1]$$

$$>qdat: = [1000,1050,1150,1250,1350,1450,1550,1650,1750,1850,2000];$$
$$qdat := [1000, 1050, 1150, 1250, 1350, 1450, 1550, 1650, 1750, 1850, 2000]$$

>inventorygas(11500,7,500,.05,20,xdat,qdat);

$$5753.039330$$
$$199862.4518, 5753.039330$$

The average cost is \$5753.04 and the inventory on hand is 199,862.4518 gallons.

9.6.2 R Applied Simulation

Algorithm:

Example 9.19 Missiles

INPUTS: N = number of F-15s

M = number of missiles fired, M = 10

P = probability the one F-15 can destroy the target, P == .9*.5 = .45

Q = probability that one air-defense missile can disable the F-15 Q = .6

X = number of F-15 disabled in the attack

Pi = probability of mission success given X = i → $1 - (1 - P)^{N-i}$

B = binomial probability given X = i $\binom{m}{i} q^i (1-q)^{m-i}, i = 0, 1, 2, \ldots, m$

S = $\sum P_i B_i$

OUTPUT   S = probability of mission success

Step 1. Initialize S = 0
Step 2. For i = 0 to M do
   Step 3. P(i) = $(1-(1-P)^{N-I}$
   Step 4. B(i) = Binomial Distribution for (m,i,q)
   Step 5. Compute S = S + P(i) * B(i)
Step 6. Output S
Step 7. Stop

R Code

```
s = 0
> n = c(10,11,12,13,14,15,16,17,18,19,20)
> for (i in 0:10) {
+ pn = 1-(1-.45)^(n-i)
+ x = rbinom(i,10,.45)
+ xs = dbinom(i,10,.6)
+ s = s+pn*xs
+ print (s)}
[1] 0.0001045920 0.0001047115 0.0001047773 0.0001048134 0.0001048333
[6] 0.0001048442 0.0001048502 0.0001048536 0.0001048554 0.0001048564
[11] 0.0001048569
[1] 0.001670212 0.001673592 0.001675450 0.001676472 0.001677034
[6] 0.001677344 0.001677514 0.001677607 0.001677659 0.001677687
[11] 0.001677703
```

[1] 0.01219815 0.01224153 0.01226539 0.01227851 0.01228573 0.01228970
[7] 0.01229188 0.01229309 0.01229375 0.01229411 0.01229431
[1] 0.05401894 0.05435326 0.05453714 0.05463827 0.05469390 0.05472449
[7] 0.05474132 0.05475057 0.05475566 0.05475846 0.05476000
[1] 0.1624099 0.1641328 0.1650804 0.1656016 0.1658883 0.1660459
[7] 0.1661326 0.1661803 0.1662066 0.1662210 0.1662289
[1] 0.3529692 0.3592366 0.3626837 0.3645796 0.3656223 0.3661958
[7] 0.3665112 0.3666847 0.3667801 0.3668326 0.3668615
[1] 0.5808401 0.5974358 0.6065634 0.6115836 0.6143447 0.6158633
[7] 0.6166986 0.6171579 0.6174106 0.6175496 0.6176260
[1] 0.7600618 0.7927536 0.8107341 0.8206234 0.8260625 0.8290540
[7] 0.8306993 0.8316042 0.8321019 0.8323757 0.8325262
[1] 0.8444121 0.8935658 0.9206004 0.9354694 0.9436473 0.9481452
[7] 0.9506190 0.9519796 0.9527280 0.9531396 0.9533659
[1] 0.8625520 0.9216826 0.9542045 0.9720915 0.9819293 0.9873402
[7] 0.9903161 0.9919529 0.9928531 0.9933482 0.9936205
[1] 0.8625520 0.9244036 0.9584220 0.9771321 0.9874226 0.9930825
[7] 0.9961954 0.9979074 0.9988491 0.9993670 0.9996519

> print (s)

[1] 0.8625520 0.9244036 0.9584220 0.9771321 0.9874226 0.9930825
[7] 0.9961954 0.9979074 0.9988491 0.9993670 0.9996519

>

N = 15 is the first time our probability of success is greater than 99% at
99.30825%.

We find that the number of F-15s equaling 15 gives us P(s) = 0.9930825.
Actually, any number of F-15 greater than 15 works to provide a result
with the probability of success we desire. We would think the 15 F-15s
yielding a P(s) = 0.9930825 would suffice. Any more would be overkill.

9.7 Exercises

1. For each generate 20 random numbers
   (a) Uniform (0,1)
   (b) Uniform (−10,10)
   (c) Exponential ($\lambda = 0.5$)
   (d) Normal(0,1)
   (e) Normal (5,0.5)

2. Modify the missile strike problem if the probability of S were only 0.95
   and the probability of an F-15 being deterred by air defense were 0.3.
   Determine the number of F-15s needed to complete the mission.

3. What if in the missile attack problem the air-defense units were modified to carry 10 missiles each? What effect does that have on the number of F-15s needed?

4. Perform sensitivity analysis on the gasoline-inventory problem by modifying the delivery to 11,450 gallons per week. What effect does this have on the average daily cost?

5. Use Monte Carlo simulation to approximate the area under the curve $f(x) = 1 + \sin x$ over the interval

$$\frac{-\pi}{2} \le x \le \frac{\pi}{2}.$$

6. Use Monte Carlo simulation to approximate the area under the curve $f(x) = x^{0.5}$ over the interval

$$\frac{1}{2} \le x \le \frac{3}{2}.$$

7. Use Monte Carlo simulation to approximate the area under the curve $f(x) = \sqrt{1-x^2}$ over the interval

$$0 \le x \le \frac{\pi}{2}.$$

8. How would you modify question 4 to obtain an approximation to π?

9. Use Monte Carlo simulation to approximate the volume under the surface $f(z) = x^2 + y^2$, the first octant.

10. Determine the area under the following non-negative curves:

    (a) $y = 0 \le x \le 1$
    (b) $y = \sqrt{4-x^2}, 0 \le x \le 2$
    (c) $y = \sin(x), 0 \le x \le pi/2 y = \sqrt{1-x^2}$
    (d) $y = x^3, 0 \le x \le 4.$

11. Find the area between the following two curves and the two axis by simulation:

    $y = 2x + 1$ and $y = -2x^2 + 4x + 8$

9.8 Projects

1. Tollbooths. Heavily traveled toll roads such as the Garden State Parkway, Interstate 95, and so forth, are multilane divided highways that are interrupted at intervals by toll plazas. Because collecting tolls is usually unpopular, it is desirable to minimize motorist annoyance by limiting the amount of traffic disruption caused by the toll plazas. Commonly, a much larger number of tollbooths are provided than the number of travel lanes entering the toll plaza. On entering the toll plaza, the flow of vehicles fans out to the larger number of tollbooths; when leaving the toll plaza, the flow of vehicles is forced to squeeze down to a number of travel lanes equal to the number of travel lanes before the toll plaza. Consequently, when traffic is heavy, congestion increases when vehicles leave the toll plaza. When traffic is very heavy, congestion also builds at the entry to the toll plaza because of the time required for each vehicle to pay the toll.

   Construct a mathematical model to help you determine the **optimal number** of tollbooths to deploy in a barrier-toll plaza. Explicitly, first consider the scenario in which there is exactly one tollbooth per incoming travel lane. Then consider multiple tollbooths per incoming lane. Under what conditions is one tollbooth per lane more or less effective than the current practice? Note that the definition of *optimal* is up to you to determine.

2. Major League Baseball. Build a simulation to model a baseball game. Use your two favorite teams or favorite all-star players to play a regulation game.

3. NBA Basketball. Build a simulation to model the NBA basketball playoffs.

4. Hospital Facilities. Build a simulation to model surgical and recovery rooms for the hospital.

5. Class Schedules. Build a simulation to model the registrar's scheduling changes for students or final exam schedules.

6. Automobile Emissions. Consider a large engineering company that performs emissions control inspections on automobiles for the state. During the peak period, cars arrive at a single location that has four lanes for inspections following exponential arrivals with a mean of 15 minutes. Service times during the same period are uniform: between [15,30] minutes. Build a simulation for the length of the queue. If cars wait more than 1 hour, the company pays a penalty of $200 per car. How much money, if any, does the company pay in penalties? Would more inspection lanes help? What costs associated with the inspection lanes need to be considered?

7. Recruiting Simulation Model

Monthly demand for recruits is provided in the table below.

Demand	Probability	CDF
300	0.05	0.05
320	0.10	0.15
340	0.20	0.35
360	0.30	0.65
380	0.25	0.90
400	0.10	1.0

Additionally, depending on conditions the average cost per recruit is between \$60 and \$80 in integer values. Returns from Higher HQ are between 20% and 30% of costs. There is a fixed cost of \$2,000/month for the office, phones, etc. Build a simulation model to determine the average monthly costs.

Assume Cost = demand* cost per recruit + fixed cost-return amount, where return = %*cost

8. Inventory Model

Demand of ammunition palette for resupply on a weekly basis is provided below.

Demand	Frequency	Probability	CDF
0	15	0.05	0.05
1	30	0.10	0.15
2	60	0.20	0.35
3	120	0.40	0.75
4	45	0.15	0.90
5	30	0.10	1.00

Assumptions:

Lead time if resupply is requires between 1 and 3 days. Currently, we have 7 palettes in stock and no orders due. Needs Order Quantity and order point to reduce Costs. Fixed cost for placing an order is \$20. The cost for holding the unused stock is \$0.02 per palette per day. Each time we cannot satisfy a demand the unit goes elsewhere and assumes a loss of \$8 to the company. We operate 24/7.

9. Simple Queuing Problem

The bank manager is trying to improve customer satisfaction by offering better service. They want the average customer to wait less than 2 minutes and the average length of the queue (line) if 2 or fewer.

The bank estimates about 150 customers per day. The existing service and arrival times are given below.

Service time	Probability	Time between arrival	Probability
1	0.25	0	0.10
2	0.20	1	0.15
3	0.40	2	0.10
4	0.15	3	0.35
		4	0.25
		5	0.05

Determine if the current servers are satisfying the goals. If not, how much improvement is needed in service to accomplish the stated goals.

10. Intelligence gathering (Information Operations) modeling Currently Intelligence reports come according to the following historical information

Time between reports	Probability
1	0.11
2	0.21
3	0.22
4	0.20
5	0.16
6	0.10

The time it takes to process these reports is given

Process time	Probability
1	0.20
2	0.19
3	0.18
4	0.17
5	0.13
6	0.10
7	0.03

Further, if we employ sensors the reports come more often

Time between reports	Probability
1	0.22
2	0.25
3	0.19
4	0.15
5	0.12
6	0.07

Advise the manager on the current system. Determine utilization and sensor satisfaction. How many report processors are needed to insure reports are processed in a timely manner.

11. *The Price Is Right.* On the popular TV game show *The Price Is Right*, at the end of each half hour, the three winning contestants face off in what is called the "Showcase Showdown." The game consists of spinning a large wheel with 20 spaces on which the pointer can land; the spaces are numbered from $0.05 to $1.00 in 5-cent increments. The contestant who has won the least amount of money at this point in the show spins first, followed by the one who has won the next most, and then by the biggest winner for that half hour.

The objective of the game is to obtain as close to $1.00 as possible without going over that amount with an allowed maximum of two spins. Naturally, if the first player does not go over, the other two will use one or both spins in their attempts to overtake the leader.

But what of the person spinning first? If he or she is an expected-value decision-maker, how high a value on the first spin does he or she need to not want to take a second spin? Remember, the person can lose if

(a) either of the other two players surpasses the player's total, or
(b) the player spins again and goes over.

12. *Let's Make a Deal.* You are "dressed to kill" in your favorite costume, and host Monte Hall picks you out of the audience. You are offered the choice of three wallets. Two wallets contain a single $50 bill, and the third contains a $1,000 bill. You choose one of the three wallets. Monte knows which wallet contains the $1,000, so he shows you one of the other two wallets – one with one of the two $50 bills inside. Monte does this on purpose because he must have at least one wallet with $50 inside. If he holds the $1,000 wallet, he shows you the other wallet, the one with $50. Otherwise, he just shows you one of his two $50 wallets. Monte then asks you if you want to trade your choice for the one he's still holding. Should you trade?

9.9 References and Suggested Further Reading

Giordano, F. R., W. P. Fox, and S. Horton (2013). *A First Course in Mathematical Modeling*. 5th ed. Cengage Publishers, Boston, MA.

Law, A., and D. Kelton (2007). *Simulation Modeling and Analysis.* 4th ed. McGraw Hill, New York.

Meerschaert, M. M. (1993). *Mathematical Modeling.* Academic Press, San Diego.

Winston, W. (1994). *Operations Research: Applications and Algorithms.* 3rd ed. Duxbury Press, Belmont, CA.

10

Modeling Decision-Making with Multi-Attribute Decision Modeling with Technology

---

Objectives

1. Understand and apply the different multi-attribute decision-making algorithms.
2. Understand the strengths and weakness of each approach.
3. Know DEA, SAW, and TOPSIS methods.
4. Know weighting methods.

10.1 Introduction

This chapter describes the elements that make up a multi-attribute decision-making (MADM) problem and provides an overview of alternative MADM methods. Section 10.2 discusses the weighting methods. Sections 10.3–10.5 provide details and example on how to use selected MADM methods, such as sum of additive weights (SAW), technique of order preference by similarity to ideal solution (TOPSIS), and data envelopment analysis (DEA).

MADM methods apply to problems where a decision-maker is choosing or ranking a finite *number of alternatives*, which are measured by two *or more relevant attributes*. The literature describing MADM theory, methods, and applications is vast. The additional readings are a good source of additional information. We provide a brief overview.

Elements of a MADM Problem

There are four elements or characteristics common to all MADM problems.

Finite (Generally Small) Set of Alternatives

MADM problems involve analysis of a finite and generally small set of discrete and predetermined options or alternatives. In this way, MADM problems are distinguished from other decision processes, which involve the *design* of a "best" alternative by considering the tradeoffs within a set of interacting design constraints.

Tradeoffs among Attributes

Problems require MADM methods if no single alternative exhibits the most preferred available value or performance for all attributes. This is often the result of an underlying tradeoff relationship among attributes. An example is the tradeoff between low desired energy costs and large glass window areas (which may raise heating and cooling costs while lowering lighting costs and enhancing aesthetics).

Incommensurable Units

The attributes in a MADM problem will generally not all be measurable in the same units. In fact, some attributes may be either impractical, impossible, or too costly to measure at all. For example, life-cycle costs are directly measured in dollars, the number and size of offices are measured in other units, and the public image of a building may not be practically measurable in any unit. If all relevant attributes characterizing alternative buildings can be expressed in terms of financial costs or benefits scheduled to occur at specifiable times, then the ranking and selection of a building does not require the application of MADM. Often we normalize these values so that they are both scaled and dimensionless.

Decision Matrix

A MADM problem can generally be characterized by a "decision matrix" as we will describe further in Sections 14.2 and 14.3. The decision matrix indicates both the set of alternatives and the set of attributes being considered in a given problem. It summarizes the "raw" data available to the decision-maker at the start of the analysis. A decision matrix has a row corresponding to each alternative being considered and a column corresponding to each attribute being considered. A problem with a total of m alternatives characterized by n attributes is described by an $m \times n$ matrix X.

MADM approaches can be viewed as alternative methods for combining the information in a problem's decision matrix together with additional information from the decision-maker in order to determine a final ranking, screening, or selection from among the alternatives. Besides the information contained in the decision matrix, all but the simplest MADA techniques require additional information from the decision-maker in order to arrive at a final ranking, screening, or selection. For example, the decision matrix provides no information about the relative importance of the different attributes to the decision-maker, nor about any minimum acceptable, maximum acceptable, or target values for particular attributes.

10.2 Weighting Methods

10.2.1 Delphi Method

The Delphi Method is a reliable way of obtaining the opinions of a group of experts on an issue by conducting several rounds of interrogative communications. This method was first developed in the U.S. Air Force in the 1950s, mainly for market research and sales forecasting (Chan et al., 2001). This method is basically a communication device that is particularly useful for achieving a consensus among experts given a complex problem. The method consists of repeated solicitations of questions from a panel of experts who are anonymous. The information and ideas of each panel member are distributed among all the panel members in the next round. They can comment on others' viewpoints and can even use new information to modify their own opinions. Panel members can change their opinions based on new information more easily than in regular group meetings and open discussion. A consensus of opinions should be ultimately achieved in this way. This method will also highlight the areas where panel members have disagreements or uncertainty in a quantitative manner. The evaluation of belief statements by the panel as a group is an explicit part of the Delphi method (Chan et al., 2001). The panel consists of a number of experts chosen based on their experience and knowledge. As mentioned previously, panel members remain anonymous to each other throughout the procedure in order to avoid the negative impacts of criticism on the innovation and creativity of panel members. The Delphi Method should be conducted by a director (facilitator) who has independent communication with each panel member. The director develops a questionnaire based on the problem at hand and sends it to each panel member. Then, the responses to the questions and all the comments are collected and evaluated by the director. The director should process the information and filter out the irrelevant content. The result of the processed information will again be distributed among the panel members. Each member receives new information and ideas and can comment on them and/or revise his own previous opinions. One can use the Delphi Method for giving weights to the short-listed critical factors. The panel members should give weights to each factor as well as their reasoning. In this way, other panel members can evaluate the weights based on the reasons given and accept, modify, or reject those reasons and weights. For example, let us assume that after application of Tier 1 by the transit agency, there still remain two or more project delivery methods as viable options. The agency has identified the following four factors to be considered in Tier 2:

- shortening the schedule,
- agency control over the project,
- project cost,
- competition among contractors.

The facilitator should ask the panel members to weight each of the factors while giving their reasons for the weights selected. The facilitator can then use the collected weights and viewpoints and establish a weight for each factor. One possible approach could be to calculate an average weight for each factor based on responses. If there are large divergences in some responses, the facilitator should study that and comment on those cases. The outcome of this analysis should be distributed to the panel again for further consideration and modification. The facilitator will decide when to stop the process of the level of consensus desired.

It might be that our panel of experts provide the weights as results for c1, c2, c3, and c4 as 0.020, 0.15, 0.40, and 0.25,

10.2.2 Rank Order Centroid (ROC)

The rank order centroid method is a simple way of giving weight to a number of items ranked according to their importance. The decision-makers usually can rank items much more easily than give weight to them. This method takes those ranks as inputs and converts them to weights for each of the items. The conversion is based on the following formula:

$$w_i = \left(\frac{1}{M}\right)\sum_{n=i}^{M}\frac{1}{n}$$

1. List objectives in order from most important to least important.
2. Use the above formulas for assigning weights.

where M is the number of items and W_i is the weight for the i item. For example, if there are 4 items, the item ranked first will be weighted $(1 + 1/2 + 1/3 + 1/4)/4 = 0.52$, the second will be weighted $(1/2 + 1/3 + 1/4)/4 = 0.27$, the third $(1/3 + 1/4)/4 = 0.15$, and the last $(1/4)/4 = 0.06$. As shown in this example, the ROC is simple and easy to follow, but it gives weights which are highly dispersed. As an example, consider the same factors to be weighted (shortening schedule, agency control over the project, project cost, and competition). If they are ranked based on their importance and influence on decision as

1. shortening schedule,
2. project cost,
3. agency control over the project,
4. competition,

their weights would be 0.52, 0.27, 0.15, and 0.06, respectively. These weights almost eliminate the effect of the fourth factor, i.e., among competitors. This could be an issue for a decision-maker. For any four criteria that are ranked order will produce these weights,

10.2.3 Ratio Method for Weights

The Ratio Method is another simple way of calculating weights for a number of critical factors. A decision-maker should first rank all the items according to their importance. The next step is giving weight to each item based on its rank. The lowest ranked item will be given a weight of 10. The weight of the rest of the items should be assigned as multiples of 10. The last step is normalizing these raw weights (Weber and Borcherding, 1993). This process is shown in the example below. Note that the weights should not necessarily jump 10 points from one item to the next. Any increase in the weight is based on the subjective judgment of the decision-maker and reflects the difference between the importance of the items. Ranking the items in the first step helps in assigning more accurate weights. Table 10.1.

Normalized weights are simply calculated by dividing the raw weight of each item over the sum of the weights for all items. For example, normalized weight for the first item (shortening schedule) is calculated as 50/(50 + 40 + 20 + 10) = 41.7%. The sum of normalized weights is equal to 100% (41.7 + 33.3 + 16.7 + 8.3 = 100). Again, any four criteria that are ranked order by importance will produce these weights,

TABLE 10.1

Ratio method

Task/Item	Shorten schedule	Project cost	Agency control	Competition
Ranking	1	2	3	4
Weighting	50	40	20	10
Normalizing	41.7%	33.3%	16.7%	8.3%

10.3 Pairwise Comparison by Saaty (AHP)

In this method, the decision-maker should compare each item with the rest of the group and give a preferential level to the item in each pairwise comparison (Chang, 2004). For example, if the item at hand is as important as the second one, the preferential level would be one. If it is much more important, its level would be 10. After conducting all of the comparisons and determining the preferential levels, the numbers will be added up and normalized. The results are the weights for each item. Table 10.2 can be used as a guide for giving a preferential level score to an item while comparing it with another one. The following example shows the application of the pairwise comparison procedure. Referring to the four critical factors identified above, let us assume that shortening the schedule, project cost, and agency control of the project are the most important parameters in the project delivery selection decision. Following the pairwise comparison, the decision-maker should pick one of these factors (e.g., shortening the schedule) and compare it with the remaining factors and give a preferential level to it. For example, shortening the schedule is more important than project cost; in this case, it will be given a level of importance of the 5. Pairwise comparisons use the information in Table 10.2.

The decision-maker should continue the pairwise comparison and give weights to each factor. The weights, which are based on the preferential levels given in each pairwise comparison, should be consistent to the extent

TABLE 10.2

Saaty's 9-point scale

Intensity of importance in pair-wise comparisons	Definition
1	Equal Importance
3	Moderate Importance
5	Strong Importance
7	Very Strong Importance
9	Extreme Importance
2,4,6,8	For comparing between the above
Reciprocals of above	In comparison of elements i and j if i is 3 compared to j, then j is $1/3$ compared to i.

TABLE 10.3

Random Consistency Index (RI)

n	1	2	3	4	5	6	7	8	9	10
RI	0	0	0.58	0.9	1.12	1.24	1.32	1.41	1.45	1.49

TABLE 10.4

Pairwise comparison example

	Shorten the schedule (1)	Project cost (2)	Agency control (3)	Competition (4)	Total (5)	Weights (6)
Shorten the schedule	1	5	5/2	8	16.5	16.5/27.225 = 0.60
Project cost	1/5	1	½	1	2.7	2.7/27/225 = 0.10
Agency control	2/5	2	1	2	5.4	5.4/27/225 = 0.20
Competition	1/8	1	½	1	2.625	2.625/27/ 225 = 0.10
				Total =	27.225	1

possible. The consistency is measured based on the matrix of preferential levels. The interested reader can find the methods and applications of consistency measurement in Temesi (2006).

Table 10.4 shows the rest of the hypothetical weights and the normalizing process, the last step in the pairwise comparison approach.

Note that Column (5) is simply the sum of the values in Columns (1) through (4). Also note that if the preferential level of factor i to factor j is n, then the preferential level of factor j to factor i is simply $1/n$. The weights calculated for this exercise are 0.6, 0.1, 0.2, and 0.1 which add up to 1.0. Note that it is possible for two factors to have the same importance and weight. Weights need to pass the consistency tests.

Let λ be the largest eigenvalue of the pairwise comparison matrix.

Saaty proved that for consistent reciprocal matrix, the largest Eigenvalue is equal to the size of comparison matrix, or $\lambda_{max} = n$. Then he gave a measure of consistency, called Consistency Index as deviation or degree of consistency using the following formula

$$CI = \frac{\lambda_{max} - n}{n-1}$$

For example, assume we have $\lambda_{max} = 3.0967$ and the size of comparison matrix is n = 3, thus the consistency index is

$$CI = \frac{\lambda_{max} - n}{n-1} = \frac{3.0967 - 3}{2} = 0.0484$$

Knowing the Consistency Index, the next question is how do we use this index? Again, Saaty proposed that we use this index by comparing it with the appropriate one. The appropriate Consistency index is called Random Consistency Index (*RI*).

He randomly generated reciprocal matrix using scale $\frac{1}{9}, \frac{1}{8}, ..., 1, ..., 8, 9$

(similar to the idea of Bootstrap) and get the random consistency index to see if it is about 10% or less. The average random consistency index of sample size 500 matrices is shown in Table 10.3.

Then, he proposed what is called Consistency Ratio, which is a comparison between Consistency Index and Random Consistency Index, or in formula

$$CR = \frac{CI}{RI}$$

If the value of Consistency Ratio is smaller or equal to 10%, the inconsistency is acceptable. If the Consistency Ratio is greater than 10%, we need to revise the subjective judgment.

For our previous example, we have CI = 0.0484 and RI for n = 3 is **0.58**, then we have $CR = \dfrac{CI}{RI} = \dfrac{0.0484}{0.58} = 8.3\% < 10\%$. Thus, we conclude our sub-jective evaluation preference is consistent.

We need to calculate this for each pair-wise matrix. The important elements is if the matrix has a $CR > 0.10$ then we must go back and revise the pair-wise matrix until the $CR < 0.10$.

10.4 Entropy Method

Shannon and Weaver (1947) proposed the entropy concept and this concept has been highlighted by Zeleny (1982) for deciding the weights of attributes. Entropy is the measure of uncertainty in the information using probability methods. It indicates that a broad distribution represents more uncertainty than does a sharply peaked distribution.

To determine the weights by the entropy method the normalized decision matrix we call R_{ij} is considered. The equation used is

$$e_j = -k \sum_{i=1}^{n} R_{ij} \ln\left(R_{ij}\right)$$

Where $k = 1/ln(n)$ is a constant that guarantees that $0 \le e_j \le 1$. The value of n refers to the number of alternatives. The degree of divergence (d_j) of the average information contained by each attribute can be calculated as:

$$d_j = 1 - e_j.$$

The more divergent the performance rating R_{ij}, for all $i \& j$, then the higher the corresponding d_j the more important the attribute B_j is considered to be.

The weights are found by the equation, $w_j = \dfrac{\left(1-e_j\right)}{\Sigma\left(1-e_j\right)}$.

Let's do an example to obtain entropy weights.

Example 10.1 Cars

(a) The data:

	Cost	Safety	Reliability	Performa	MPG City	MPG HW	Interior/ style
a1	27.8	9.4	3	7.5	44	40	8.7
a2	28.5	9.6	4	8.4	47	47	8.1
a3	38.668	9.6	3	8.2	35	40	6.3
a4	25.5	9.4	5	7.8	43	39	7.5
a5	27.5	9.6	5	7.6	36	40	8.3
a6	36.2	9.4	3	8.1	40	40	8

(b) Sum the columns

Sums	184.168	57	23	47.6	245	246	46.9

(c) Normalize the data. Divide each data element in a column by the sum of the column.

0.150949	0.164912	0.13043478	0.157563	0.17959184	0.162602	0.185501066
0.15475	0.168421	0.17391304	0.176471	0.19183673	0.191057	0.172707889
0.20996	0.168421	0.13043478	0.172269	0.14285714	0.162602	0.134328358
0.138461	0.164912	0.2173913	0.163866	0.1755102	0.158537	0.159914712
0.14932	0.168421	0.2173913	0.159664	0.14693878	0.162602	0.176972281
0.19656	0.164912	0.13043478	0.170168	0.16326531	0.162602	0.170575693

(d) Use the entropy formula, where in the case $k = 6$.

$$e_j = -k\sum_{i=1}^{n} R_{ij} \ln\left(R_{ij}\right)$$

e1	e2	e3	e4	e5	e6	e7	k = 0.558111
-0.28542	-0.29723	-0.2656803	-0.29117	-0.3083715	-0.29536	-0.31251265	
-0.28875	-0.30001	-0.3042087	-0.30611	-0.31674367	-0.31623	-0.30330158	
-0.32771	-0.30001	-0.2656803	-0.30297	-0.27798716	-0.29536	-0.26965989	
-0.27376	-0.29723	-0.3317514	-0.29639	-0.30539795	-0.29199	-0.293142	
-0.28396	-0.30001	-0.3317514	-0.29293	-0.28179026	-0.29536	-0.3064739	
-0.31976	-0.29723	-0.2656803	-0.30136	-0.29589857	-0.29536	-0.3016761	

(e)   Find $e_{j'}$

0.993081	0.999969	0.98492694	0.999532	0.99689113	0.998825	0.997213162

(f)   Compute weights by formula,

	0.006919	3.09E–05	0.01507306	0.000468	0.00310887	0.001175	0.002786838	0.029561
w	0.234044	0.001046	0.50989363	0.015834	0.1051674	0.039742	0.094273553

(g)   Check that weights sum to 1, as they did above.

(h)   Interpret weights and rankings.

(i)   Use these weights in further analysis.

10.5 Simple Additive Weights (SAW) Method

Description and Uses

The simple additive weights method is also called the weighted sum method (Fishburn, 1967) and is the simplest, and still one of the widest used of the MADM methods. Its simplistic approach makes it easy to use. Depending on the type relational data used, we might either want the larger average or the smaller average.

Methodology

Here, each criterion (attribute) is given a weight, and the sum of all weights must be equal to one. Each alternative is assessed with regard to every criterion (attribute). The overall or composite performance score of an alternative is given simply by Equation 10.1 with m criteria.

$$P_i = \left(\sum_{j=1}^{m} w_j m_{ij} \right) / m \tag{10.1}$$

Previously, it was argued that SAW should be used only when the decision criteria can be expressed in identical units of measure (e.g., only dollars, only pounds, only seconds, etc.). However, if all the elements of the decision table are normalized, then this procedure can be used for any type and any number of criteria. In that case, Equation 10.1 will take the following form still with m criteria shown as Equation 10.2:

$$P_i = \left(\sum_{j=1}^{m} w_j m_{ijNormalized} \right) / m \tag{10.2}$$

where $(m_{ijNormalized})$ represents the normalized value of m_{ij}, and P_i is the overall or composite score of the alternative A_i. The alternative with the highest value of P_i is considered the best alternative.

Strengths and Limitations

The strengths are the ease of use and the normalized data allow for comparison across many differing criteria. Limitations include larger is always better or smaller is always better. There is not the flexibility in this method to state which criterion should be larger or smaller to achieve better performance. This makes gathering useful data of the same relational value scheme (larger or smaller) essential.

Sensitivity Analysis

Sensitivity analysis should be applied to the weighting scheme employed to determine how sensitive the model is to the weights. Weighting can be arbitrary for a decision-maker or in order to obtain weights you might choose to use a scheme to perform pairwise comparison as we show in AHP that we discuss later. Whenever subjectivity enters into the process for finding weights, then sensitivity analysis is recommended. Please see later sections for a suggested scheme for dealing with sensitivity analysis for individual criteria weights.

Illustrative Examples of SAW

Example 10.2 Manufacturing Example

We have data (Table 10.5) for our three alternatives over four criteria.

We want to normalize the data in the table for the alternatives. We normalize by first summing the data in each column and then dividing each data elements by its column sum (Figure 10.1).

Our company has ranked order alternative 3 first, followed by alternative 2, then alternative 1.

TABLE 10.5

Manufacturing weights and alternatives

	C_1	C_2	C_3	C_4
Alts./Weights	0.2	0.15	0.4	0.25
A_1	25	20	15	30
A_2	10	30	20	30
A_3	30	10	30	10

Alts / Weights	c1	c2	c3	c4
	0.2	0.15	0.4	0.25
a1	25	20	15	30
a2	10	30	20	30
a3	30	10	30	10
Total	65	60	65	70

	c1	c2	c3	c4
Normalized	0.2	0.15	0.4	0.25
a1	0.3846	0.3333	0.2308	0.4286
a2	0.1538	0.5000	0.3077	0.4286
a3	0.4615	0.1667	0.4615	0.1429

					Sum	Rank
a1	0.0769	0.0500	0.0923	0.1071	0.3264	2
a2	0.0592	0.0750	0.1231	0.1071	0.3360	1
a3	0.0710	0.0250	0.1846	0.0357	0.3376	3

FIGURE 10.1
Excel screenshot of simple additives weight method.

TABLE 10.6

Raw data

Cars	Cost ($000)	MPG City	MPG HW	Performance	Interior & Style	Safety	Reliability
Prius	27.8	44	40	7.5	8.7	9.4	3
Fusion	28.5	47	47	8.4	8.1	9.6	4
Volt	38.668	35	40	8.2	6.3	9.6	3
Camry	25.5	43	39	7.8	7.5	9.4	5
Sonata	27.5	36	40	7.6	8.3	9.6	5
Leaf	36.2	40	40	8.1	8.0	9.4	3

Example 10.3 Car Selection (data from Consumer's Report and US News and World Report on-line data)

We are considering six cars: Ford Fusion, Toyota Prius, Toyota Camry, Nissan Leaf, Chevy Volt, and Hyundai Sonata. For each car we have data on seven criteria that were extracted from Consumer's Report and US News and World Report data sources. They are *cost, mpg city, mpg highway, performance, interior & style, safety,* and *reliability*. We provide the extracted information in Table 10.6.

Initially, we might assume all weights are equal to obtain a baseline ranking. We substitute the rank orders (1st to 6th) for the actual data. We compute the average rank attempting to find the best ranking (smaller is better).We find our rank ordering is Fusion, Sonata, Camry, Prius, Volt, and Leaf.

We use pairwise comparisons to obtain weights. The weights are

Cost (0.3112), MPG City (0.1336), MPG Highway (0.0.0958), Performance (0.0551), Interior & Style (0.0499), Safety (0.1294), Reliability (0.2250)

We compute the expected value E[X] for each alternative.

	Expected values	Ranks
Prius	0.164133	4
Fusion	0.177761	3
Volt	0.142215	6
Camry	0.186879	1
Sonata	0.18009	2
Leaf	0.148922	5

We decide on the Camry as our best alternative since it has the largest expected value.

10.6 Technique of Order Preference by Similarity to the Ideal Solution (TOPSIS)

The principle behind TOPSIS is simple: The chosen alternative should be as close to the ideal solution as possible and as far from the negative-ideal solution as possible. The ideal solution is formed as a composite of the best performance values exhibited (in the decision matrix) by any alternative for each attribute. The negative-ideal solution is the composite of the worst performance values. Proximity to each of these performance poles is measured in the Euclidean sense (the square root of the sum of the squared distances along each axis in the "attribute space").

Introduction

TOPSIS was developed in 1980 by K. Yoon and H. Ching-Lai at Kansas State University. Their premise was that the alternative chosen should have the shortest distance from the ideal solution and the farthest distance from the negative-ideal solution. In the late 1980s the Department of Defense was using the TOPSIS program to rank order defense acquisitions across all branches for the POM and the budget cycle.

To begin with, assume we have m alternatives and n attributes. Further assume that either the decision-maker provides attributes weights or that we use something like Saaty's Analytical Hierarchy Process (AHP) to find the decision weights as applied to the attributes.

There are three key assumptions:

(1) Each attribute in the decision matrix takes on either monotonically increasing or monotonically decreasing utility.
(2) Weights for the attributes are either provided or can solved for with a procedure such as AHP.
(3) Any outcomes which is expressed in anon-numerical way, should be quantified through some appropriate scaling technique. Again Saaty's 9 point scale works well.

Step in TOPSIS

1. Create the *m x n* matrix with alternatives as rows and attributes as columns using the appropriate scaling techniques.
2. Transform to a normalized decision matrix which provided a non-dimensional value and allows comparison across attributes.
3. Multiply the decision weights for the attributes by each value in the column.
4. Determine for each column both the max and min values.
5. Calculate the distance measures:
6. Calculate the relative closeness metric to the ideal solution.
7. Rank order the results in descending order.

TOPSIS Methodology

The TOPSIS process is carried out as follows:

Step 1

Create an evaluation matrix consisting of *m* alternatives and *n* criteria, with the intersection of each alternative and criteria given as x_{ij}, giving us a matrix $(X_{ij})_{mxn}$.

$$
D = \begin{array}{c} \\ A_1 \\ A_2 \\ A_3 \\ \cdot \\ \cdot \\ \cdot \\ A_m \end{array}
\begin{array}{ccccc}
x_1 & x_2 & x_3 & \cdots & x_n \\
\left[\begin{array}{ccccc}
x_{11} & x_{12} & x_{13} & \cdots & x_{1n} \\
x_{21} & x_{22} & x_{23} & \cdots & x_{2n} \\
x_{31} & x_{32} & x_{33} & \cdots & x_{3n} \\
\cdot & \cdot & \cdot & & \cdot \\
\cdot & \cdot & \cdot & & \cdot \\
\cdot & \cdot & \cdot & & \cdot \\
x_{m1} & x_{m2} & x_{m3} & \cdots & x_{mn}
\end{array}\right]
\end{array}
$$

Step 2

The matrix shown as D above then normalized to form the matrix $R = (R_{ij})_{mxn}$, using the normalization method

$$
r_{ij} = \frac{x_{ij}}{\sqrt{\sum x_{ij}^2}}
$$

for $i = 1,2\ldots,m; j = 1,2,\ldots n.$

Step 3

Calculate the weighted normalized decision matrix. First we need the weights. Weights can come from either the decision-maker or by computation.

Step 3a

Use either the decision-maker's weights for the attributes $x_1, x_2 \ldots x_n$ or compute the weights through the use Saaty's (1980) AHP's decision-maker weights method to obtain the weights as the eigenvector to the attributes versus attribute pair-wise comparison matrix.

$$
\sum_{j=1}^{n} w_j = 1
$$

The sum of the weights over all attributes must equal 1 regardless of the method used.

Step 3b

Multiply the weights to each of the column entries in the matrix from *Step 2* to obtain the matrix, T.

$$
T = (t_{ij})_{mxn} = (w_j r_{ij})_{mxn}, i = 1, 2, \ldots, m
$$

Step 4

Determine the worst alternative (A_w) and the best alternative (A_b): Examine each attribute's column and select the largest and smallest values appropriately. If the values imply larger is better (profit) then the best alternatives are the largest values and if the values imply smaller is better (such as cost) then the best alternative is the smallest value.

$$A_w = \left\{ \left\langle \max(t_{ij} | i = 1, 2, \ldots, m \,|\, j \in J_-) \right\rangle, \left\langle \min(t_{ij} | i = 1, 2, \ldots, m) \,|\, j \in J_+ \right\rangle \right\}$$
$$\equiv \left\{ t_{wj} | j = 1, 2, \ldots, n \right\},$$

$$A_{wb} = \left\{ \left\langle \min(t_{ij} | i = 1, 2, \ldots, m \,|\, j \in J_-) \right\rangle, \left\langle \max(t_{ij} | i = 1, 2, \ldots, m) \,|\, j \in J_+ \right\rangle \right\}$$
$$\equiv \left\{ t_{bj} | j = 1, 2, \ldots, n \right\},$$

where,

$J_+ = \{j = 1, 2, \ldots n | j)$ associated with the criteria having a positive impact, and
$J_- = \{j = 1, 2, \ldots n | j)$ associated with the criteria having a negative impact.

We suggest that if possible make all entry values in terms of positive impacts.

Step 5

Calculate the L2-distance between the target alternative i and the worst condition A_w

$$d_{iw} = \sqrt{\sum_{j=1}^{n} (t_{ij} - t_{wj})^2}, i = 1, 2, \ldots m$$

and the distance between the alternative i and the best condition A_b

$$d_{ib} = \sqrt{\sum_{j=1}^{n} (t_{ij} - t_{bj})^2}, i = 1, 2, \ldots m$$

where d_{iw} and d_{ib} are L2-norm distances from the target alternative i to the worst and best conditions, respectively.

Step 6

Calculate the similarity to the worst condition:

$$s_{iw} = \frac{d_{iw}}{(d_{iw} + d_{ib})}, 0 \le s_{iw} \le 1, i = 1, 2, \ldots, m$$

$S_{iw} = 1$ if and only if the alternative solution has the worst condition; and
$S_{iw} = 0$ if and only if the alternative solution has the best condition.

Step 7

Rank the alternatives according to their value from $S_{iw}(i = 1,2,...,m)$.

Normalization

Two methods of normalization that have been used to deal with incongruous criteria dimensions are linear normalization and vector normalization.

Linear normalization can be calculated as in *Step 2* of the TOPSIS process above. Vector normalization was incorporated with the original development of the TOPSIS method (Hwang et al., 1987), and is calculated using the following formula:

$$r_{ij} = \frac{x_{ij}}{\sqrt{\sum x_{ij}^2}}$$

for $i = 1,2...,m; j = 1,2,...n.$

In using vector normalization, the nonlinear distances between single dimension scores and ratios should produce smoother trade-offs (Huang et al., 2011).

Example 10.4 Selecting a New Car

We want to choose from among four cars: Civic, Saturn, Ford, and Mazda. We considered four criteria: body style, reliability, fuel economy, and cost.

Step 1. Obtain the decision weights. Choose any method. We use the AHP method to obtain the weights as 0.054931, 0.4144559, 0.408604, 0.121906 as shown below (Figure 10.2).

Step 2. We create a matrix of the alternatives rated in each criterion. We used the 1–9 scale as before (Figure 10.3).

Step 3. Square the entries, sum and take square root to the totals (Figure 10.4).

Step 4. Normalize the entries in Step 2 by dividing by the square root of the column total (Figure 10.5).

Step 5. Multiply all entries in the normalized matrix by the criterion weights found in step 1 (Figure 10.6).

Style	Reliability	Fuel	Cost
1	0.1667	0.1111	0.3333
6	1.0000	1.0000	4.0000
9	1.0000	1.0000	3.0000
3	0.2500	0.3333	1.0000
19	2.4167	2.4444	8.3333
0.0526	0.0690	0.0455	0.04
0.3158	0.4138	0.4091	0.48
0.4737	0.4138	0.4091	0.36
0.1579	0.1034	0.1364	0.12

Style	Reliability	Fuel	Cost
0	1	0	0
0.0690	0.4138	0.4138	0.1034
0.0551	0.4119	0.4104	0.1225
0.0549	0.4146	0.4086	0.1220
0.0549	0.4146	0.4086	0.1219
0.0549	0.4146	0.4086	0.1219
0.0549	0.4146	0.4086	0.1219
0.0549	0.4146	0.4086	0.1219
0.0549	0.4146	0.4086	0.1219
0.0549	0.4146	0.4086	0.1219
0.0549	0.4146	0.4086	0.1219
0.0549	0.4146	0.4086	0.1219

FIGURE 10.2
Excel screenshot of Step 1 AHP method.

		Criterion		
Cars	Style	Reliability	Fuel Econ	Cost
Civic	7	9	9	8
Saturn	8	7	8	7
Ford	9	6	8	9
Mazda	6	7	8	6

FIGURE 10.3
Excel screenshot of Step 2 AHP method.

Step 6. Pick the largest and smallest value in each column. We use max(column values) and min(column values) to obtain (Figure 10.7):

Step 7. Compute the separation from the max ideal (Figure 10.8).

$$(Value\text{-}maximum\ value)^2.$$

	49	81	81	64
	64	49	64	49
	81	36	64	81
	36	49	64	36
	230	215	273	230
	15.16575089	14.6628783	16.5227116	15.16575089

FIGURE 10.4
Excel screenshot of Step 3 AHP method.

0.461566331	0.613794906	0.54470478	0.527504379
0.527504379	0.477396038	0.48418203	0.461566331
0.593442426	0.409196604	0.48418203	0.593442426
0.395628284	0.477396038	0.48418203	0.395628284

FIGURE 10.5
Excel screenshot of Step 4 AHP method.

Multiply by weights.			
Assume weights ate			
0.054930536	0.414559405	0.40860391	0.121906151
0.025354086	0.254454451	0.2225685	0.064306028
0.028976098	0.197909017	0.19783867	0.056267775
0.03259811	0.1696363	0.19783867	0.072344282
0.021732074	0.197909017	0.19783867	0.048229521

FIGURE 10.6
Excel screenshot of Step 5 AHP method.

Pick largest and smallest in each column			
0.03259811	0.254454451	0.2225685	0.072344282
0.021732074	0.1696363	0.19783867	0.048229521

FIGURE 10.7
Excel screenshot of Step 6 AHP method.

Separation from max ideal.

5.24759E-05	0	0	0.000258454	0.000311	0.017633
1.3119E-05	0.003197386	0.00061156	6.46135E-05	0.003887	0.062343
0	0.007194119	0.00061156	0.000581522	0.008387	0.091582
0.000118071	0.003197386	0.00061156	0	0.003927	0.062666

FIGURE 10.8
Excel screenshot of Step 7 AHP method.

Separation from minimum ideal

1.3119E-05	0.007194119	0.00061156	6.46135E-05	0.007883	0.088789
5.24759E-05	0.000799347	0	0.000258454	0.00111	0.033321
0.000118071	0	0	0	0.000118	0.010866
0	0.000799347	0	0.000581522	0.001381	0.03716

FIGURE 10.9
Excel screenshot of Step 8 AHP method.

Relative Closeness	
	0.834308335
	0.348310581
	0.106064211
	0.372248112

FIGURE 10.10
Excel screenshot of Step 9 AHP method.

Step 8. Compute the separation from the min ideal (Figure 10.9).

$$(Value-minimum\ value)^2.$$

Step 9. Calculate the relative closeness to the ideal solution $C_i^* = S'_i / (S_i^* + S'_i)$ (Figure 10.10)

Civic: 0.8343
Saturn: 0.3483
Ford: 0.1066
Mazda: 0.3722

Our ranking have Civic 1st, followed by the Mazda, the Saturn, and the Ford.

10.7 Modeling of Ranking Units Using Data Envelopment Analysis (DEA) with Linear Programming

Introduction

Data envelopment analysis (DEA), occasionally called frontier analysis, was first put forward by Charnes, Cooper, and Rhodes in 1978. It is a performance measurement technique which, as we shall see, can be used for evaluating the *relative efficiency* of *decision-making units* (*DMUs*) in organizations. Here a DMU is a distinct unit within an organization that has flexibility with respect to some of the decisions it makes, but not necessarily complete freedom with respect to these decisions.

Examples of such units to which DEA has been applied are: banks, police stations, hospitals, tax offices, prisons, defense bases (army, navy, air force), schools, and university departments. Note here that one advantage of DEA is that it can be applied to non-profit making organizations.

Since the technique was first proposed much theoretical and empirical work has been done. Many studies have been published dealing with applying DEA in real-world situations. Obviously there are many more unpublished studies, e.g. done internally by companies or by external consultants.

We will initially illustrate DEA by means of a small example. More about DEA can be found on line using Google: "Data Envelopment Analysis". Note here that much of what you will see below is a graphical (pictorial) approach to DEA. This is very useful if you are attempting to explain DEA to those less technically qualified (such as many you might meet in the military or management world). There is a mathematical approach to DEA that can be adopted however. We will present the single measure first to demonstrate the idea and then move to multiple measures and use linear programming methodology from our course.

Example 10.5 Banks

Consider a number of bank branches. For each branch we have a single output measure (number of personal transactions completed) and a single input measure (number of staff).

The data (Table 10.7) we have is as follows:

For example, for the Branch 2 in one year, there were 44,000 transactions relating to personal accounts and 16 staff members were employed.

How then can we compare these branches and measure their performance using this data?

TABLE 10.7

Bank transactions and staff

Branch	Personal transactions ('000s)	Number of staff
1	125	18
2	44	16
3	80	17
4	23	11

TABLE 10.8

Personal transactions

Branch	Personal transactions ('000s)
1	6.94
2	2.75
3	4.71
4	2.09

Ratios

A commonly used method is *ratios*. Typically we take some output measure and divide it by some input measure. Note the terminology here, we view branches as taking *inputs* and converting them (with varying degrees of efficiency, as we shall see below) into *outputs*.

For our bank branch example, we have a single input measure, the number of staff, and a single output measure, the number of personal transactions. Hence we have (Table 10.8):

Here we can see that Branch1 has the highest ratio of personal transactions per staff member, whereas Branch 4 has the lowest ratio of personal transactions per staff member.

As Branch 1 has the highest ratio of 6.94 we can compare all other branches to it and calculate their *relative efficiency* with respect to Branch 1. To do this we divide the ratio for any branch by 6.94 (the value for Croydon) and multiply by 100 to convert to a percentage. This gives (Table 10.9):

The other branches do not compare well with Branch 1, so are presumably performing less well. That is, they are relatively less efficient at using their given input resource (staff members) to produce output (number of personal transactions).

We could, if we wish, use this comparison with Branch 1 to set *targets* for the other branches.

For example, we could set a target for Branch 4 of continuing to process the same level of output but with one less member of staff. This is an example of an ***input target*** as it deals with an input measure.

An example of an ***output target*** would be for Branch 4 to increase the number of personal transactions by 10% (e.g. by obtaining new accounts).

Plainly, in practice, we might well set a branch a mix of input and output targets which we want it to achieve. We can use linear programming.

Illustrative Examples

Example 10.6 Bank Example with Linear Programming

Typically we have more than one input and one output. For the bank branch example suppose now that we have two output measures (number of personal transactions completed and number of business transactions completed) and the same single input measure (number of staff) as before.

The data (Table 10.10) we have is as follows:

We start be scaling (via ratios) the inputs and outputs to reflect the ratio of 1 unit (Table 10.11).

Pick a DMU to maximize: *E1, E2, E3,* or *E4*

TABLE 10.9

Relative efficiency

Branch	Relative efficiency
1	100 (6.94/6.94) = 100%
2	100 (2.75/6.94) = 40%
3	100 (4.71/6.94) = 68%
4	100 (2.09/6.94) = 30%

TABLE 10.10

Bank Example 10.6 for linear programming

Branch	Personal transactions ('000s)	Business transactions ('000s)	Number of staff
1	125	50	18
2	44	20	16
3	80	55	17
4	23	12	11

TABLE 10.11

Bank Example 10.6 scaling

Branch	Personal Transactions ('000s)	Business Transactions ('000s)	Number of Staff
1	125/18 = 6.94	50/18 = 2.78	18/18 = 1
2	44/16 = 2.75	20/16 = 1.25	16/16 = 1
3	80/17 = 4.71	55/17 = 3.24	17/17 = 1
4	23/11 = 2.09	12/11 = 109	11/11 = 1

Let $W1$ and $W2$ be the personal and business transactions at each branch. In this example, we choose to maximize branch one, $E1$.
The LP formulation is:

$$\text{Maximize } E1$$

Subject to

$$E1 = 6.94\ W1 + 2.78\ W2$$

$$E2 = 2.75\ W1 + 1.25\ W2$$

$$E3 = 4.71\ W1 + 3.24\ W2$$

$$E4 = 2.09\ W1 + 1.09\ W2$$

$$E1 \leq 1$$

$$E2 \leq 1$$

$$E3 \leq 1$$

$$E4 \leq 1$$

$$Non\text{-}negativity$$

Here is the Excel output (Figure 10.11):

maximize E1			E1	1
	0.431809		E2	0.431809
			E3	1
			E4	0.361777
			w1	0.04877
subject to			w2	0.238152
	1	1	w3	0
	0.431809	0.431809		
	1	1		
	0.361777	0.361777		
	1	1		
	0.431809	1		
	1	1		
	0.361777	1		

FIGURE 10.11

Excel screenshot of LP formulation.

Microsoft Excel 11.0 Answer Report
Worksheet: [DEA example for class.xls]Sheet1
Report Created: 5/13/2009 12:41:28 PM

Target Cell (Max)

Cell	Name	Original Value	Final Value
$M$4	w2	0	0.431808681

Adjustable Cells

Cell	Name	Original Value	Final Value
$P$3	E1	0	1
$P$4	E2	0	0.431808681
$P$5	E3	0	1
$P$6	E4	0	0.361776759
$P$7	w1	0	0.048770292
$P$8	w2	0	0.238152303
$P$9	w3	0	0

Constraints

Cell	Name	Cell Value	Formula	Status	Slack
$M$9		1	$M$9 = $N$9	Not Binding	0
$M$10		0.431808681	$M$10 = $N$10	Not Binding	0
$M$11		1	$M$11 = $N$11	Not Binding	0
$M$12		0.361776759	$M$12 = $N$12	Not Binding	0
$M$13		1	$M$13< = 1	Binding	0
$M$14		0.431808681	$M$14< = 1	Not Binding	0.568191319
$M$15		1	$M$15< = 1	Binding	0
$M$16		0.361776759	$M$16< = 1	Not Binding	0.638223241

FIGURE 10.12

Excel screenshot of LP formulation solution.

The answer report from Excel (Figure 10.12):

The sensitivity report from Excel (Figure 10.13):

Now, what did we learn from this? If we rank-ordered the branches on efficiency performance of our inputs and outputs, we find

Branch 1	100%
Branch 3	100%
Branch 2	43.2%
Branch 4	36.2%

We know we need to improve on Branch 2 and Branch 4 performance while not losing our efficiency in Branches 1 and 3. A better interpretation

Microsoft Excel 11.0 Sensitivity Report
Worksheet: [DEA example for class.xls]Sheet1
Report Created: 5/13/2009 12:41:28 PM

Adjustable Cells

	Cell	Name	Final Value	Reduced Cost	Objective Coefficient	Allowable Increase	Allowable Decrease
	$P$3	E1	1	0	0	1E+30	0.321361388
1.1	$P$4	E2	0.431808681	0	1	1E+30	1
1.2	$P$5	E3	1	0	0	1E+30	0.110447293
1.3	$P$6	E4	0.361776759	0	0	1E+30	0.587754985
1.4	$P$7	w1	0.048770292	0	0	0.373	0.931818182
1.5	$P$8	w2	0.238152303	0	0	0.640625	0.149295549
1.6	$P$9	w3	0	0	0	0	1E+30
1.7	1.8	1.9	1.10	1.11	1.12	1.13	1.14

Constraints

	Cell	Name	Final Value	Shadow Price	Constraint R.H. Side	Allowable Increase	Allowable Decrease
	$M$9		1	0.321361388	0	0.47475	0.141414141
1.15	$M$10		0.431808681	-1	0	0.431808681	0.568191319
1.16	$M$11		1	0.11044723	0	0.164705882	0.321918969
1.17	$M$12		0.361776759	0	0	0.361776759	0.638223241
1.18	$M$13		1	0.321361388	1	0.47475	0.141414141
1.19	$M$14		0.431808681	0	1	1E+30	0.568191319
1.20	$M$15		1	0.110447293	1	0.164705882	0.321918969
1.21	$M$16		0.361776759	0	1	1E+30	0.638223241
1.22	1.23	1.24	1.25	1.26	1.27	1.28	1.29

FIGURE 10.13

Excel screenshot of LP formulation sensitivity analysis.

could be that the practices and procedures used by the other branches were to be adopted by Branch 4, they could improve their performance.

For multiple inputs and outputs, we recommend the formulations by Winston (1995) in Equation (10.3).

For any DMU_0, let X_i be the inputs and Y_i be the outputs. Let X_0 and Y_0 be the DMU being modeled.

$$Min \ \theta$$

Subject to

$$\sum \lambda_i X_i \leq \theta X_0 \qquad (10.3)$$
$$\sum \lambda_i Y_i \leq Y_0$$
$$\lambda i \geq 0$$

Non-negativity

Example 10.7 Manufacturing

Consider the following process (Trick, Chapter 12; Winston, 1995) where we have three DMUs where each has 2 inputs and 3 outputs as shown in the data table.

DMU	Input #1	Input #2	Output #1	Output #2	Output #3
1	5	14	9	4	16
2	8	15	5	7	10
3	7	12	4	9	13

We define the following decision variables:

t_i = value of a single unit of output of *DMU i*, for $i = 1,2,3$
w_i = cost or weights for one unit of inputs of *DMU i*, for $i = 1,2$
$efficiency_i$ = (total value of i's outputs)/(total cost of i's inputs), for $i = 1,2,3$.

The following modeling assumptions are made:

a. No unit will have an efficiency more than 100%.

b. If any efficiency is less than 1, then it is inefficient.

c. We should scale the costs as the costs of the inputs equals 1 for each linear program. For example, we will use $5w_1 + 14w_2 = 1$ in our program for DMU 1.

d. All values and weights must be strictly positive, so we use a constant such as 0.0001 in lieu of 0.

To calculate the efficiency of unit 1, we define the linear program

$$Maximize\ 9t_1 + 4t_2 + 16t_3$$

Subject to

$$-9t_1 - 4t_2 - 16t_3 + 5w_1 + 14w_2 \geq 0$$
$$-5t_1 - 7t_2 - 10t_3 + 8w_1 + 15w_2 \geq 0$$
$$-4t_1 - 9t_2 - 13t_3 + 7w_1 + 12w_2 \geq 0$$
$$5w_1 + 14w_2 = 1$$
$$t_i \geq 0.0001, i = 1,2,3$$
$$w_i \geq 0.0001, i = 1,2$$
$$Non\text{-}negativity$$

To calculate the efficiency of unit 2, we define the linear program as

$$Maximize\ 5t_1 + 7t_2 + 10t_3$$

Subject to

$$-9t_1 - 4t_2 - 16t_3 + 5w_1 + 14w_2 \geq 0$$
$$-5t_1 - 7t_2 - 10t_3 + 8w_1 + 15w_2 \geq 0$$
$$-4t_1 - 9t_2 - 13t_3 + 7w_1 + 12w_2 \geq 0$$
$$8w_1 + 15w_2 = 1$$
$$t_i \geq 0.0001, i = 1,2,3$$
$$w_i \geq 0.0001, i = 1,2$$

Non-negativity

To calculate the efficiency of unit 3, we define the linear program as

$$\text{Maximize } 4t_1 + 9t_2 + 13t_3$$

Subject to

$$-9t_1 - 4t_2 - 16t_3 + 5w_1 + 14w_2 \geq 0$$
$$-5t_1 - 7t_2 - 10t_3 + 8w_1 + 15w_2 \geq 0$$
$$-4t_1 - 9t_2 - 13t_3 + 7w_1 + 12w_2 \geq 0$$
$$7w_1 + 12w_2 = 1$$
$$t_i \geq 0.0001, i = 1,2,3$$
$$w_i \geq 0.0001, i = 1,2$$

Non-negativity

The linear programming solutions show the efficiencies as $u_1 = u_3 = i$, $u_2 = 0.77303$.

   Interpretation: u_2 is operating at 77.303% of the efficiency of u_1 and u_3. You could concentrate some improvements or best practices from u_1 or u_3 for u_2. An examination of the dual prices for the linear program of DMU 2 yields $\lambda_1 = 0.261538$, $\lambda_2 = 0$, and $\lambda_3 = 0.661538$. The average output vector for DMU 2 can be written as:

$$0.261538 \begin{bmatrix} 9 \\ 4 \\ 16 \end{bmatrix} + 0.661538 \begin{bmatrix} 4 \\ 9 \\ 13 \end{bmatrix} = \begin{bmatrix} 5 \\ 7 \\ 12.785 \end{bmatrix}$$

And the average input vector can be written as

$$0.261538 \begin{bmatrix} 5 \\ 14 \end{bmatrix} + 0.661538 \begin{bmatrix} 7 \\ 12 \end{bmatrix} = \begin{bmatrix} 5.938 \\ 11.6 \end{bmatrix}$$

We may clearly see the inefficiency is output 3 where $12.785 - 10 = 2.785$ more of output 3. This helps focus on treating the inefficiency found for units.

Sensitivity Analysis: This is also called "what if" analysis. So let's assume that without management engaging some training for u_2 that u_2 production dips of output #2 dips 9 units of output while the input hours increases from 15 to 16 hours.

We find that changes in technology coefficients are easily handled in resolving the LPs.

Since u_2 is affected, we might only modify and solve the LP concerning u_2. We find with these changes that u_2 is now only 74% as effective as u_1 and u_3.

10.8 Technology for Multi-Attribute Decision-Making

10.8.1 Excel and Multi-Attribute Decision-Making

We illustrated our example is using Excel. We have created used friendly templates for SAW and TOPSIS with all the weighting methods. They are available upon request or from our website.

We illustrate the template in Figure 10.14 and Figure 10.15 and will cover in more detail in Chapter 11.

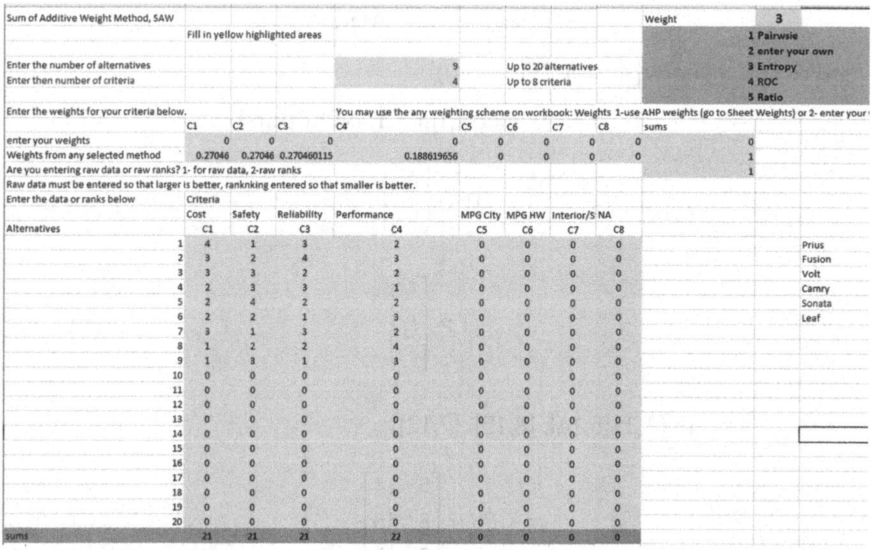

FIGURE 10.14
Screenshot of SAW template.

weights used		0.27046	0.27046	0.270460115		0.188619656	0	0	0	0	sums of alternative	Rank	
	1	0.051516	0.012879	0.038637159		0.017147241	0	0	0	0	0.120179666	2	1
	2	0.038637	0.025758	0.051516212		0.025720862	0	0	0	0	0.14163234	1	2
	3	0.038637	0.038637	0.025758106		0.017147241	0	0	0	0	0.120179666	2	3
	4	0.025758	0.038637	0.038637159		0.008573621	0	0	0	0	0.111606045	5	4
	5	0.025758	0.051516	0.025758106		0.017147241	0	0	0	0	0.120179666	2	5
	6	0.025758	0.025758	0.012879053		0.025720862	0	0	0	0	0.090116128	8	6
	7	0.038637	0.012879	0.038637159		0.017147241	0	0	0	0	0.107300613	6	7
	8	0.012879	0.025758	0.025758106		0.034294483	0	0	0	0	0.098689748	7	8
	9	0.012879	0.038637	0.012879053		0.025720862	0	0	0	0	0.090116128	8	9

FIGURE 10.15
Screenshot of results.

10.8.2 Maple and Multi-Attribute Decision-Making

We may use the template that we wrote provided in Maple to solve this problem.

DEA Analysis with INPUTS and Outputs

   (1) Enter number of dmu(s), inputs, outputs
   (2) Enter input matrix as IM and output matrix as OutM

>ndmu: = 3; dmu: = ndmu;

$$ndmu: = 3$$

$$dmu: = 3$$

> NInputs: = 2;

$$NInputs: = 2;$$

>NOutputs: = 3;

$$NOutputs: = 3*400$$

>W: = '<,>(seq(w[i],i = 1..NInputs));

$$T: = '<,>(seq(t[i],i = 1..NOutputs));$$

$$W := \begin{bmatrix} w_1 \\ w_2 \end{bmatrix}$$

$$T := \begin{bmatrix} t_1 \\ t_2 \\ t_3 \end{bmatrix}$$

> IM: = Matrix (3,2,[[5,14], [8,15], [7,12]]);

$$IM := \begin{bmatrix} 5 & 14 \\ 8 & 15 \\ 7 & 12 \end{bmatrix}$$

>OutM: = Matrix (3,3,[[9,4,16], [5,7,10], [4,9,13]]);

$$OutM := \begin{bmatrix} 9 & 4 & 16 \\ 5 & 7 & 10 \\ 4 & 9 & 13 \end{bmatrix}$$

Putting together the constraints and objective functions

> INM: = IM.W;

$$INM := \begin{bmatrix} 5w_1 + 14w_2 \\ 8w_1 + 15w_2 \\ 7w_1 + 12w_2 \end{bmatrix}$$

> OMC: = OutM.T;

$$OMC := \begin{bmatrix} 9t_1 + 4t_2 + 16t_3 \\ 5t_1 + 7t_2 + 10t_3 \\ 4t_1 + 9t_2 + 13t_3 \end{bmatrix}$$

> cl: = seq(((IM.W)[jj] − (OMC[jj]))≥0,jj = 1.. ndmu);

$$cl := 0 \le 5w_1 + 14w_2 - 9t_1 - 4t_2 - 16t_3, 0 \le 8w_1 + 15w_2 - 5t_1$$
$$- 7t_2 - 10t_3, 0 \le 7w_1 + 12w_2 - 4t_1 - 9t_2 - 13t_3$$

> obj: = seq(OMC[ii], ii = 1..ndmu);

$$obj := 9t_1 + 4t_2 + 16t_3, 5t_1 + 7t_2 + 10t_3, 4t_1 + 9t_2 + 13t_3$$

> c: = seq(INM[jjj], jjj = 1..ndmu);

$$c := 5w_1 + 14w_2, 8w_1 + 15w_2, 7w_1 + 12w_2$$

Solutions

>print(solutions): **forkk from 1 to** ndmu **do**

s: = LPSolve(obj[kk], {cl, c[kk] = 1}, assume = nonnegative, maximize); printf ("%g", kk, s); **end do;**

solutions

$$s := [1.00000000000000, [t_1 = 0., t_2 = 0., t_3 = 0.0625000000000000, w_1 = 0., w_2 = 0.0714285714285714]]$$

1

$$s := [0.773333333333333, [t_1 = 0.800000000000000, t_2$$
$$= 0.0533333333333333, t_3 = 0., w_1 = 0., w_2$$
$$= 0.0666666666666667]]$$

2

$$s := [1., [t_1 = 0., t_2 = 0.0090579710144278, t_3$$
$$= 0.0706521739130435, w_1 = 0., w_2 = 0.0833333333333333]]$$

3

Finding the Shadow Prices

SHADOW PRICES

>

>obj[2];

$$5t_1 + 7t_2 + 10t_3$$

>cl;

$$0 \le 5w_1 + 14w_2 - 9t_1 - 4t_2 - 16t_3, 0 \le 8w_1 + 15w_2 - 5t_1 - 7t_2$$
$$- 10t_3, 0 \le 7w_1 + 12w_2 - 4t_1 - 9t_2 - 13t_3$$

>

> with(simplex):

> d: = dual(obj[2], {c[2]> = 1,c[2]< = 1,c1}, z);

$d; = -z1 + z2, \{0 \le -15z1 + 15z2 - 15z3 - 12z4 - 14z5, 0 \le$
$- 8z1 + 8z2 - 8z3 - 7z4 - 5z5, 5 \le 5z3 + 4z4 + 9z5, 7 \le 7z3$
$+ 9z4 + 4z5, 10 \le 10z3 + 13z4 + 16z5$

>LPSolved(d, assume = nonnegative);

$[0.773333333264493, [z1 = 0., z2 = 0.773333333264493, z3 = 0., z4$
$= 0.661538461538462, z5 = 0.261538461538461]]$

Maple TOPSIS and SAW
SAW

    SAW: = **proc** *(A, C, RA, RC, k, Dweights, PCM, AltM)*

    local *v1, v2, v3, v4, v5, v6, v7, v8, s1, s2, s3, s4, s5, s6, s7, s8, sp1, sp2, sp3, sp4, sp5, sp6, sp7, sp8:*

    local *vp1, vp2, vp3, vp4, vp5, vp6, vp7, vp8:*

    v1: = Vector [column](1..A, [seq(AltM[i,1],i = 1..A)]):

    v2: = Vector [column](1..A, [seq(AltM[i,2],i = 1..A)]):

    v3: = Vector [column](1..A, [seq(AltM[i,3],i = 1..A)]):

    v4: = Vector [column](1..A, [seq(AltM[i,4],i = 1..A)]):

    v5: = Vector [column](1..A, [seq(AltM[i,5],i = 1..A)]):

    v6: = Vector [column](1..A, [seq(AltM[i,6],i = 1..A)]):

    v7: = Vector [column](1..A, [seq(AltM[i,7],i = 1..A)]):

    v8: = Vector [column](1..A, [seq(AltM[i,8],i = 1..A)]):

    s1: = add(v1[i],i = 1..A): s2: = add(v2[i],i = 1..A):

    s3: = add(v3[i],i = 1..A): s4: = add(v4[i],i = 1..A):

    s5: = add(v5[i],i = 1..A): s6: = add(v6[i],i = 1..A):

    s7: = add(v7[i],i = 1..A): s8: = add(v8[i],i = 1..A):

    if *s1 = 0* **then** *s1 = 1* **end if;ifs2** *= 0* **then** *s2 = 1;* **end if; if** *s3 = 0* **then** *s3 = 1*

    end if;ifs4 *= 0* **then** *s4 = 1* **end if;ifs5** *= 0* **then** *s5 = 1;* **end if;**

    if *s6 = 0* **then** *s6 = 1:* **end if;ifs7** *= 0* **then** *s7 = 1;* **end if;ifs8** *= 0* **then** *s8 = 1;* **end if;**

    vp1: = Vector [column](1..A, [seq(PCM[i,1],i = 1..A)]):vp2 : = Vector [column](1..A, [seq(PCM[i,1],i = 2..A)]):

    vp3: = Vector [column](1..A, [seq(PCM[i,1],i = 3..A)]):vp4 : = Vector [column](1..A, [seq(PCM[i,1],i = 4..A)]):

    vp5: = Vector [column](1..A, [seq(PCM[i,1],i = 5..A)]):vp6 : = Vector [column](1..A, [seq(PCM[i,1],i = 6..A)]):

    vp7: = Vector [column](1..A, [seq(PCM[i,1],i = 7..A)]):vp8 : = Vector [column](1..A, [seq(PCM[i,1],i = 8..A)]):

    sp1: = add(vp1[i],i = 1..A):

$sp2: = add(vp2[i], i = 1..A): sp3: = add(vp3[i], i = 1..A): sp4$
$: = add(vp4[i], i = 1..A):$

$sp5: = add(vp5[i], i = 1..A): sp6: = add(vp6[i], i = 1..A):sp7$
$: = add(vp7[i], i = 1..A)$
$: sp8: = add(vp8[i], i = 1..A):$

$$vsp1 := evalf\left(Multiply\left(vp1, \frac{1}{sp1}\right)\right): vsp2 := evalf\left(Multiply\left(vp2, \frac{1}{sp2}\right)\right):$$

$$vsp3 := evalf\left(Multiply\left(vp3, \frac{1}{sp3}\right)\right): vsp4 := evalf\left(Multiply\left(vp4, \frac{1}{sp4}\right)\right):$$

$$vsp5 := evalf\left(Multiply\left(vp5, \frac{1}{sp5}\right)\right): vsp6 := evalf\left(Multiply\left(vp6\frac{1}{sp6}\right)\right):$$

$$vsp7 := evalf\left(Multiply\left(vp7, \frac{1}{sp7}\right)\right): vsp8 := evalf\left(Multiply\left(vp8, \frac{1}{sp8}\right)\right):$$

$DM: = Matrix(A,C,[vsp1, vsp2, vsp3, vsp4 vsp5, vsp6, vsp7, vsp8]):$
$DMW: = DM^{10};$

$DMWeights: = Transpose (DMW): DMWV: = Vector[row](1..C,$
$[seq(DMWeights[1,i], i = 1..C)]):$

$Numera: = PCM.DMWV:numerb: = DMWV.numera:DMWVC$
$: = Vector[column](DMWV):$

$$Den := DMWV.DMWVC : lamnda := \frac{numerb}{den} : CI := \frac{(lambda - RC)}{(RC - 1)} :$$

$RI: = Vector([0,0,.52,.89, 1.1, 1.24, 1.35, 1.4, 1.45, 1.49]): CR$

$$:= \frac{CI}{RI(RC)} :$$

$DMWR: = Vector [row] (DMWV):$

$$vs1 := evalf\left(Multiply\left(v1, \frac{DMWV[1]}{s1}\right)\right):$$

$$vs2 := evalf\left(Multiply\left(v2, \frac{DMWV[2]}{s2}\right)\right):$$

$$vs3 := evalf\left(Multiply\left(v3, \frac{DMWV[3]}{s3}\right)\right):$$

$$vs4 := evalf\left(Multiply\left(v4, \frac{DMWV[4]}{s4}\right)\right):$$

$$vs5 := evalf\left(Multiply\left(v5, \frac{DMWV[5]}{s5} \right) \right):$$

$$vs6 := evalf\left(Multiply\left(v6, \frac{DMWV[6]}{s6} \right) \right):$$

$$vs7 := evalf\left(Multiply\left(v7, \frac{DMWV[7]}{s7} \right) \right):$$

$$vs8 := evalf\left(Multiply\left(v8, \frac{DMWV[8]}{s8} \right) \right):$$

DMR: = Matrix9A, C, [vs1,vs2,vs3,vs4,vs5,vs6,vs7,vs8]):

DMR1: = Transpose(DMR):

r1: = add(DMR1(i,1), i = 1..C): r2: = add(DMR1(i,2), i = 1..C):
r3: = add(DMR1(i,3), i = 1..C):r4: = add(DMR1(I,4), i = 1..C):
r5: = add(DMR1(i,5), i = 1..C):
r6: = add(DMR1(i,6), i = 1..C):
r7: = add(DMR1(i,7), i = 1..C):
r8: = add(DMR1(i,8), i = 1..C):
ranks: = Vector([r1, r2, r3, r4, r5, r6, r7, r8]);
print("Consistency ratio = ", CR);
print("Decision Weights are"DMWV, "Ranks by SAW are"ranks);
end;

Example 10.8 Cars Using SAW

=

>*SAW(8,8,6,6,1,Dweights,PCM,AltM);*

"Consistency ration = ",0.01907648081

"Decision weights are"[0.347448273536262 0.334526073040970 0.131896506942109 0.100435989552961 0.0477395587995010 0.0379535991686066 0.0.],"Ranks are"

$$\begin{vmatrix} 0.1401711563694767 \\ 0.105036380080595 \\ 0.2381717460459063 \\ 0.157711383739355 \\ 0.161315773138818 \\ 0.197593439927820 \\ 0. \\ 0. \end{vmatrix} =$$

>

=

>

Next, we apply a scheme to the weights and still use the ranks 1–6 as before. Perhaps we apply a technique similar to the pairwise comparison that we will discuss in the AHP section 4. Using the pairwise comparison to obtain new weights, we obtain a new ordering:

Camry, Sonata, Fusion, Prius, Leaf, and Volt. The changes in results of the rank ordering differ from using equal weights shows the sensitivity that the model has to the given criteria weights. We assume the criteria in order of importance are: cost, reliability, MPG City, safety, MPG HW, performance, interior,and style.

TOPSIS

Example 10.9 Car Selection Revisited Using TOPSIS

We might assume that our decision-maker weights from the AHP section are still valid for our use so we can use the same PCM matrix as before. We must place the input data for the alternatives in the same order as the prioritized criteria. The input data look like this 8 x 8 matrix:

$$AltM := \begin{vmatrix} 27.8 & 9.4 & 3 & 7.5 & 44 & 40 & 8.7 & 0 \\ 28.5 & 9.6 & 4 & 8.4 & 47 & 47 & 8.1 & 0 \\ 38.668 & 9.6 & 3 & 8.2 & 35 & 40 & 6.3 & 0 \\ 25.5 & 9.4 & 5 & 7.8 & 43 & 39 & 7.5 & 0 \\ 27.5 & 9.6 & 5 & 7.6 & 36 & 40 & 8.3 & 0 \\ 36.2 & 9.4 & 3 & 8.1 & 40 & 40 & 8 & 0 \\ 0 & 0 & 0 & 0 & 0 & 0 & 0 & 0 \\ 0 & 0 & 0 & 0 & 0 & 0 & 0 & 0 \end{vmatrix}$$

As before our criteria weights are:

CR: = 0.02149944074

DMWR: = [0.361233126487974, 0.209324398363431,
　　　　0.144589995206757, 0.116672945567780, 0.0801478041928341,
　　　　0.0529870560310582, 0.0350446742839783,0.]

and the CR is less than 0.1.

　　>part1weights (k, R, A, RA, RC, AltM, PCM, Dweights, lsize);
　　"CR and weights = ", 0.02149944074, [0.361233126487974,
　　　　0.209324398363431, 0.144589995206757, 0.116672945567780,
　　　　0.0801478041928341, 0.0529870560310582,
　　　　0.0350446742839783, 0.]

The TOPSIS values for our alternatives:

$$Ranks := \begin{vmatrix} 0.614967288202874 \\ 0.715258858184150 \\ 0.0614818245926484 \\ 0.908784041886248 \\ 0.810338906822239 \\ 0.182468079847254 \end{vmatrix}$$

The order of alternatives are: Camry (0.9087), Sonata (0.8103), Fusion (0.7152), Prius (0.6149), Leaf (0.1824), and Volt (0.06148).

10.8.3 R and Multi-Attribute Decision-Making

Example 10.10 We revisit Example 10.2 in Section 10.5 using R

```
w = matrix(c(.2,.15,.4,.25),ncol = 1)
> A = matrix(c(25,20,15,30,10,30,20,30,30,10,30,10),ncol = 4,nrow = 3)
> A
```

```
[,1] [,2] [,3] [,4]
[1,] 25 30 20 10
[2,] 20 10 30 30
[3,] 15 30 30 10
```

```
> A = matrix(c(25,10,30,20,30,10,15,20,30,30,30,10),ncol = 4,nrow = 3)
> A
```

```
[,1] [,2] [,3] [,4]
[1,] 25 20 15 30
[2,] 10 30 20 30
[3,] 30 10 30 10
```

```
>colSums (A, na.rm = FALSE, dims = 1)
[1] 65 60 65 70
```

```
> A[,1] = A[,1]/65
> A[,2] = A[,2]/60
> A[,3] = A[,3]/65
> A[,4] = A[,4]/70
> A
```

```
         [,1]      [,2]      [,3]      [,4]
[1,] 0.3846154 0.3333333 0.2307692 0.4285714
[2,] 0.1538462 0.5000000 0.3076923 0.4285714
[3,] 0.4615385 0.1666667 0.4615385 0.1428571
```

> R = A%*%w
> R

```
      [,1]
[1,] 0.3263736
[2,] 0.3359890
[3,] 0.3376374
```

The ranking above for Alternative 3> Alternative 2 > Alternative 1.

We also point out the R had a MADAM package but it does not install on R version 4.0.

10.9 Exercises

In each problem, use SAW and then TOPSIS to find the ranking under these weighted conditions:

    (a) All weights are equal.
    (b) Choose and state your weights.

1. For a given hospital, rank order the procedure using the data below.

	Procedure			
	1	2	3	4
Profit	$200	$150	$100	$80
X-Ray times	6	5	4	3
Laboratory Time	5	4	3	2

2. For a given hospital, rank order the procedure using the data below.

	Procedure			
	1	2	3	4
Profit	$190	$150	$110	980
X-Ray times	6	5	5	3
Laboratory Time	5	4	3	3

3. Rank order the following threats:

Threat Alternatives/ Criterion	Reliability of threat assessment	Approximate associated deaths (000)	Cost to fix damages in (Millions)	Location density in millions	Destructive psychological Influence	Number of intelligence related tips
Dirty Bomb Threat	0.40	10	150	4.5	9	3
Anthrax-Bio Terror Threat	0.45	.8	10	3.2	7.5	12
DC-Road & Bridge network threat	0.35	0.005	300	.85	6	8
NY subway threat	0.73	12	200	6.3	7	5
DC Metro Threat	0.69	11	200	2.5	7	5
Major bank robbery	0.81	0.0002	10	.57	2	16
FAA Threat	0.70	0.001	5	.15	4.5	15

4. Consider a scenario where we want to move, rank the cities.

City	Affordability of housing (average home cost in hundreds of thousands)	Cultural opportunities-events per month	Crime rate –number of reported # crimes per month (in hundreds)	Quality of Schools on average (quality rating between [0,1])
1	250	5	10	.75
2	325	4	12	.6
3	676	6	9	.81
4	1,020	10	6	.8
5	275	3	11	.35
6	290	4	13	.41
7	425	6	12	.62
8	500	7	10	.73
9	300	8	9	.79

5. Consider rating departments at a college.

The following table is provided:

DMU departments	Inputs # Faculty	Outputs Student credit hours	Outputs Number of students	Outputs Total degrees
Unit1	25	18,341	9,086	63
Unit2	15	8,190	4,049	23
Unit3	10	2,857	1,255	31
Unit4	33	22,277	6,102	31
Unit5	12	6,830	2,910	19

Formulate and solve the DEA model and rank order the five departments.

6. Consider ranking companies within a larger organization. For simplification reasons we will consider only 6 companies.

Companies	Inputs # Size of Unit	Output #1	Output #2	Output #3
Unit1	120	18,341	9,086	63
Unit2	110	8,190	4,049	23
Unit3	100	2,857	1,255	31
Unit4	135	22,277	6,102	31
Unit5	120	6,830	2,910	19
Unit 6	95	5,050	1835	12

7. Given the input output table below for three hospitals where inputs are number of beds and labor hours in thousands per month and outputs, all measured in hundreds, are patient-days for patients under 14, patient-days for patients between 14 and 65, and patient-days for patients over 65. Determine the efficiency of the three hospitals.

	Inputs		Outputs		
Hospital	1	2	1	2	3
1	5	14	9	4	16
2	8	15	5	7	10
3	7	12	4	9	13

8. Resolve problem 3 with the following inputs and outputs.

	Inputs		Outputs		
Hospital	1	2	1	2	3
1	4	16	6	5	15
2	9	13	10	6	9
3	5	11	5	10	12

9. Consider ranking 4 bank branches in a particular city. The inputs are:

Input 1 = labor hours in hundred per month
Input 2 = space used for tellers in hundreds of square feet
Input 3 = supplies used in dollars per month
Output 1 = loan applications per month
Output 2 = deposits made in thousands of dollars per month

Output 3 = checks processed thousands of dollars per month
The following data table is for the bank branches.

Branches	Input 1	Input 2	Input 3	Output 1	Output 2	Output 3
1	15	20	50	200	15	35
2	14	23	51	220	18	45
3	16	19	51	210	17	20
4	13	18	49	199	21	35

10. What "best practices" might you suggest to the branches that are less efficient in problem 9?

10.10 References and Suggested Further Reading

Alinezhad, A. and A. Amini (2011). Sensitivity analysis of TOPSIS technique: The results of change in the weight of one attribute on the final ranking of alternatives. *Journal of Optimization in Industrial Engineering*, 7(2011): 23–28.

Baker, T. and Z. Zabinsky (2011). A multicriteria decision making model for re verse logistics using Analytical Hierarchy Process. *Omega* (39): 558–573.

Burden, R. and D. Faires (2013). *Numerical Analysis*, 9th ed., Cengage Publishers, Boston, MA.

Butler, J., J. Jia, and J. Dyer (1997). Simulation techniques for the sensitivity analysis of multi-criteria decision models. *European Journal of Operations Research*,103, 531–546.

Callen, J. (1991). Data envelopment analysis: practical survey and managerial accounting applications. *Journal of Management Accounting Research*, 3 (1991): 35–57.

Carley, K. M. (2011). *Organizational Risk Analyzer (ORA)*. Pittsburgh, PA: Center for Computational Analysis of Social and Organizational Systems (CASOS): Carnegie Mellon University.

Chan, A., E. Yung, P. Lam, C. Tam, and S. Cheung (2001). Application of Delphi method in selection of procurement systems for construction projects. *Construction Management and Economics*, 19(7): 699–718, DOI: 10.1080/01446190110066128.

Chang, Y, S. Chou, and C. Shen (2008). A fuzzy simple additive weighting system under group decision-making for facility location selection with objective/subjective attributes. *European Journal of Operational Research*, 189: 132–145.

Charnes, A., W. Cooper, and E. Rhodes (1978). Measuring the efficiency of decision making units. *European Journal of Operations Research*, 2 (1978): 429–444.

Chen, H. and D. Kocaoglu (2008). A sensitivity analysis algorithm for hierarchical decision models. *European Journal of Operations Research*, 185(1): 266–288.

Cooper, W., L. Seiford, and K. Tone (2000). *Data Envelopment Analysis*. Kluwer Academic Press, London, UK.

Cooper, W., S. Li, L. Seiford, R. M. Thrall, and J. Zhu (2001). Sensitivity and stability analysis in DEA: Some recent developments. *Journal of Productivity Analysis,* 15(3): 217–246.

Figueroa, S. (2014). Improving recruiting in the 6th Recruiting Brigade through statistical analysis and efficiency measures. Master's Thesis, NPS, December.

Fishburn, P. C. (1967). Additive utilities with incomplete product set: Applications to priorities and assignments. *Operations Research Society of America (ORSA),* 15: 537–542.

Fox. W.P (2012). Mathematical modeling of the analytical hierarchy process using discrete dynamical systems in decision analysis. *Computers in Education Journal,* July–Sept.: 27–34.

Fox, W. P. (2013). *Mathematical Modeling with Maple.* Cengage Publishing, Boston, MA.

Fox, W. P. (2018). *Mathematical Modeling for Business Analytics.* Taylor & Francis Publishers, Boca Raton, FL.

Fox, W. P. and W. Bauldry (2020). *Problem Solving with Maple.* Taylor & Francis Publishers, Boca Raton, FL.

Fox, W. P. and S. Everton (2013). Mathematical modeling in social network analysis: Using TOPSIS to find node influences in a social network. *Journal of Mathematics and System Science,* 3(10): 531–541.

Fox, W. P. and S. Everton (2014). Mathematical modeling in social network analysis: Using data envelopment analysis and analytical hierarchy process to find node influences in a social network. *Journal of Defense Modeling and Simulation,* 2 (2014): 1–9.

Giordano, F. R., W. Fox, and S. Horton (2014). *A First Course in Mathematical Modeling,* 5th ed., Brooks-Cole Publishers, Boston, MA.

Hartwich, F. (1999). *Weighting of Agricultural Research Results: Strength and Limitations of the Analytic Hierarchy Process (AHP),* Universität Hohenheim. Retrieved from https://entwicklungspolitik.uni-hohenheim.de/uploads/media/DP_09_1999_Hartwich_02.pdf

Huang, I., J. Keisler, and I. Linkov (2011). Multi-criteria decision analysis in environmental sciences: Ten years of applications and trends. *Science of The Total Environment,* 409(19): 3578–3594.

Hurly, W. J. (2001). The analytical hierarchy process: A note on an approach to sensitivity which preserves rank order. Computers and Operations Research, 28: 185–188.

Hwang, C. L. and M. Lin (1987). *Group Decision Making under Multiple Criteria.* Springer-Verlag, New York.

Hwang, C. L. and K. Yoon (1981). *Multiple Attribute Decision Making: Methods and Applications.* Springer-Verlag, New York.

Hwang, C. L., Y. Lai, and T. Y. Liu (1993). A new approach for multiple objective decision making. *Computers and Operational Research,* 20: 889–899.

Krackhardt, D. (1990). Assessing the political landscape: Structure, cognition, and power in organizations. *Admin. Science Quarterly,* 35: 342–369.

Leonelli, R. (2012). Enhancing a decision support tool with sensitivity analysis. Thesis, University of Manchester.

Neralic, L. (1998). Sensitivity analysis in models of data envelopment analysis. *Mathematical Communications,* 3: 41–59.

Saaty, T. (1980). *The Analytical Hierarchy Process.* McGraw Hill, New York.

Shannon, C. and W. Weaver. (1947). *The Mathematical Theory of Communication*. The University of Illinois Press, Urbana.

Temesi, J. (2006). Consistency of the decision-maker in pair-wise comparisons. *International Journal of Management and Decision Making*, 3(2–3).

Thanassoulis, E. (2011). *Introduction to the Theory and Application of Data Envelopment Analysis: A Foundation Text with Integrated Software*. Kluwer Academic Press, London, UK.

Trick, M. A. (1996). Multiple Criteria Decision Making for Consultants, http://mat.gsia.cmu.edu/classes/mstc/multiple/multiple.html. Accessed April 2014.

Trick, M. A. (2014). Data Envelopment Analysis, Chapter 12, http://mat.gsia.cmu.edu/classes/QUANT/NOTES/chap12.pdf. Accessed April 2014.

Weber, M. and K. Borcherding (1993). Behavioral influences on weight judgments in multiattribute decision making. *European Journal of Operational Research*, 67: 1–12.

Winston, W. (1995). *Introduction to Mathematical Programming*. Duxbury Press, Belmont, CA, pp. 322–325.

Yoon, K. (1987). A reconciliation among discrete compromise situations. *Journal of Operational Research* Society, 38: 277–286.

Zeleny, M. (1982). *Multiple Criteria Decision Making*. McGraw Hill, New York.

Zhenhua, G. (2009).The application of DEA/AHP method to supplier selection. *2009 International Conference on Information Management, Innovation Management and Industrial Engineering*, 449–451.

11

Modeling with Game Theory

Objectives

1. Understand the assumption for using game theory.
2. Understand the solution techniques.

Consider the following total conflict game between two players Houston's premier 2019 pitcher, Gerrit Cole and New York Yankees star outfielder Aaron Judge. We have historical statistics on the two players versus each other over the past seasons. We put this information into a payoff matrix for the players where we assume success is a measure of correct guessing. The information collected is the batting average of Aaron in each set of strategies guess and executed by the pitcher: {fastball, fastball; fastball, slider; slider, fastball; slider, slider shown in Table 11.1}.Given the information, what decision should each player make when they face each other in the big game?

Games can have several features; a few of the most common are listed here.

- Number of players: Each person who makes a choice in a game or who receives a payoff from the outcome of those choices is a player. A two-person game has two players. A three or more person is referred to as an N-person game.
- Strategies per player: Each player chooses from a set of possible actions, known as strategies. In a two-person game we allow the row player to have up to *m* strategies and the column player to have up to n strategies. The choice of a particular strategy by each player determines the pay-off to each player
- Pure strategy solution. If a player should always choose one strategy over all other strategies to obtain their best outcome in a game, then that strategy represents a pure strategy solution. Otherwise if strategies should be played randomly then the solution is a mixed strategy solution.

TABLE 11.1

Cole versus Judge

		Gerrit Cole		
	Guess/Pitch	Fastball	Slider	
Aaron Judge	Fastball	.373	.150	
	Slider	.109	.250	

- Nash equilibrium: A Nash equilibrium is a set of strategies that represents mutual best responses to the other player's strategies. In other words, if every player is playing their part of Nash equilibrium, no player has an incentive to unilaterally change his or her strategy. Considering only situations where players play a single strategy without randomizing (a pure strategy) a game can have any number of Nash equilibrium.

- Sequential game: A game is sequential if one player performs her/his actions after another; otherwise the game is a simultaneous game.

- Simultaneous game: A game is simultaneous if the players each choose their strategy for the game and implement them at the same time.

- Perfect information: A game has perfect information if either in a sequential game every player knows the strategies chosen by the players who preceded them or in a simultaneous game each player knows the other players strategies and outcomes in advance.

- Constant sum or zero-sum: A game is constant sum if the sums of the payoffs are the same for every set of strategies and zeros-sum if the sum is always equal to zero. In these games one player gains if and only if another player loses otherwise, we have variable sum game.

- Extensive form presents the game in a tree diagram while normative form presents the game in a payoff matrix. In this chapter we only present the normative form and its associated solution methodologies.

- Outcomes: an outcome is a set of strategies taken by the players, or it is their payoffs resulting from the actions or strategies taken by all players.

- **Total conflict** games are games between players where the sums of the outcomes for all strategy pairs are either the same constant or zero. Games whose outcome sums are variable are known as **partial conflict** games.

The study of game theory has provided many classical and standard games that provide insights into the playing of games. Table 11.2 provides a short summary of some of these games. A complete list maybe viewed at the following website: http://en.wikipedia.org/wiki/List_of_games_in_game_theory.

TABLE 11.2

Summary of classical games in game theory (from http://en.wikipedia.org/wiki/List_of_games_in_game_theory)

Game	Players	Strategies per player	Number of pure strategy Nash equilibrium	Sequential	Perfect information	Zerosum
Battle of the Sexes	2	2	2	No	No	No
Biottogames	2	variable	variable	No	No	Yes
Chicken (aka hawk-dove)	2	2	2	No	No	No
Matching Pennies	2	2	0	No	No	Yes
Nash bargaining game	2	infinite	infinite	No	No	No
Prisoner's dilemma	2	2	1	No	No	No
Rock, Paper, Scissors	2	3	0	No	No	Yes
Staghunt	2	2	2	No	No	No
Trust game	2	infinite	1	Yes	Yes	No

This chapter is primarily concerned with two-person games. The irreconcilable, conflicting interests between the two players in a game resemble parlor games and military encounters between enemy states. Giordano et al. (2014) explain the two-person game in a context of modeling. Players make moves and countermoves, until the rules of engagement declare the game is ended. The rules of engagement determine what each player can or must do at each stage (the available and/or required moves given the circumstances of the game at this stage) as the game unfolds. For example, in the game rock, paper, scissors both players simultaneously make one move, with rock beating scissors beating paper beating rock. While this game consists of only one move, games like chess require many moves to resolve the conflict.

Outcomes or payoffs used in a game are determined as a result of playing a pair of strategies by the players. These outcomes may come from calculated values or expected values, ordinal ranking, cardinal values developed from a lottery system (Von Neumann and Morgenstern, 2004; Straffin, 2004) or cardinal values from pairwise comparisons; Fox, 2014). Here we will assume that we have the cardinal (interval or ratio data) outcomes or payoffs for our games since cardinal outcomes allow us to do mathematical calculations.

11.1 Introduction to Total Conflict (Zero-Sum) Games

Game theory is a branch of applied mathematics that is used in the social sciences (most notably economics), biology, decision sciences, engineering, political science, international relations, operations research, computer science, and philosophy. Game theory attempts to mathematically capture behavior in strategic situations where an individual's success in making choices depends on the choices of others. We say these players are in conflict. Game theory describes how players should play games if both the players are rational. By rational, we imply that each player desires the best outcomes that they can achieve playing the game. Here we analyze competitions in which one individual does better at another's expense. If one player wins \$1 the other player loses \$1, we call these game total conflict games or zero sum and constant sum games.

Traditional applications of game theory provide techniques to find an equilibrium value in zero-sum games. John F. Nash Jr. proved that every game has at least one Nash equilibrium (1950). At equilibrium, each player has adopted a strategy that they are unlikely to change. This equilibrium point (or points) is called the Nash equilibrium.

Our plan in this chapter is to illustrate a game and provide the short cut methods to solve 2 x 2 games. When games are larger than 2 x 2 then we will use linear programming to obtain the solution to both players.

In summary, the current methods to find the Nash equilibrium include Dominance, Minimax, Maximin, equalizing strategies, and William's Method (Williams, 1986; Straffin, 2004). These techniques are found in many modern game theory textbooks such as Straffin (2004) and Barron (2013). Again Straffin (2004) only comments about linear programming but does not use linear programming as one of his techniques.

11.1.1 Total Conflict (Zero-Sum) Games

Using the same conventions as in the Straffin text, we call the row player, Rose, and the column player, Colin. Let's define the zero-sum game with the following payoff matrix that has components for both Rose and Colin where Rose has m strategies and Colin has n strategies:

$$(M,N) = \begin{bmatrix} (M_{1,1}, N_{1,1}) & (M_{1,2}, N_{1,2}) & \cdots & (M_{1,n}, N_{1,n}) \\ (M_{2,1}, N_{2,1}) & (M_{2,2}, N_{2,2}) & \cdots & (M_{2,n}, N_{2,n}) \\ \cdot & \cdot & \cdots & \cdot \\ \cdot & \cdot & \cdots & \cdot \\ \cdot & \cdot & \cdots & \cdot \\ (M_{m,1}, N_{m,1}) & (M_{m,2}, N_{m,2}) & \cdots & (M_{m,n}, N_{m,n}) \end{bmatrix}$$

In the special case of zero-sum games, each pair sums to zero. For example, one such pair is $M_{11} + N_{11} = 0$. In the special case of the constant sum game all pairs sum to the same constant, C. For example the sum of all $M_{ij} + N_{ij} = C$.

Our knowledge of the zero-sum game and the Primal–Dual relations suggest formulating Rose's game and finding Colin's solution through the dual solution. This works well if you have a single problem to solve. But what if you have many games to consider and you only have Excel. How best to construct a technology assistant?

Treating the zero-sum game as above translates into two linear programming formulations, one for each maximizing player. We combine the formulation into a single formulation shown in Equation (11.1) where the solution provides the values of the game and the probabilities that the players should play their strategies. Further, if any pairs $(M_{mn}, N_{m,n})$ are negative then there is a chance that the game solution can be negative. Since the game solution will be a decision variable in our formulation, we must account for that possibility. Our best recommendation is to use the method suggested by Winston [1995, pp.172–178] to replace any variable that could take on negative values with the difference in two positive variables, $x_j - x'_j$. We assume that the value of the game could be positive or negative. The other values we are looking for are probabilities that are always between 0 and 1. Since this occurs only in the value of the game, we use as a substitute variable, $V = v_1 - v_2$.

$$\text{Maximize } v_1 - v_2 \qquad (11.1)$$

Subject to:

$$M_{1,1}x_1 + M_{2,1}x_2 + \ldots + M_{m,1}x_n - v_1 + v_2 \geq 0$$
$$M_{1,2}x_1 + M_{2,2}x_2 + \ldots + M_{m,2}x_n - v_1 + v_2 \geq 0$$
$$\ldots$$
$$M_{1,m}x_1 + M_{2,m}x_2 + \ldots + M_{m,n}x_n - v_1 + v_2 \geq 0$$
$$x_1 + x_2 + \ldots + x_n = 1$$
$$\textit{Nonnegativity}$$

where the weights x_i yields Rose strategy and the value of V is the value of the game to Rose.

$$\text{Maximize } v_3 - v_4$$

Subject to:

$$N_{1,1}y_1 + N_{1,2}y_2 + \ldots + N_{1,m}y_n - v_3 + v_4 \geq 0$$
$$N_{2,1}y_1 + N_{2,2}y_2 + \ldots + N_{2,m}y_n - v_3 + v_4 \geq 0$$
$$\ldots$$
$$N_{m,1}y_1 + N_{m,2}y_2 + \ldots + N_{m,n}y_n - v_3 + v_4 \geq 0$$
$$y_1 + y_2 + \ldots + y_n = 1$$
$$\textit{Nonnegativity}$$

where the weights y_i yield Colin's strategy and the value of $v_3 - v_4$ is the value of the game to Colin.

To accomplish this as one formulation, we combine as

$$\text{Maximize } v_1 - v_2 + v_3 - v_4 \tag{11.2}$$

Subject to:

$$M_{1,1}x_1 + M_{2,1}x_2 + ... + M_{m,1}x_n - v_1 + v_2 \geq 0$$
$$M_{1,2}x_1 + M_{2,2}x_2 + ... + M_{m,2}x_n - v_1 + v_2 \geq 0$$
$$...$$
$$M_{1,m}x_1 + M_{2,m}x_2 + ... + M_{m,n}x_n - v_1 + v_2 \geq 0$$
$$x_1 + x_2 + ... + x_n = 1$$
$$Nonnegativity$$

$$N_{1,1}y_1 + N_{1,2}y_2 + ... + N_{1,m}y_n - v_3 + v_4 \geq 0$$
$$N_{2,1}y_1 + N_{2,2}y_2 + ... + N_{2,m}y_n - v_3 + v_4 \geq 0$$
$$...$$
$$N_{m,1}y_1 + N_{m,2}y_2 + ... + N_{m,n}y_n - v_3 + v_4 \geq 0$$
$$y_1 + y_2 + ... + y_n = 1$$
$$Nonnegativity$$

The algorithm for the LP template solver, saved as macro-enabled, is described as follows:

Step 1. Input the payoff matrix for Rose only.

Step 2. Make all decision variables $\{x_1, x_2, ...,x_{10}, y_1, y_2, ...,y_{10}, v_1, v_2, v_3, v_4\}$ initially set to 0.

Step 3. Highlight the objective function cell, open the Solver, insure it has non-negativity for variables checked, and is set for SimplexLP. Then hit solve.

Step 4. The answers are updated automatically.

Step 5. Extract and interpret the answers.

Template Algorithm

The algorithm for building the template for up to 10 strategies for Rose and Colin required several steps using *Logical If* statements that allow for including or not including constraints based upon the number of variables in the problem. The user input for the numbers of active rows and active columns is a critical user input. These changes in the formulation are essential because of the additional generic decision variables. In creating the constraints, the constraints that exceed the number of rows or column need to be zeroed out. *Logical If* statements are used to handle that condition within the constraints. Recall the template is setup to accept up to ten rows and ten columns. If our game theory problem has less than ten rows and/or less than ten columns, then the used number of decision variables are eliminated if these exceed the number of row or columns as well. Again, a *Logical If* statement has been added to handle that eventuality. Figure 11.2a–11.2d show screenshots of the Solver Excel template.

The template uses linear programming in the Excel Solver (called *SimplexLP*) to solve for the game's solution and to identify the strategies to be played by each player to achieve the "best" solution either by pure or mixed strategies.

How do we deal with alternate optimal solution identification? If a pure strategy solution has alternate solutions this is identified by examining the Sensitivity Report. A basic decision variable, a variable whose value is 1, having either an increase or decrease of 0 in the report identifies a possible alternative optimal solution. To find the other solution, add a constraint that requires the optimal solution to equal the current value and change the objective function to Maximize another decision variable whose final value is 0 and also has an increase or decrease of 0. This requires some additional linear programming work beyond what is contained in this template.

Generic Solver Formulation from the Template with 43 Constraints (Figure 11.1)

The remaining constraints are shown here (Figure 11.2b):

Finally (Figure 11.2c – 11.2d),

In the template created for student use, Figure 11.3, the student enters only what is highlighted in within the template. This consists of the number of rows (cell F2), the number of columns (cell F3), the payoff matrix for Rose (cells B2:K16), and initial values for the decision variables usually set at 0 (cells B32: B55). The cells L2 and L3 are the values of the game and cells L18:L27 and B28:K28 are the strategies played by Rose and Colin, respectively. We illustrate the template with two examples one with pure strategy solution and the other with mixed strategy solutions.

Example 11.1 Pure Strategy 4 x 4 Game

Consider the following zero-sum game displayed in the payoff matrix for Rose.

There is a pure strategy solution at R1C3 and R3C3, with value of 2 to Rose. Since this is a zero-sum game we also know Colin's game value of –2. We may use the template to insert this game and obtain a solution (Figure 11.3–11.4).

Used	Used		RHS
=IF(F32<=F$3,L32-B$42+B$43,	=G32		0
=IF(F33<=F$3,L33-B$42+B$43,	=G33		0
=IF(F34<=F$3,L34-B$42+B$43,	=G34		0
=IF(F35<=F$3,L35-B$42+B$43,	=G35		0
=IF(F36<=F$3,L36-B$42+B$43,	=G36		0
=IF(F37<=F$3,L37-B$42+B$43,	=G37		0
=IF(F38<=F$3,L38-B$42+B$43,	=G38		0
=IF(F39<=F$3,L39-B$42+B$43,	=G39		0
=IF(F40<=F$3,L40-B$42+B$43,	=G40		0
=IF(F41<=F$3,L41-B$42+B$43,	=G41		0
=H42		=B32+B33+B34+B35+B36+B37+B38+B39+B4(	1
=IF(F43<=F$2,L42-B$54+B$55,	=G43		0
=IF(F44<=F$2,L43-B$54+B$55,	=G44		0
=IF(F45<=F$2,L44-B$54+B$55,	=G45		0
=IF(F46<=F$2,L45-B$54+B$55,	=G46		0
=IF(F47<=F$2,L46-B$54+B$55,	=G47		0
=IF(F48<=F$2,L47-B$54+B$55,	=G48		0
=IF(F49<=F$2,L48-B$54+B$55,	=G49		0
=IF(F50<=F$2,L49-B$54+B$55,	=G50		0
=IF(F51<=F$2,L50-B$54+B$55,	=G51		0
=IF(F52<=F$2,L51-B$54+B$55,	=G52		0
=B44+B45+B46+B47+B48+B49	=G53		1
=B42-B43+B54-B55	=G54		0
=B32			=IF(F55<=F$2,1,0)
=B33			=IF(F56<=F$2,1,0)
=B34			=IF(F57<=F$2,1,0)
=B35			=IF(F58<=F$2,1,0)
=B36			=IF(F59<=F$2,1,0)
=B37			=IF(F60<=F$2,1,0)
=B38			=IF(F61<=F$2,1,0)
=B39			=IF(F62<=F$2,1,0)
=B40			=IF(F63<=F$2,1,0)
=B41			=IF(F64<=F$2,1,0)
=B44			=IF(F65<=F$3,1,0)
=B45			=IF(F66<=F$3,1,0)
=B46			=IF(F67<=F$3,1,0)
=B47			=IF(F68<=F$3,1,0)
=B48			=IF(F69<=F$3,1,0)
=B49			=IF(F70<=F$3,1,0)
=B50			=IF(F71<=F$3,1,0)
=B51			=IF(F72<=F$3,1,0)
=B52			=IF(F73<=F$3,1,0)
=B53			=IF(F74<=F$3,1,0)

FIGURE 11.1

Excel screenshot of generic solve formulation with 43 constraints.

TABLE 11.3

Pure strategy 4 x 4 game

		C1	C2	C3	C4
	R1	4	3	2	5
Rose	R2	−10	2	0	−1
	R3	7	5	2	3
	R4	0	0	−4	−5

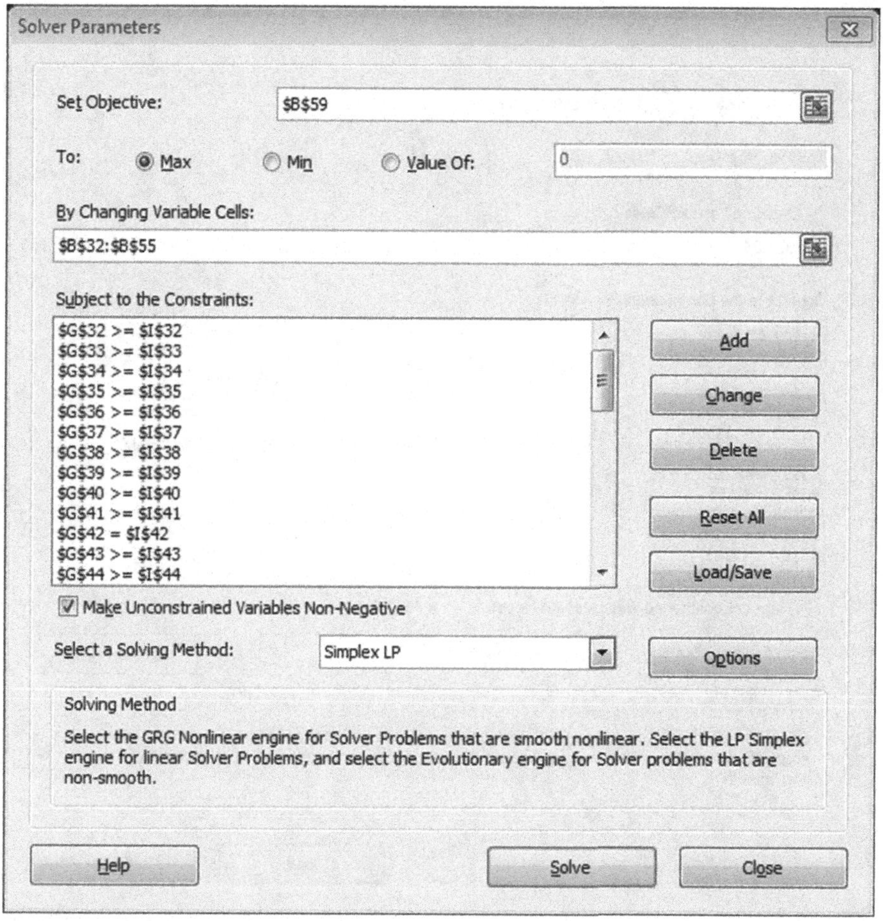

FIGURE 11.2a
Excel screenshot of Solver with constraints.

With the template we have found the pure strategy solution at R1C3. So how do we know if there are alternate optimal solutions? We present this later using the steps provided by Fox (2014) to find an alternate optimal solution, our solution is now at R3C3.

Let's go back to hitter–pitcher duel.

Example 11.2 Hitter–Pitcher Duel between Aaron Judge and Gerrit Cole

We enter the game previously presented into the template where R1 = guesses fastball, R2 = guesses slider, C1 = pitches fastball, C2 = pitches slider.

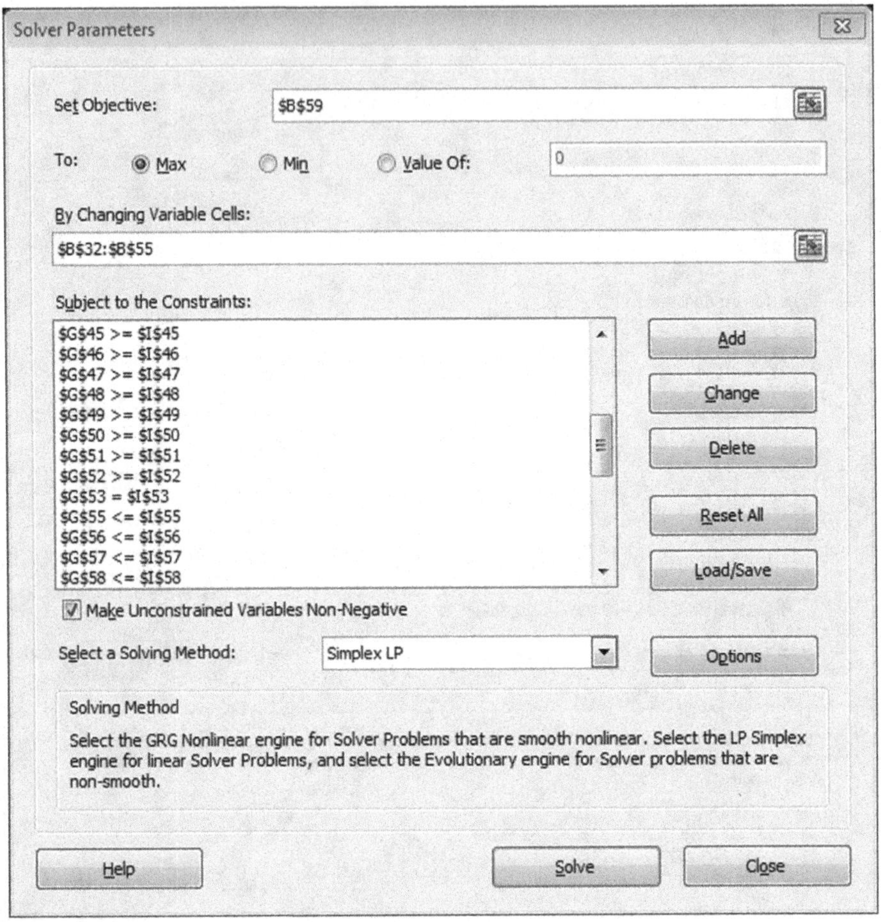

FIGURE 11.2b
Excel screenshot of Solver with constraints.

		Gerrit Cole	
	Guess/Pitch	Fastball	Slider
Aaron Judge	Fastball	0.373	0.150
	Slider	0.109	0.250

There is no saddle point solution. You will be asked to show that as an exercise. We know a solution exists so we move to method of oddments (see Straffin, 2004 for more details). In the method of oddments, all subtractions are higher value–lower value. We switch the position to calculate the appropriate probabilities (Figure 11.5).

The solution, by the methods of oddments, is the Aaron guesses fastball 38.7% of the time and guesses slider 61.3% of the time while Gerrit

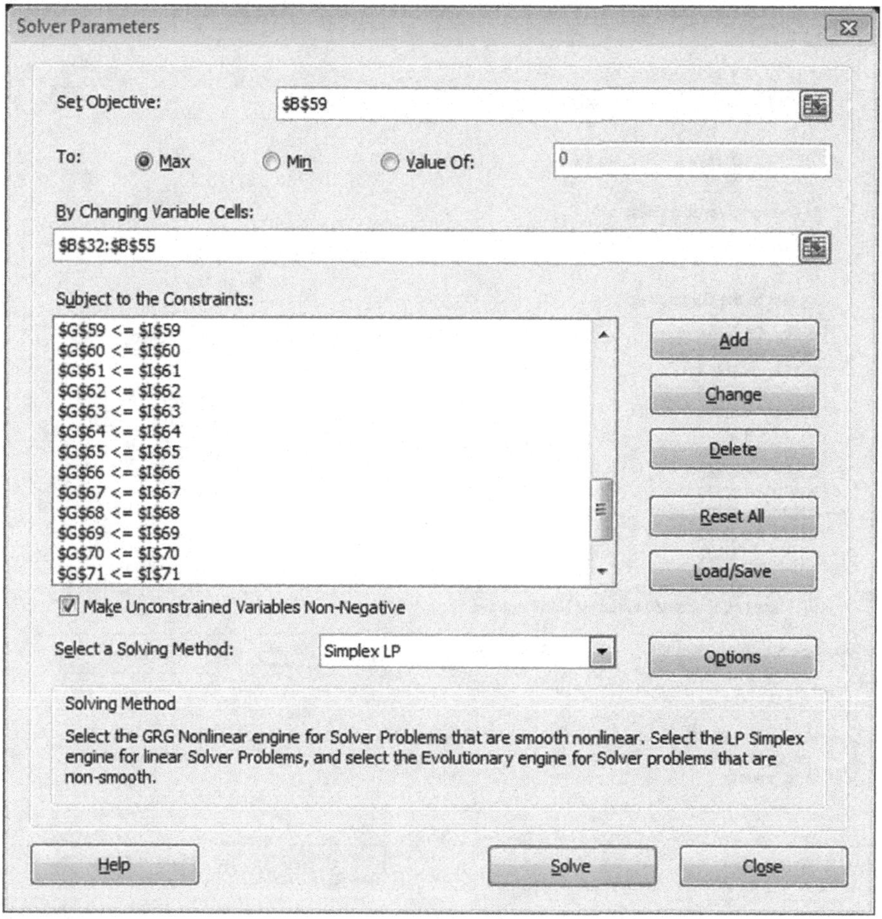

FIGURE 11.2c
Excel screenshot of Solver with constraints.

throws the fastball 27.5% of the time and throws the slider 72.5% of the time. This makes Aaron's batting average versus Gerrit as:

$$BA = 0.373 (0.3870(0.725)+0.15(0.387)(0.275) + 0.109(/0.613)$$
$$(0.725)+0.25(0.6130(0.275) = 0.211$$

Any deviations from these percentages result in a higher average for Aaron.

As we said previous, any games larger than 2 x 2 are better solved using linear programming. Equations (11.1) and (11.2) illustrate the LP formulations.

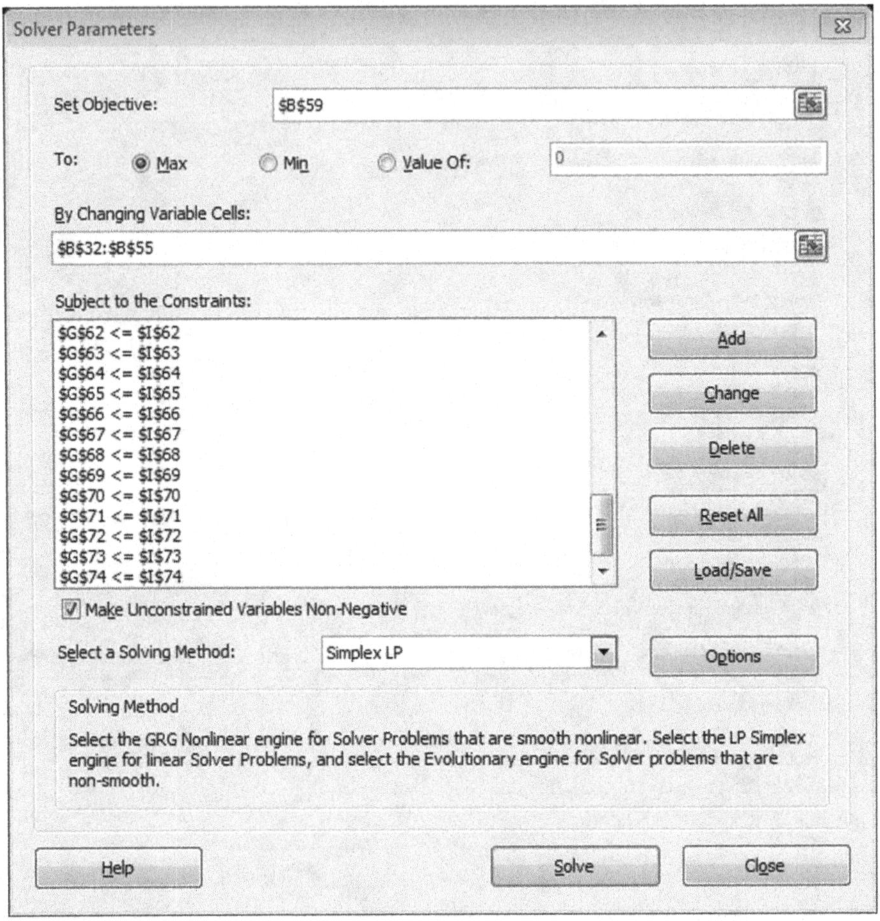

FIGURE 11.2d
Excel screenshot of Solver with constraints.

Example 11.3 Mixed Strategy *3 x 6* Game

Consider the following zero-sum game displayed in the payoff matrix for Rose.

This game does not have a pure strategy solution, so we need to find the mixed strategy solution. Our linear programming template will find either pure or mixed strategy solutions. We put the payoff matrix into the template and let the template know there are 3 rows and 6 columns as asked in cells F2:F3.

	A	B	C	D	E	F	G	H	I	J	K	L
1	This template will allow you to solve up to 10 strategies for each player in a two-person zero-sum game										Game Values	
2	Enter the number of Strategies for Rose					4					Rose	2
3	Enter the number of Strategies for Colin					4					Colin	-2
4												
5	Enter the payoff to the Row player only											
6	R/C	1	2	3	4	5	6	7	8	9	10	
7	1	4	3	2	5	0	0	0	0	0	0	
8	2	-10	2	0	-1	0	0	0	0	0	0	
9	3	7	5	2	3	0	0	0	0	0	0	
10	4	0	0	-4	-5	0	0	0	0	0	0	
11	5	0	0	0	0	0	0	0	0	0	1	
12	6	0	0	0	0	0	0	0	0	0	0	
13	7	0	0	0	0	0	0	0	0	0	0	
14	8	0	0	0	0	0	0	0	0	0	0	
15	9	0	0	0	0	0	0	0	0	0	0	
16	10	0	0	0	0	0	0	0	0	0	0	
17	R/C	1	2	3	4	5	6	7	8	9	10	Rose's strategies
18	1	4	3	2	5	0	0	0	0	0	0	1.0000
19	2	-10	2	0	-1	0	0	0	0	0	0	0.0000
20	3	7	5	2	3	0	0	0	0	0	0	0.0000
21	4	0	0	-4	-5	0	0	0	0	0	0	0.0000
22	5	0	0	0	0	0	0	0	0	0	0	0.0000
23	6	0	0	0	0	0	0	0	0	0	0	0.0000
24	7	0	0	0	0	0	0	0	0	0	0	0.0000
25	8	0	0	0	0	0	0	0	0	0	0	0.0000
26	9	0	0	0	0	0	0	0	0	0	0	0.0000
27	10	0	0	0	0	0	0	0	0	0	0	0.0000
28	Colin's	0.0000	0.0000	1.0000	0.0000	0.0000	0.0000	0.0000	0.0000	0.0000	0.0000	
29	Strategies											

FIGURE 11.3
Screen capture of a proportion of the student template.

	A	B	C	D	E	F	G	H	I	J	K	L
1	This template will allow you to solve up to 10 strategies for each player in a two-person zero-sum game										Game Values	
2	Enter the number of Strategies for Rose					4					Rose	2
3	Enter the number of Strategies for Colin					4					Colin	-2
4												
5	Enter the payoff to the Row player only											
6	R/C	1	2	3	4	5	6	7	8	9	10	
7	1	4	3	2	5	0	0	0	0	0	0	
8	2	-10	2	0	-1	0	0	0	0	0	0	
9	3	7	5	2	3	0	0	0	0	0	0	
10	4	0	0	-4	-5	0	0	0	0	0	0	
11	5	0	0	0	0	0	0	0	0	0	1	
12	6	0	0	0	0	0	0	0	0	0	0	
13	7	0	0	0	0	0	0	0	0	0	0	
14	8	0	0	0	0	0	0	0	0	0	0	
15	9	0	0	0	0	0	0	0	0	0	0	
16	10	0	0	0	0	0	0	0	0	0	0	
17	R/C	1	2	3	4	5	6	7	8	9	10	Rose's strategies
18	1	4	3	2	5	0	0	0	0	0	0	0.0000
19	2	-10	2	0	-1	0	0	0	0	0	0	0.0000
20	3	7	5	2	3	0	0	0	0	0	0	1.0000
21	4	0	0	-4	-5	0	0	0	0	0	0	0.0000
22	5	0	0	0	0	0	0	0	0	0	0	0.0000
23	6	0	0	0	0	0	0	0	0	0	0	0.0000
24	7	0	0	0	0	0	0	0	0	0	0	0.0000
25	8	0	0	0	0	0	0	0	0	0	0	0.0000
26	9	0	0	0	0	0	0	0	0	0	0	0.0000
27	10	0	0	0	0	0	0	0	0	0	0	0.0000
28	Colin's	0.0000	0.0000	1.0000	0.0000	0.0000	0.0000	0.0000	0.0000	0.0000	0.0000	
29	Strategies											

FIGURE 11.4
Screen capture of a solution of the 4 x 4 game.

	Guess/Pitch	Gerrit Cole Fastball	Slider		
Aaron Judge	Fastball	.373	.150	.373 -.150=.223	.141/.364 =0.387
	Slider	.109	.250	.250 -.109=.141	.223/.364= 0.613
		.373 - .109=.264	.250 -.150 -.1	.223+.141=.364 = .264+1.00	
		.1/.364=.275	.264/.364=.725		

FIGURE 11.5

Hitter–pitcher duel between Aaron Judge and Gerrit Cole

TABLE 11.4

Pure strategy 3 x 6 game

		C1	C2	C3	C4	C5	C6
	R1	4	−4	3	−2	−3	3
Rose	R2	−1	−1	−2	0	0	4
	R3	−1	2	1	−1	2	−3

The template output solution states that the values of the game are (Figure 11.6):

Rose   -0.71428571
Colin   0.71428571

Rose plays R1 with 23.8095238%, R2 with 21.4285714%, and R3 with 54.7619048% to achieve her best solution and Colin never plays C1, C4, or C6 and plays C2 with 35.7143%, C3 with 57.14286%, and C5 with 7.1429% to achieve his best solution.

11.2 Finding Alternate Optimal Solutions in a Two Person Zero-Sum Game

Often in game theory there are ties in the value of the game by playing the strategies that produce the "best" or optimal result. In this chapter, we provide several two-person zero-sum game theory examples that produce alternate optimal solutions to playing the game. Further, we show how MS-Excel can be used to find these alternate optimal solutions.

Introduction and Background to Mathematical Modeling

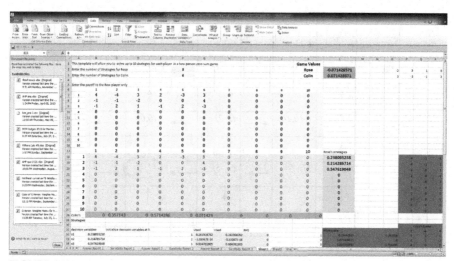

FIGURE 11.6
Screen capture of 3 x 6 game.

Pure Strategy Solutions

Most texts (Straffin, 2004; Williams, 1986; Giordano et al., 2014) suggest using movement diagrams or the saddle point method (minimax = maximin) as the methods to examine a payoff matrix for pure strategy solutions for a specific class of zero-sum game problems. We quickly describe these methods.

Movement Diagrams

In a movement diagram, we draw vertical arrows from the smaller values in a row to the corresponding larger value in another row and then draw horizontal arrows from the other players smaller to larger values in columns. We tell the students to follow the arrows after they have drawn them in the payoff matrix. If following the arrows leads into a point but no arrow leaves that point, then that point is an equilibrium. We might find multiple equilibria or no equilibrium may exist using this procedure. We illustrate with an example where Rose has four strategies {R1,R2,R3,R4} and Colin has four strategies {C1,C2,C3, C4}.

> **Example 11.4 Consider the following game between Rose and Colin where each has four strategies.**
>
> We draw the arrows as described. By following the arrows in Figure 11.7a, we find the arrows all *point in* and no arrow departs from either R1C3 (2, –2) and R3C3 (2, –2). We have found two pure strategy solutions (Figure 11.7a–11.7b).

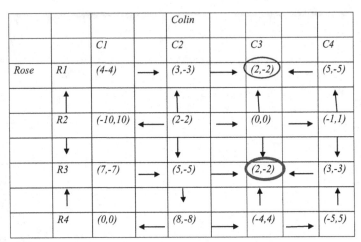

FIGURE 11.7a
Movement diagram for Example 11.4.

		Colin				Row minimums	Maximin
		C1	C2	C3	C4		
Rose	R1	4	3	2	5	2	②
	R2	-10	2	0	-1	-10	
	R3	7	5	2	3	2	②
	R4	0	8	-4	-5	-5	
	Column maximum	7	8	2	5			
	Minimax			②			

FIGURE 11.7b
Movement diagram solution for Example 11.4.

Saddle Point Method

The following definition of a saddle point is taken from Straffin [5],

> An outcome in a matrix game (with payoff to the row player) is
> called a saddle point if the entry at that outcome is both less than or
> equal to any entry in its row and greater than or equal to any entry
> in its column. (p. 9)

We illustrate the same example using this method using only the row player's payoffs.

We find a tie at R1C2 = 2, and R3C3 = 2 as before. The value of the game for Rose is this saddle point, $V = 2$.

Linear Programming Method

Linear programming is a method that can always be used in solving all zero sum games. We present a generic formulation for Rose's payoff matrix from Fox (2010).

$$\text{Maximize } v_1 - v_2$$

Subject to:

$$M_{1,1}x_1 + M_{2,1}x_2 + ... + M_{m,1}x_n - v_1 + v_2 \geq 0$$
$$M_{1,2}x_1 + M_{2,2}x_2 + ... + M_{2,m}x_n - v_1 + v_2 \geq 0$$
$$...$$
$$M_{1,m}x_1 + M_{2,m}x_2 + ... + M_{m,n}x_n - v_1 + v_2 \geq 0$$
$$x_1 + x_2 + ... + x_n = 1$$
$$Nonnegativity$$

where the weights x_i yield Rose strategies and the value of game, $V = v_1 - v_2$, is the value of the game to Rose. We use $V = v_1 - v_2$ because the negatives in the payoff matrix might yield a negative value of the game, since all variables are non-negative this change in variables allows for negative values of the game.

Continuing our example with payoff matrix as shown in Figure 11.7b, we formulate the linear program as follows:

$$\text{Maximize } v_1 - v_2$$

$$4x_1 - 10x_2 + 7x_3 - v_1 + v_2 \geq 0$$
$$3x_1 + 2x_2 + 5x_3 + 8x_4 - v_1 + v_2 \geq 0$$
$$2x_1 + 2x_3 - 4x_4 - v_1 + v_2 \geq 0$$
$$5x_1 - x_2 + 3x_3 - 5x_4 - v_1 + v_2 \geq 0$$
$$x_1 + x_2 + x_3 + x_4 = 1$$
$$x_i, v_j \geq 0 \; i = 1,2,3,4, \, j = 1,2.$$

We use a template developed for class use to obtain a solution. We enter the number of rows and columns in cells F2:F3, Rose's payoff's in cells B7:E10, initial the decision variables in cells B34:B55. We open the Solver dialog box, and press Solve.

R/C	1	2	3	4	5	6	7	8	9	10	
	This template will allow you to solve up to 10 strategies for each player in a two-person zero-sum game									Game Values	
Enter the number of Strategies for Rose				4						Rose	2
Enter the number of Strategies for Colin				4						Colin	-2
Enter the payoff to the Row player only											
R/C	1	2	3	4	5	6	7	8	9	10	
1	4	3	2	5	0	0	0	0	0	0	
2	-10	2	0	-1	0	0	0	0	0	0	
3	7	5	2	3	0	0	0	0	0	0	
4	0	8	-4	-5	0	0	0	0	0	0	
5	0	0	0	0	0	0	0	0	0	1	
6	0	0	0	0	0	0	0	0	0	0	
7	0	0	0	0	0	0	0	0	0	0	
8	0	0	0	0	0	0	0	0	0	0	
9	0	0	0	0	0	0	0	0	0	0	
10	0	0	0	0	0	0	0	0	0	0	
R/C	1	2	3	4	5	6	7	8	9	10	Rose's strategies
1	4	3	2	5	0	0	0	0	0	0	1.0000
2	-10	2	0	-1	0	0	0	0	0	0	0.0000
3	7	5	2	3	0	0	0	0	0	0	0.0000
4	0	8	-4	-5	0	0	0	0	0	0	0.0000
5	0	0	0	0	0	0	0	0	0	0	0.0000
6	0	0	0	0	0	0	0	0	0	0	0.0000
7	0	0	0	0	0	0	0	0	0	0	0.0000
8	0	0	0	0	0	0	0	0	0	0	0.0000
9	0	0	0	0	0	0	0	0	0	0	0.0000
10	0	0	0	0	0	0	0	0	0	0	0.0000
Colin's	0.0000	0.0000	1.0000	0.0000	0.0000	0.0000	0.0000	0.0000	0.0000	0.0000	
Strategies											

FIGURE 11.8
Excel screenshottemplate output for Example 11.4.

We interpret the output in cells L2:L3, cells L18:L26 and cells B28:K28. The solution shown here is R1C3 with values 2 for Rose and −2 for Colin. This is displayed in Figure 11.8.

We save both the *Answer* and *Sensitivity* reports in Figure 11.8 and 11.9 respectively.

11.2.1 Excel and Game Theory

The Solver's main solution is $x1 = 1$ corresponding to R1 and $y3 = 1$ corresponding to C3 with a game value for Rose is $v_1 - v_2 = 2 - 0 = 2$ and for Colin $v_3 - v_4 = 0 - 2 = -2$.

We examined the sensitivity report to look for the possibility of alternate optimal solutions to a zero-sum game. From the Sensitivity Report, we find the indicators of alternate optimal solutions (Figure 11.10).

What about Alternate Solutions?

The indicators are found in the variable cells of the Sensitivity Report as highlighted here. Variable $x1$ currently at 1 can decrease by 0 and $x3$ can increase by 0. This indicates possible alternate optimal solution in Excel. So how do we find the alternate solution, if one exists?

First, we change the objective function from maximize $v_1 - v_2$ to maximize x_3. We maximize x_3 because it is currently at value 0 and can replace x_1, which

	A	B	C	D	E	F	G	H	I	J
8		Iterations: 36 Subproblems: 0								
9	**Solver Options**									
10		Max Time Unlimited, Iterations Unlimited, Precision 0.000001, Use Automatic Scaling								
11		Max Subproblems Unlimited, Max Integer Sols Unlimited, Integer Tolerance 1%, Assume NonNegative								
12										
13										
14	Objective Cell (Max)									
15		Cell	Name	Original Value	Final Value					
16		$B$59		0	0					
17										
18										
19	Variable Cells									
20		Cell	Name	Original Value	Final Value	Integer				
21		$B$32 x1		0.000000	1.000000 Contin					
22		$B$33 x2		0.000000	0.000000 Contin					
23		$B$34 x3		0.00000	0.00000 Contin					
24		$B$35 x4		0.00000	0.00000 Contin					
25		$B$36 x5		0.00000	0.00000 Contin					
26		$B$37 x6		0.00000	0.00000 Contin					
27		$B$38 x7		0.00000	0.00000 Contin					
28		$B$39 x8		0.00000	0.00000 Contin					
29		$B$40 x9		0.00000	0.00000 Contin					
30		$B$41 x10		0.00000	0.00000 Contin					
31		$B$42 v1		0.00000	2.00000 Contin					
32		$B$43 v2		0.00000	0.00000 Contin					
33		$B$44 y1		0.00000	0.00000 Contin					
34		$B$45 y2		0.00000	0.00000 Contin					
35		$B$46 y3		0.00000	1.00000 Contin					
36		$B$47 y4		0.00000	0.00000 Contin					
37		$B$48 y5		0.00000	0.00000 Contin					
38		$B$49 y6		0.00000	0.00000 Contin					
39		$B$50 y7		0.00000	0.00000 Contin					
40		$B$51 y8		0.00000	0.00000 Contin					
41		$B$52 y9		0.00000	0.00000 Contin					
42		$B$53 y10		0.00000	0.00000 Contin					
43		$B$54 v3		0.00000	0.00000 Contin					
44		$B$55 v4		0.00000	2.00000 Contin					
45										

FIGURE 11.9
Excel screenshot answer report for Example 11.4.

is currently at value 1. Then we add a new constraint that states the objective function remain at 2, $v_1 - v_2 = 2$. After making those two modifications with the Solver's dialog box, we press Solve.

After making these two alterations to the formulation, we can obtain the alternate solution, $x3 = 1, y3 = 1$, $V = 2$, corresponding to strategies $R3C3$ obtaining a value of 2.

Since there are no other variables with increases or decreases allowable to 0 in the original or this solution's sensitivity report we have found all the alternate solutions, Figure 11.11.

	Cell	Name	Final Value	Reduced Cost	Objective Coefficient	Allowable Increase	Allowable Decrease
	$B$32	x1	1	0	0	1E+30	0
	$B$33	x2	0	-2	0	2	1E+30
	$B$34	x3	2.22045E-16	0	0	0	2
	$B$35	x4	0	-6	0	6	1E+30
	$B$36	x5	0	-2	0	2	1E+30
	$B$37	x6	0	-2	0	2	1E+30
	$B$38	x7	0	-2	0	2	1E+30
	$B$39	x8	0	-2	0	2	1E+30
	$B$40	x9	0	-2	0	2	1E+30
	$B$41	x10	0	-2	0	2	1E+30
	$B$42	v1	2	0	1	0	1
	$B$43	v2	0	0	-1	0	1E+30
	$B$44	y1	0	-2	0	2	1E+30
	$B$45	y2	0	-1	0	1	1E+30
	$B$46	y3	1	0	0	2	1
	$B$47	y4	0	-3	0	3	1E+30
	$B$48	y5	0	0	0	1E+30	2
	$B$49	y6	0	0	0	1E+30	2
	$B$50	y7	0	0	0	1E+30	2
	$B$51	y8	0	0	0	1E+30	2
	$B$52	y9	0	0	0	1E+30	2
	$B$53	y10	0	0	0	1E+30	2
	$B$54	v3	0	0	1	0	1E+30
	$B$55	v4	2	0	-1	0	1E+30

Allowable decrease of 0.

Allowable increase of 0 and currently has value 0.

FIGURE 11.10

Excel screenshot sensitivity report for Example 11.4.

FIGURE 11.11

Excel alternate solutions for Example 11.4.

Mixed Solutions

The typical solution methods for mixed solution, algebra, or the method of oddments do not support finding alternate solutions. Thus, we turn to linear programming using the same formulation and template as before.

Example 11.5 A 4×5 Game Payoff Matrix from Straffin

In the following game (Straffin (2004), exercise 2, chapter 2), Rose has four strategies and Colin has five strategies. Using the techniques from Straffin's solution, we use dominating strategies and then oddments to find as the solution: $V = v_1 - v_2 = 4/3$ when $x_1 = 2/3$, $x_2 = 0$, $x_3 = 1/3$ corresponding to play R1 with probability 2/3, to never play R2, and to play R3 with probability 1/3 while playing C3 with probability 2/3, to play C5 with probability 1/3, and to never playing strategies C1, C2, or C4.

We employ linear programming to solve this more complex problem quickly and to look for possible alternate optimal solutions.

$$\text{Maximize } v_1 - v_2$$

Subject to

$$x1 + 2x2 + 2x3 + 2x4 - v1 + v2 > 0$$
$$x1 + x2 + 2x3 + 2x4 - v1 + v2 > 0$$
$$x1 + x2 + x3 + 2x4 - v1 + v2 > 0$$
$$2x1 + x2 + x3 + x4 - v1 + v2 > 0$$
$$2x1 + 2x2 + 2x3 - v1 + v2 > 0$$
$$x1 + x2 + x3 + x4 = 1$$
$$xi, vj > 0 \text{ for } i = 1,2,3,4, j = 1,2$$

Using our linear programming template and Solving on Excel, we obtain the following solution, Figure 11.12.

We note that, using linear programming, our solution is to play R1 with probability 1/3, R2 with probability 1/3, and R4 with probability 1/3 while playing C2 with probability 1/3, C3 with probability 1/3, and C5 with probability 1/3 yielding a solution to the game of $V = v_1 - v_2 = 4/3$.

Our applying linear programming yielded an alternate optimal solution to a previous method shown in Straffin (2004).

We examine the sensitivity report. Further analysis of the sensitivity report shows that x_3 might replace either x_2 or x_4 as a basic variable in the solution and y_1 might replace either affect y_2 or y_3. Our queues are as before.

R/C	1	2	3	4	5	6	7	8	9	10	
	Enter the number of Strategies for Rose			4					Rose	1.333333333	
	Enter the number of Strategies for Colin			5					Colin	-1.333333333	
	Enter the payoff to the Row player only										
R/C	1	2	3	4	5	6	7	8	9	10	
1	1	1	1	2	2	0	0	0	0	0	
2	2	1	1	1	2	0	0	0	0	0	
3	2	2	1	1	1	0	0	0	0	0	
4	2	2	2	1	0	0	0	0	0	0	
5	0	0	0	0	0	0	0	0	0	1	
6	0	0	0	0	0	0	0	0	0	0	
7	0	0	0	0	0	0	0	0	0	0	
8	0	0	0	0	0	0	0	0	0	0	
9	0	0	0	0	0	0	0	0	0	0	
10	0	0	0	0	0	0	0	0	0	0	
R/C	1	2	3	4	5	6	7	8	9	10	Rose's strategies
1	1	1	1	2	2	0	0	0	0	0	0.3333
2	2	1	1	1	2	0	0	0	0	0	0.3333
3	2	2	1	1	1	0	0	0	0	0	0.0000
4	2	2	2	1	0	0	0	0	0	0	0.3333
5	0	0	0	0	0	0	0	0	0	0	0.0000
6	0	0	0	0	0	0	0	0	0	0	0.0000
7	0	0	0	0	0	0	0	0	0	0	0.0000
8	0	0	0	0	0	0	0	0	0	0	0.0000
9	0	0	0	0	0	0	0	0	0	0	0.0000
10	0	0	0	0	0	0	0	0	0	0	0.0000
Colin's	0.0000	0.3333	0.3333	0.0000	0.3333	0.0000	0.0000	0.0000	0.0000	0.0000	
Strategies											

FIGURE 11.12
Excel screenshot of solution for 4 x 5 game.

In a systemic way, we look for the other solutions. Not only can we obtain the solution as described above but also $V = v_1 - v_2 = 4/3$ when $x1 = 2/3$, $x2 = 0$, $x3 = 1/3$ corresponding to play $R1$ with probability $2/3$ never play $R2$ and play $R3$ with probability $1/3$ while playing $C3$ with probability $2/3$ and $C5$ with probability $1/3$ and never playing strategies $C1$, $C2$, or $C4$,but we can find the following solutions as well – all yielding a value of the game of $V = 4/3$, $R2 = 1/3$, $R3 = 2/3$, $R1 = R4 = 0$; $C1 = C4 = 0$, $C2 = C3 = C5 = 1/3$, Figure 11.13.

Ultimately, we have found 15 alternate solutions to this game and suspect that there might be more. We would not have searched at all if we had used linear programming and noticed the conditions for alternate optimal solution existed.

11.2.2 Maple and Game Theory

In Maple we will use with(Optimization) and with(Simplex) to assist us for the total conflict games. We resolve the hitter–pitcher duel of Example 11.2.

>*with(Optimization)*:

=

>*objfunc: = V.*

$$objfunc: = V$$

	A	B	C	D	E	F	G	H	I	J
1		Microsoft Excel 14.0 Sensitivity Report								
2		Worksheet: [LP Template for Zero Sum Game Theory feb 5 2014 4 x 5 alt op example 2 mixed Dec 2.xlsm]ZERO SUM Template								
3		Report Created: 12/2/2014 11:49:02 AM								
4										
5										
6		Variable Cells								
7				Final	Reduced	Objective	Allowable	Allowable		
8		Cell	Name	Value	Cost	Coefficient	Increase	Decrease		
9		$B$32	x1	0.333333333	0	1	1E+30	1		
10		$B$33	x2	0.333333333	0	0	0	1E+30		
11		$B$34	x3	0	0	0	0	0		
12		$B$35	x4	0.333333333	0	0	1E+30	0		
13		$B$36	x5	0	1	0	1E+30	1		
14		$B$37	x6	0	1	0	1E+30	1		
15		$B$38	x7	0	1	0	1E+30	1		
16		$B$39	x8	0	1	0	1E+30	1		
17		$B$40	x9	0	1	0	1E+30	1		
18		$B$41	x10	0	1	0	1E+30	1		
19		$B$42	v1	2	0	0	1E+30	0		
20		$B$43	v2	0.666666667	0	0	1	0		
21		$B$44	y1	1.11022E-16	0	0	0	0.333333333		
22		$B$45	y2	0.333333333	0	0	0	0		
23		$B$46	y3	0.333333333	0	0	0.5	0		
24		$B$47	y4	0	0.333333333	0	1E+30	0.333333333		
25		$B$48	y5	0.333333333	0	0	0.5	1		
26		$B$49	y6	0	0	0	1.333333333	1E+30		
27		$B$50	y7	0	0	0	1.333333333	1E+30		
28		$B$51	y8	0	0	0	1.333333333	1E+30		
29		$B$52	y9	0	0	0	1.333333333	1E+30		
30		$B$53	y10	0	0	0	1.333333333	1E+30		
31		$B$54	v3	0	0	0	1E+30	0		
32		$B$55	v4	1.333333333	0	0	1E+30	0		
33										

FIGURE 11.13

Excel screenshot of alternate optimal solution to Example 11.5.

TABLE 11.5

4 x 5 payoff matrix alternative solutions

R1	R2	R3	R4	C1	C2	C3	C4	C5	V
2/3	0	1/3	0	0	0	2/3	0	1/3	4/3
2/3	0	1/3	0	0	1/3	1/3	0	1/3	4/3
2/3	0	1/3	0	0	1	0	0	0	4/3
2/3	0	1/3	0	0	0	1	0	0	4/3
2/3	0	1/3	0	0	0	0	0	1	4/3
2/3	0	1/3	0	1	0	0	0	0	4/3
1/3	2/3	0	0	0	0	0	1	0	4/3
1/3	1/3	0	1/3	0	0	0	0	1	4/3
2/3	0	0	1/3	0	0	1	0	0	4/3
0	2/3	1/3	0	0	0	1	0	0	4/3
0	0	2/3	1/3	0	0	1	0	0	4/3
0	2/3	0	1/3	0	0	2/3	0	1/3	4/3

TABLE 11.5

4 x 5 payoff matrix

		Colin				
		C1	C2	C3	C4	C5
Rose	R1	1	1	1	2	2
	R2	2	1	1	1	2
	R3	2	2	i	1	1
	R4	2	2	2	1	0

=

>*constr:* = {*.373·x1 + .109·x2 – V ≥ 0,.150·x1 + .250·x2–V ≥ 0,x1 + x2 ≤ 1,x1 + x2 ≥ 1, x1 ≥ 0, x2 ≥ 0*};

constr: = {0 ≤ x1, 0 ≤ x2, 0 ≤ 0.150 x1 + 0.250x2 – V, 0 ≤ 0.373x1 + 0.190x2 – V, 1 ≤ x1 + x2,x1 + x2 ≤1}

=

>

=

>*LPSolve(V, {.373·x1 + .109·x2 – V ≥ 0,.150·x1 + .250·x2 – V≥0,x1 + x2 ≤ 1,x1 + x2 ≥ 1, x1 ≥ 0,x2 ≥ 0},maximize);*

[0.211263736765567,[*V* = 0.211263736765567, *x1* = 0.387362634925820, *x2* = 0.612637366106778]]

=

>*with (simplex):*

=

>*with (simplex):*

=

>*dual (V, {.373·x1 + .109·x2 – V ≥ 0,.150·x1 + .250·x2–V ≥ 0, x1 + x2 ≤ 1, x1 + x2 ≥ 1,x1 ≥ 0,x2 ≥ 0},z)*

–z3 + z4,{0 ≤ –z1 – z3 + z4 – 0.373z5 – 0.150z6, 0 ≤ –z2 – z3 + z4 –0.109z5 – 0.250z6, 1 ≤ z5 + z6

=

>*LPSolve (–z3 + z4, {0 ≤ –z1 – z3 + z4 – 0.373z5 – 0.150z6, 0 ≤ – z2 – z3 + z4 – 0.109z5 – 0.250z6,1 ≤ z5 + z6), assume = nonnegative):*

[0.211263735863746, [*z1* = 0.,*z2* = 0.,*z3* = 0., *z4* = 0.2112637358638746, *z5* = 0.274725277562082, *z6* = 0.7252747224379188]]

=

>

We obtain the same solutions as before, but we need to understand duality to know where to find these solutions within the output.

11.2.3 R and Game Theory

Solving total conflict games in R using linear programming. Using LP as described in Chapter 6, we input the problems and obtain our output as:

```
>library(lpSolve)
>obf<-c(0,0,1)
>constr<-matrix(c(.373,.15,-1,.109,.250,-1,1,1,0),ncol=3,byrow=TRUE)
>constr.dir<-c(">=",">=","=")
>rhs<-c(0,0,1)
>prod.sol<-lp("max",obf,constr,constr.dir,rhs,compute.sens = TRUE)
>prod.sol$objval
[1] 0.2112637
>prod.sol$solution
[1] 0.2747253 0.7252747 0.2112637
>prod.sol$duals
[1] -0.3873626 -0.6126374 0.2112637 0.0000000 0.0000000 0.0000000
```

Again, we must know how to extract the primary and dual solutions. Because of the signs of the duals (a convention of the R algorithm), we must multiply the values by –1 to obtain our solution to the column player.

11.3 The Partial Conflict Game Analysis without Communication

Let's first define a partial conflict game. As opposed to a total conflict game where if a player wins x his opponent loses x, in a partial conflict game the players are not strictly opposed, so it is possible for both players to win or lose some value. In a partial sum game the sum of the values for the two players do not sum to zero.

> **Example 11.6 Consider the following game where the sums of the outcomes do not all sum to zero or the same constant.**

> In Figure 11.14, we note that a plot of the payoffs to each player do not lie in a line, indicating that the game is a partial conflict game because total conflict game values lie in a straight line.

TABLE 11.6

2 × 2 partial conflict game

		Player II	
		C_1	C_2
Player I	R_1	(2,4)	(1,0)
	R_2	(3,1)	(0,4)

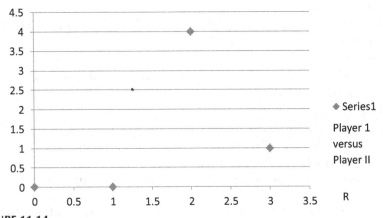

FIGURE 11.14

Payoffs in a partial conflict game do not lie on a line.

What are the objectives of the players in a partial conflict game? In total conflict, each player attempts to maximize his payoffs and necessarily minimizes the other player in the process. But in a partial conflict game, a player may have any of the following objectives from Giordano et al. (2014).

Maximize his payoffs. Each player chooses a strategy in an attempt to maximize his payoff. While he reasons what the other player's response will be he does not have the objective of insuring the other player gets a "fair" outcome. Instead, he "selfishly" maximizes his payoff.

Find a stable outcome. Quite often players have an interest in finding a stable outcome. *A Nash equilibrium outcome is an outcome from which neither player can unilaterally improve*, and therefore represents a stable situation. For example, we may be interested in determining whether two species in a habitat will find equilibrium and coexist, or will one species dominate and drive the other to extinction? The Nash equilibrium is named in honor of John Nash (1950) who proved that every two-person game has at least one equilibrium in either pure strategies or mixed strategies.

Minimize the opposing player. Suppose we have two corporations whose marketing of products interact with each other, but not in total conflict. Each may begin with the objective of maximizing his payoffs. But, if dissatisfied with the outcome, one, or both corporations, may turn hostile and choose the objective of minimizing the other player. That is, a player may forego their long-term goal of maximizing their own profits and choose the short-term goal of minimizing the opposing player's profits. For example, consider a large, successful corporation attempting to bankrupt a "start-up venture" in order to drive him out of business, or perhaps motivate him to agree to an arbitrated "fair" solution.

Find a "mutually fair" outcome, perhaps with the aid of an arbiter. Both players may be dissatisfied with the current situation. Perhaps, both have a poor outcome as a result of minimizing each other. Or perhaps one has executed a "threat" as we study below, causing both players to suffer. In such cases the players may agree to abide by the decision of an arbiter who must then determine a "fair" solution.

In this introduction to partial conflict games, we will assume that both players have the objective of maximizing their payoffs. Next we must determine if the game is played without communication or with communication. "Without communication" indicates that the players must choose their strategies without knowing the choice of the opposing player. For example, perhaps they choose their strategies simultaneously. The term "with communication" indicates that perhaps one player can move first and make his move known to the other player, or that the players can talk to one another before they move. We assume that our games do not allow communication and are played simultaneously.

Further we assume our players are rational, attempting to obtain their best outcomes and that games are repetitive.

One method to find a pure strategy solution is the movement diagram. We define the movement diagram as follows:

Movement Diagram: For Player one, examine the first value in the coordinate and compare R_1 to R_2. For each C_1 and C_2 draw an arrow from the smaller to larger values between R_1 and R_2. For Player two examine the second value in the coordinate and compare C_1 to C_2. For each R_1 and R_2 draw an arrow from the smaller to larger values between C_1 and C_2.

For example, under C_1, we draw an arrow from 2 to 3 and under C_2 from 0 to 1. Under R_1 we draw the arrow from 0 to 4 and under R_2 from 1 to 4. We show this in Figure 11.15.

		Player II	
	C_1		C_2
R_1	(2, 4)	←	(1, 0)
Player I	↓		↑
R_2	(3, 1)	→	(0, 4)

FIGURE 11.15
Movement diagram.

For Solving a 2 x 2 game for Equalizing Strategies
Step 1. Enter Rose's and Colin's Values into the appropriate cells

			Colin				
		C1				C2	
	R1	2	4 <--			1	0
Rose		down_|/			UP_/|\		
	R2	3	1 -->			0	4
Follow the arrows:		FALSE					0
		FALSE					0

FIGURE 11.16
Screenshot of Excel template for movement diagram.

Using the Excel template, Figure 11.16, the arrows indicate "false" in all directions so there is no pure strategy.

We follow the arrows. If the arrows lead us to a value or values where no arrows point out, then we have a pure strategy solution. If the arrows move in a clockwise or counterclockwise direction, then we have no pure strategy solution. Here we move counterclockwise and have no pure strategy solution. As Nash proved all games have a solution either by pure or mixed strategies. As a matter of fact others (Barron [1]; Houseman and Gillman[5]) have shown that some partial conflict games have both a pure and mixed (equalizing) strategy.

We start here by defining the mixed (equalizing) strategy for a partial conflict game.

Rose's game: Rose maximizing, Colin "equalizing" is a total conflict game that yields Colin's equalizing strategy.

Colin's game: Colin maximizing, Rose "equalizing" is a total conflict game that yields Rose's equalizing strategy.

Note: If either side plays its equalizing strategy, then the other side "unilaterally" cannot improve its own situation (it stymies the other player).

We will call this strategy, an equalizing strategy. Each player is restricting what his opponent can obtain by insuring no matter what they do that his opponent always gets the identical solution (Straffin, 2004).

11.4 Methods to Obtain the Equalizing Strategies

We present two methods to obtain equalizing strategies and we will apply these methods to our previous example. The two methods are: linear programming and nonlinear programming. We state here that linear programming works only because each player has only two strategies.

Linear Programming with Two Players and Two Strategies Each

This translates into two maximizing linear programming formulations as shown in Equations (11.3) and (11.4). Formulation (11.3) provides the Nash equalizing solution for Colin with strategies played by Rose while formulation (11.4) provides the Nash equalizing solution for Rose and strategies played by Colin. The two constraints representing strategies are implicitly equal to each other per this formulation (see Fox, 2010).

$$\text{Maximize } V$$

Subject to:

$$N_{1,1}x_1 + N_{2,1}x_2 - V \geq 0$$
$$N_{1,2}x_1 + N_{2,2}x_2 - V \geq 0$$
$$(N_{1,1} - N_{1,2})x_1 + (N_{2,1} - N_{2,2})x_2 = 0 \tag{11.3}$$
$$x_1 + x_2 = 1$$
$$\textit{Nonnegativity}$$

$$\text{Maximize } v$$

Subject to:

$$M_{1,1}y_1 + M_{1,2}y_2 - v \geq 0$$
$$M_{2,1}y_1 + M_{2,2}y_2 - v \geq 0$$
$$(M_{1,1} - M_{2,1})y_1 + (M_{1,2} - M_{2,2})y_2 = 0 \tag{11.4}$$
$$y_1 + y_2 = 1$$
$$\textit{Nonnegativity}$$

With our example, we obtain the following formulation

$$\text{Maximize } V$$

Subject to:

$$4x_1 + x_2 - V \geq 0$$
$$0x_1 + 4x_2 - V \geq 0$$
$$4x_1 - 3x_2 = 0$$
$$x_1 + x_2 = 1$$
$$\textit{Nonnegativity}$$

and

$$\text{Maximize } v$$

Subject to:

$$2y_1 + y_2 - v \geq 0$$
$$3y_1 + 0y_2 - v \geq 0$$
$$-y_1 + y_2 = 0$$
$$y_1 + y_2 = 1$$
$$\text{Nonnegativity}$$

11.4.1 Excel to Obtain Equalizing Strategies

We will now use Excel's solver to address the two player and two strategy problem (Figure 11.17).

$3/7\, x_1, 4/7\, x_2$ corresponding to $3/7\, R_1, 4/7\, R_2$ and $\frac{1}{2}y_1, \frac{1}{2}y_2$ corresponding to $\frac{1}{2}\, C1, \frac{1}{2}C_2$. The Nash equilibrium is $(3/2, 16/7)$.

	Linear Programming				
Decision Variables					
x1	0.571429				
x2	0.428571				
vc	1.714286				
y1	0.5				
y2	0.5				
vr	1.5				
OBJ	3.214286				
Constraints			0	0	2y1+y2-vr>=0
			0	0	3*y1-vc>=0
			1	1	y1+y2=1
			1	0	4x1+x2-vc>=0
			0	0	4x2-vc>=0
			1	1	x1+x2=1
			0	0	-y1+y2=0'
			0	0	3x1-4x2=0

FIGURE 11.17
Screenshot of Excel for two player strategy.

Nonlinear Programming Approach for Two or More Strategies for Each Player

For games with two players and more than two strategies each, we present the nonlinear optimization approach by Barron [1]. Consider a two-person game with a payoff matrix as before. Let's separate the payoff matrix into two matrices M and N for players I and II. We solve the following nonlinear optimization formulation in expanded form, in Equation (11.5).

$$Maximiz \sum_{i=1}^{n} \sum_{j=1}^{m} x_i a_{ij} y_j + \sum_{i=1}^{n} \sum_{j=1}^{m} x_i b_{ij} y_j + -p - q$$

Subject to

$$\sum_{j=1}^{m} a_{ij} y_j \le p. \quad i = 1, 2, \ldots, n,$$

$$\sum_{i=1}^{n} x_i b_{ij} \le q, \quad j = 1, 2, \ldots, m, \qquad (11.5)$$

$$\sum_{i=1}^{n} x_i = \sum_{j=1}^{m} y_j = 1$$

$$x_i \ge 0, y_j \ge 0$$

We return to our previous example. We define M and N as:

$$M = \begin{bmatrix} 2 & 1 \\ 3 & 0 \end{bmatrix} \text{ and } N = \begin{bmatrix} 4 & 0 \\ 1 & 4 \end{bmatrix}$$

We define x_1, x_2, y_1, y_2 as the probabilities for players playing their respective strategies.

By substitution and simplification, we obtain (Figure 11.18)

Maximize

$$6y_1 x_1 + 4y_1 x_2 + x_1 y_2 + 4x_2 y_2 - p - q$$

Subject to:

$$x1 + x2 = 1$$
$$y1 + y2 = 1$$

A				
	2	1		
	3	0		
B	4	0		
	1	4		
dv				
x1	0.428572			
x2	0.571429			
p	1.500001			
y1	0.5			
y2	0.5			
q	2.285715			

FIGURE 11.18
Excel screenshot of nonlinear approach for two player strategy.

$$4x2 - q < 0$$
$$4x1 + x2 - q < 0$$
$$2y1 + y2 - p < 0$$
$$3y1 - p < 0$$
$$Non\text{-}negativity$$

We find the exact same solution as before with the larger screenshot (Figure 11.19).

Finding a Solution

According to Straffin [8], a Nash equilibrium is a solution if and only if it is unique and Pareto Optimal. Pareto optimal refers to the northeast region of a payoff polygon where the payoff polygon is found as the convex setformed by the outcome coordinates, Figure 11.20.

We see in the figure that the Nash equilibrium (1.5, 2.28) is not Pareto optimal and not the solution that we should seek.

At this point, we might try to allow communication and try strategic moves which we do not describe here but can be reviewed in Giordano et al. (2013). Further, we might want to show the method of Nash arbitration although we do not illustrate that here.

	A	B	C	D	E	F	G
	Decision Variables						
	x1	0.428571				Objective Function	
	x2	0.571429				z	0
	x3	0					
	x4	0					
	x5	0			X^TAY^T	1.5	
	x6	0			X^TBY^T	2.285714	
	x7	0					
	x8	0					
	x9	0					
	x10	0					
	P	1.5					
	v2	0					
	y1	0.5					
	y2	0.5					
	y3	0					
	y4	0					
	y5	0					
	y6	0					
	y7	0					
	y8	0					
	y9	0					
	y10	0					
	q	2.285714					
	v4	0					

FIGURE 11.19
Excel screenshot of nonlinear approach for two player strategy.

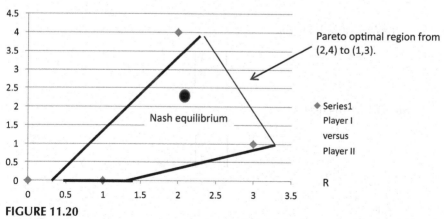

FIGURE 11.20
Payoff polygon and Pareto optimal region.

11.4.2 Maple to Obtain Equalizing Strategies

Partial Conflict Games in Maple. We point out the Barron's equation (2013) for solving pure and equalizing strategies is a nonlinear quadratic programming problem.

>\# Example 11.7 Pure Strategy

=

>$A: = Matrix([[-1,0,0],[2,1,0],[0,1,2,]]);$

$$A := \begin{bmatrix} -1 & 0 & 0 \\ 2 & 1 & 0 \\ 0 & 1 & 2 \end{bmatrix}$$

>$B: = Matrix([[1,2,2],[1,-1,0],[0,1,2,]]);$

$$B := \begin{bmatrix} -1 & 2 & 2 \\ 1 & -1 & 0 \\ 0 & 1 & 2 \end{bmatrix}$$

=

>$ar: = 3: ac: = 3:$
>$PartialConflictGame(ar,ac,A,B,10,10);$

$$[0.,[p \quad = 2.,q = 1.,x_1 = 0.,x_2 = 1.,x_3 = 0.,y_1 = 1.,y_2 = 0.,y_3 = 0.]]$$

=

>\# Example 11.8 Equaling Strategy (no Pure strategy

=

>$A: = Matrix([[2,1],[3,0]]);$

$$A := \begin{bmatrix} 2 & 1 \\ 3 & 0 \end{bmatrix}$$

>$B: = Matrix([[4,0],[1,4]]);$

$$B := \begin{bmatrix} 4 & 0 \\ 1 & 4 \end{bmatrix}$$

=

>$ar: = 2: ac: = 2:$
>$PartialConflictGame(ar,ac,A,B,10,10);$
$$[0.,[p = 1.50000000000000, q = 2.28571428571429, x_1 = 0.428571428571429,$$

$x_2 = 0.571428571428571, y_1 = 0.500000000000000, y_2 = 0.500000000000000]]$

=

>

-

We extract the solution as $p = 1.5$, $q = 2.28571428571429$ *when* $x_1 =$ 0.42857142857149, $x_2 = 0.571428571428571$, $y_1 = 0.5$, $y_2 = 0.5$.

We wrote a small Maple program to solve for these solutions.

-

>PartialConflictGame: = **proc** *(ar, ac, A, B, ip,iq)*

with(LinearAlgebra): with(Optimization):

X: = '<,>'(seq(x[i], i = 1..ar));
Y: = '<,>'(seq(y[i], i = 1..ac));
Cnst: = {seq ((A.Y)[i]≤p,i = 1..ar),seq((Transpose(X)B[i]≤q,i = 1..ac),
add(x[i], i = 1..ar) = 1.add(y[i], i = 1..ac) = 1}; objective: = expand
(Transpose(X) A.Y + Transpose(X) B.Y-p-q);
QPSolve(objective, Cnst,assume = nonnegative, maximize, initialpoint
= {p = ip,q = iq});

11.4.3 R to Obtain Equalizing Strategies

We provide only the code to solve the QP problem. You must load MASS from the CRAN library and then open the library. The following commands should prove the results. Unfortunately, my computer gets an error installing the MASS package,

```
library(MASS)
objfun = function(x){return(6*x[1]*y[1]+4*y[1]*x[2]+
    4*x[2]*y[2]=x[1]*y[2]–p–q}
functionconfun = function(x){
f=NULL
f=rbind(f,x[1]+x[2]–1)
f=rbind(f,y[1]+y[2]–1)
f=rbind(4*x[2]–q)
f=rbind(4*x[1]+x[2]–q)
f=rbind(2*y[1]+y[2]–p)
f=rbind(3*y[1]–p)
return(list(ceq=f,c=NULL))}
x0=c(1,1,1,1,1,1)
solnl(x0,objfun=objfun,confun=confun)
```

The solution for the variables, x_1, x_2, y_1, y_2, p, and q are:

[1] 0.4285714 x0.5714857 0.5000000, 0.5000000 1.0000000 2.2857714.

Another method is using quadratics programming. Please see Lay (2003) for how to formulate a quadratic form. You must load the "quadprog" from the CRAN library.

```
Dmat<-matrix(c(0,0,3,.5,0,0,0,0,2,2,0,0,3,2,0,0,0,0,.5,2,0,0,0,0,0,0,0,0,0,0,0,0,0
    ,0,0,0), nrow=6,byrow=TRUE)[[[I_03124]]]
dvec<-c(0,0,0,0,-1,-1)
A<-matrix(c(1,1,0,0,0,0,0,0,1,1,0,0,4,1,0,0,0,-1,0,0,2,1,-1,0,0,0,3,0,-1,[[[I_
    03125]]]
0),ncol=6,byrow=TRUE)
bvec<-c(1,1,0,0,0,0)
Amat<-t(A)
sol<-solve.QP(Dmat, dvec, Amat, bvec, meq = 0)
```

$solution
[1] 0.4285714 x0.5714857 0.5000000, 0.5000000 1.0000000 2.2857714

$value
[1] 0.0000000

11.5 Nash Arbitration Method

When we have not achieved an acceptable Pareto Optimal solution by other methods that is acceptable by the players, then a game might move to arbitration. The Nash arbitration theorem (1950) states that

> there is one and only arbitration scheme which satisfies rationality, linear invariance, symmetry, and independence of irrelevant alternatives. It is this: if the status quo (SQ) point is (x_0, y_0), then the arbitrated solution point N is the point (x, y) in the polygon with $x > x_0$, $y > y_0$ which maximizes the product of $(x - x_0)$ $(y - y_0)$.

There are a few terms, strategies, and methods we must discuss prior to illustrating this entire process.

Using this security level as our status quo point we can now formulate the Nash arbitration scheme.

There are four axioms that are required to be meet using the arbitration scheme.

- Axiom 1: Rationality. The solution should be in the negotiation set.
- Axiom 2: Linear Invariance. If either Rose's or Colin's utilities are transformed by a positive linear function, the solution point should be transformed by the same function.
- Axiom 3. Symmetry. If the polygon happens to be symmetric about the line of slope±1 through the status quo point, then the solution should be on this line.
- Axiom 4: Independence of Irrelevant Alternatives. Suppose N is the solution point for a polygon, P with status quo point SQ. Suppose Q is another polygon which contains both SQ and N, and is totally contained in P. Then N should also be the solution point to Q with status quo point SQ.
- First, we will define the SQ point as either the security levels found through Prudential Strategies or the threat level found by communications methods. We will use only the security levels here.

Finding the Prudential Strategy (Security Levels)

The security levels are the payoffs to the players in a partial conflict game where each player attempt to maximize their own payoff. We can solve for these payoffs using a separate linear program for each security level.

We have Rose in Rose's game and Colin in Colin's game. So, the LP formulation are:

$$\text{Maximize } V$$

Subject to:

$$N_{1,1}y_1 + N_{1,2}y_2 + ... + N_{1,m}y_n - V \geq 0$$
$$N_{2,1}y_1 + N_{2,2}y_2 + ... + N_{2,m}y_n - V \geq 0$$
$$...$$
$$N_{m,1}x_1 + N_{m,2}x_2 + ... + N_{m,n}x_n - V \geq 0$$
$$y_1 + y_2 + ... + y_n = 1$$
$$y_j \leq 1 \quad \text{for} \quad j = 1,...,n$$
$$\text{Nonnegativity}$$

where the weights y_i yield Colin's prudential strategy and the value of V is the security level for Colin.

$$\text{Maximize } v$$

Subject to:

$$M_{1,1}x_1 + M_{2,1}x_2 + \ldots + M_{n,1}x_n - v \geq 0$$
$$M_{1,2}x_1 + M_{2,2}x_2 + \ldots + M_{n,2}x_n - v \geq 0$$
$$\ldots$$
$$M_{1,m}y_1 + M_{2,m}y_2 + \ldots + M_{m,n}y_n - v \geq 0$$
$$x_1 + x_2 + \ldots + x_m = 1$$
$$x_i \leq 1 \ for \ i = 1, \ldots, m$$
$$Nonnegativity$$

Example 11.9 Given the following payoff matrix from Example 11.7, Section 11.4

Let's return to Section 11.4, Example 11.7 to illustrate finding the security levels. Let *SLR* and *SLC* represent the security levels for Rose and Colin, respectively (Figure 11.21). We use linear programming to find these values using the following formulations:

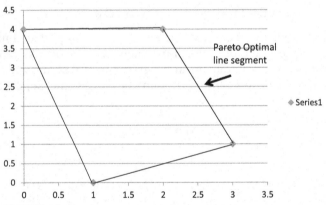

FIGURE 11.21
Payoff polygon and Pareto optimal region.

TABLE 11.7

Payoff matrix

		Player II	
		C_1	C_2
Player I	R_1	(2, 4)	(1, 0)
	R_2	(3, 1)	(0, 4)

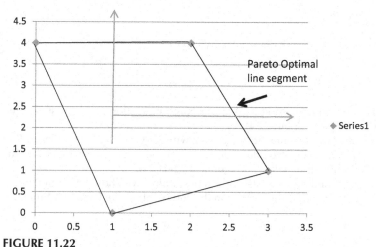

FIGURE 11.22
Payoff polygon and Pareto optimal region.

MaxSLR

$$2\,x1 + 3\,x2 - SLR > 0$$

$$1x1 + 0x2 - SLR > 0$$

$$x1 + x2 = 1$$

Non-negativity

Max SLC

$$4y1 - SLC > 0$$

$$1y1 + 4y2 - SLC > 0$$

$$y1 + y2 = 1$$

Non-negativity

The solution yields both how the game with Prudential Strategies is played and the security levels. Player 1 always plays R1 and Player 2 plays 4/7 C1 and 3/7 C2. The security level, the values of each player Prudential Strategy is (1, 16/7) = (1, 2.286).

Using this security level as our status quo point we can now formulate the Nash arbitration scheme (Figure 11.22).

We apply this theorem in a nonlinear optimization model framework,

Maximize

$$(x-1)(y-2.286)$$

Subject to

$$3x + y = 10$$
$$x \geq 1$$
$$y \geq 2.286.$$

We find, in this example, that our Nash arbitration point for status quo (1, 2.286) is the point (2,4).

Here, we recommend the Solver with GRG nonlinear routine to find the solution. We provide screen shots of the input to the solver and the resulting output solution (Figures 11.23–11.24).

After finding the Nash arbitration point, we should also determine how this point is obtained from the end points of the Pareto Optimal line

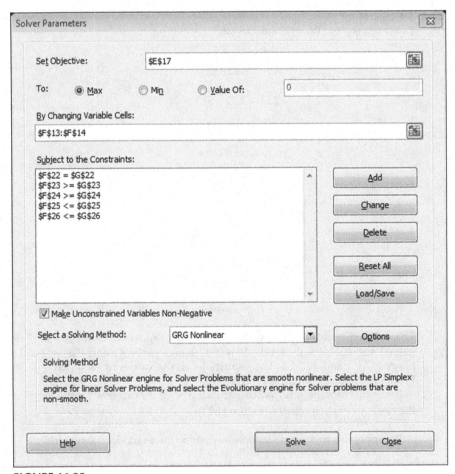

FIGURE 11.23
Excel Solverscreen for Nash arbitration.

Decision Variables					
x	2				
y	4				

Objective Function					
1.714					

Constraints					
	Used	RHS			
	10	10	On the Pareto Optimal		
	2	1			
	4	2.286			
	2	3			
	4	4			

FIGURE 11.24
Excel screenshot of solution for Nash arbitration.

segment. We will solve for the probabilities to play (2,4) and to play (3,1). Here is the formulation we will use:

$$2p_1 + 3p_2 = 2(x \text{ coordinate of Nash arbitration point})$$
$$4p_1 + 1p_2 = 4 \text{ (y coordinate of Nash arbitration point)}$$

This solves with $p_1 = 1, p_2 = 0$. The arbitrators hold on to the strategies that yield (2,4) always.

Example 11.10 Consider the payoff matrix (Table 11.8).

The Nash equilibrium is found through equalizing strategies as (34/9, 28/5). It is not Pareto Optimal, so Rose is assumed to be unhappy with

TABLE 11.8

Payoff matrix

		Colin	
Rose		C1	C2
	R1	(2,6)	(10,5)
	R2	(4,4)	(3,8)

Decision Variables		
x	6.18889	
y	6.63333	

Objective Function
2.49148

Constraints

	Used	RHS	
	9.28572	9.28571	On the Pareto Optimal Line
	6.18889	3.78	
	6.63333	5.6	
	6.18889	10	
	6.63333	8	

FIGURE 11.25
Excel screenshot of solution for Nash arbitration.

her outcome and sets out for arbitration. We find the security levels by linear programming as (34/9, 28/5) with Prudential strategies 1/9 R1, 8/9 R2, and 3/5 c1, 2/5 C2.

Using this security level as our status quo we can formulate the Nash arbitration as (Figure 11.25)

Maximize

$$(x - 34/9)(y - 28/5)$$

Subject to

$$3/7x + y = 65/7$$

$$x \geq 34/9$$

$$y \geq 28/5$$

$$x < 10$$

$$y < 8$$

non-negativity

Our Nash arbitrated solution is (6.1889, 6.6333) which is better for Rose than the equilibrium value of 3.78 and better for Colin than his equilibrium value of 5.6. Both players are better off with arbitration. The scheme to achieve this output is found from the solution to the following systems of equations:

$$10p_1 + 3p_2 = 6.18889$$

$$5p_1 + 8p_2 = 6.63333.$$

We find p_1 = 45.5% and p_2 = 54.5%. Further since both (10,5) and (3,8) are in C2, Colin always plays C2 while Rose plays 45.5 R1 and 54.5% R2.

Example 11.11 Writer's Guild Strike and Nash Arbitration

We examine strategic moves. The writers move first, and their best results is again (2,4). If management moves first the best result is (2,4). First moves keep us at the Nash equilibrium. The writers consider a threat and tell management that if they choose SQ that they will strike putting us at (1,3). This result is indeed a threat as it is worse for both the writers and management.

However, the options for management under IN are both worse than (1,3) so they do not accept the threat. The writers do not have a promise. At this point we might involve an arbiter using the method as suggested earlier.

Writers, management security levels are found from prudential strategies and the following

LP models. The security levels are (2, 3). We show this in Figure 11.26. The Nash arbitration formulation is

Maximize

$$(x-2)(y-3)$$

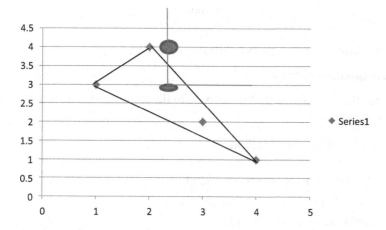

FIGURE 11.26
Payoff polygon for Writers Guild strike.

subject to

$$3/2x + y = 7$$
$$x > 2$$
$$y > 3.$$

The Nash equilibrium value, (2,4), lies along the Pareto Optimal line segment. But the Writer's want to do better by going on strike and forcing arbitration, which is what they did. In this example, we consider "binding arbitration" where the players have a third party work out the outcomes that best meets their desires and is acceptable to all players. Nash found that this outcome can be obtained by:

The status quo point is the security levels of each side. We find these values using prudential strategies as (2,3). The function for the Nash Arbitration scheme is *Maximize* $(x - 2)(y - 3)$.

Using technology, we find the desired solution to our NLP as

$$x = 2.3333$$
$$y = 3.5.$$

We have the x and y coordinate (2.3333, 3.5) as our arbitrated solution. We can also determine how the arbiters should proceed. We solve the following simultaneous equations

$$2p_1 + 4p_2 = 2.3333$$
$$4p_1 + p_2 = 3.5$$

We find that the probabilities to be played are 5/6 and 1/6, Further we see that player I, the writers, always play R_2 so management arbiter plays 5/6 C_1 and 1/6 C_2 during the arbitration.

11.5.1 Maple and the Nash Arbitration Method

Nash Arbitration in Maple

Here are the commands that we choose to illustrate as we wrote a Nash arbitration procedure.

Nash Arbitration Method
-

```
>NashArbitration: = proc(slr, slc, x1,y1, x2, y2)
    with(LinearAlgebra): with(Optimization):
    objective: = (x–slr)·(y–slc);
```

$$in1: = (y - y1) - \frac{(y1 - y2)}{(x1 - x2)}(x - x1);$$

```
    NLPSolve(objective, {in1 = 0, x ≥ slr, y ≥ slc, x ≤ x2, y ≤ y1},maximize);
```

An example with our writer's guild problem,

```
=
>NashArbitration(2, 3, 2, 4, 4, 1);
     [0.166666666666666741, [x = 2.33333333333333,
        y = 3.50000000000000]]
=
>
-
```

You will also still need to solve a set of simultaneous equations to get the percentages to play each set of strategies.

```
=
>solve({2p + 4q = 2.333, 4p + q = 3.5}, {p,q});
     {p = 0.8333571429, q = 0.1665714286}
=
=
```

The solutions are as before.

11.5.2 R and the Nash Arbitration Method

The CRAN library routines for game theory that are beyond our coverage.
   However, we provide some R routines to find the Nash Arbitration point.
   First, we need to have solved for the security values and have the function to maximize.
   We return to our writer's guild example.

Maximize

$$(x - 2)(y - 3)$$

subject to

$$3/2x + y = 7$$

$$x > 2$$

$$y > 3.$$

Recall, by default R solves the minimum so we will use $-1 * (x - 2)*(y - 3)$ and the optimization methods described in Chapters 7 and 8.
   First, we convert to a single variable problem and use Golden Section as described in Chapter 7.

```
f=function(x)
+ {
+ –1*(x – 2)*(4 – 1.5*x)
+}
>
>golden.section.search(f,2,4,0.05)
```

Iteration # 1
f1=0.1114562
f2=1.055728
f2>f1
New Upper Bound=3.236068
New Lower Bound=2
New Upper Test Point=2.763932
New Lower Test Point=2.472136

Iteration # 2
f1=–0.1377674
f2=0.1114562
f2>f1
New Upper Bound=2.763932
New Lower Bound=2
New Upper Test Point=2.472136
New Lower Test Point=2.291796

Iteration # 3
f1=–0.1640786
f2=–0.1377674
f2>f1
New Upper Bound=2.472136
New Lower Bound=2
New Upper Test Point=2.291796
New Lower Test Point=2.18034

Iteration # 4
f1=–0.1315562
f2=–0.1640786

f2<f1
New Upper Bound=2.472136
New Lower Bound=2.18034
New Lower Test Point=2.291796
New Upper Test Point=2.36068

Iteration # 5
f1=−0.1640786
f2=−0.1655449
f2<f1
New Upper Bound=2.472136
New Lower Bound=2.291796
New Lower Test Point=2.36068
New Upper Test Point=2.403252

Iteration # 6
f1=−0.1655449
f2=−0.1593337
f2>f1
New Upper Bound=2.403252
New Lower Bound=2.291796
New Upper Test Point=2.36068
New Lower Test Point=2.334369

Iteration # 7
f1=−0.1666651
f2=−0.1655449
f2>f1
New Upper Bound=2.36068
New Lower Bound=2.291796
New Upper Test Point=2.334369
New Lower Test Point=2.318107

Iteration # 8
f1=−0.1663189
f2=−0.1666651
f2<f1

New Upper Bound=2.36068

New Lower Bound=2.318107

New Lower Test Point=2.334369

New Upper Test Point=2.344419

We accept our x value as the midpoint of 2.334369 and 2.344419. This value is 2.339394.

We back solve to find y, since the solution falls on the line, $3/2x + y = 7$. We compute y as $y = 3.4909$.

Our Nash arbitration point, using R, is (2.339394, 3.4909).The value of the function $f(2.339394, 3.4909) = 0.16666$ (remember to change the sign if you take the functional value directly from R).

We still solve the following simultaneous equations

$$2p_1 + 4p_2 = 2.3333$$

$$4p_1 + p_2 = 3.5.$$

We find that the probabilities to be played are 5/6 and 1/6.

11.6 Illustrative Modeling Examples of Zero-Sum Games

In this section we present some illustrative examples of game theory. We present the scenario, discuss the outcomes used in the payoff matrix, and present a possible solution for the game. In most game theory problems, the solution suggests insights in how to play the game rather than a definitive methodology to "winning" the game. We present only total conflict game as illustrative examples in this section.

Example 11.12 Penalty Kicks in Soccer

This example is adapted from an article by Chiappori, Levitt, and Groseclose (see additional readings). Let's consider a penalty kick in soccer between two players, the kicker and the opponent's goalie. The kicker basically has two alternatives of strategies that we will consider: he might kick the ball left or kick it right. The goalie will also have two strategies: the goal can prepare to dive left or right to block the kick. We will start very simply in the payoff matrix by awarding a 1 to the player that is successful and a –1 to the player that is unsuccessful. The payoff matrix would simply be as follows:

TABLE 11.9

Payoff matrix for Example 11.12

		Goalie	
		Dive Left	**Dive Right**
Kicker	Kick left	(–1,1)	(1,–1)
	Kick right	(1,–1)	(–1,1)

TABLE 11.10

Payoff matrix from kicker perspective

		Goalie	
		Dive Left	**Dive Right**
Kicker	Kick left	-1	1
	Kick right	1	-1

TABLE 11.11

Historic datafor Example 11.12

		Goalie	
		Dive Left	**Dive Right**
Kicker	Kick left	0.58,-0.58	0.95,-0.95
	Kick right	0.93,-0.93	0.70,-0.70

or as just from the kicker's prospective.

There is no pure strategy. We find a mixed strategy to the zero-sum game using either linear programming or the method of oddments. The mixed strategy results are that the kicker randomly kicks 50% left and 50% right while the goalie randomly dives 50% left and 50% right. The value of the game to each player is 0.

Let's refine the game using real data. A study was done in the Italian Football league in 2002 by Ignacio Palacios-Huerta. As he observed the kicker can aim the ball to the left or to the right of the goalie, and the goalie can dive either left or right as well. The ball is kicked with enough speed that the decisions of the kicker and goalie are affectively made simultaneously. Based on these decisions the kicker is likely to score or not score. The structure of the game is remarkably similar to our simplified game. If the goalie dives in the direction that the ball is kicked then he has a good chance of stopping the goal, and if he dives in the wrong direction then the kicker is likely to get a goal.

Based upon an analysis of roughly 1,400 penalty kicks, Palacios-Huerta determined the empirical probabilities of scoring for each of four outcomes: the kicker kicks left or right and the goalie dives left or right. His results led to the following payoff matrix (Table 11.11).

TABLE 11.12

Oddments determined for Example 11.12

		Goalie			
		Dive Left	Dive Right	Oddments	Probabilities
Kicker	Kick left	0.58	0.95	.37	.23/.60 = 0.383
	Kick right	0.93	0.70	.23	.37/.60 = 0.6166
Oddments		0.35	0.25		
Probabilities		0.25/0.60 = 0.416	0.35/0.60 = 0.5833		

There is no pure strategy equilibrium as expected. The kicker and the goalie must use a mixture of strategies since the game is played over and over. Neither player wants to reveal a pattern to their decision. We apply the method of oddments to determine the mixed strategies for each player based on this data (Table 11.12).

We find the mixed strategy for the kicker is 38.3% kicking left and 61.7% kicking right while the goalie dives right 58.3% and dives left 41.7%. If we merely count percentages from the data that was collected by Palacios-Huerta in his study from 459 penalty kicks over 5 years of data, we find the kicker did 40% kicking left and 60% kicking right while the goalie dove left 42% and right 58%. Since our model closely approximates our data, our game theory approach adequately models the penalty kick.

Example 11.13 Batter–Pitcher Duel

In view of the use of technology in sports today, we present an example of the hitter-pitcher duel. First in this example we extend the strategies for each player in our model. We consider a batter-pitcher duel between Ryan Howard of the Philadelphia Phillies and various pitchers in the national league where the pitcher throws a fastball, a split-finger fastball, a curve ball, and a change-up. The batter, aware of these pitches, must prepare appropriately for the pitch. Data is available from many web sites that we might use. In this example, we obtained the data from the internet, www.STATS.com. We consider both a right-handed pitcher and a left-handed pitcher separately in this analysis.

For a National League right-handed pitcher (RHP) versus Ryan Howard, we have compiled the following data. Let FB = fastball, CB = curveball CH = change up, SF = split-fingered fastball.

Howard/RHP	FB	CB	CH	SF
FB	0.337	0.246	0.220	0.200
CB	0.283	0.571	0.339	0.303
CH	0.188	0.347	0.714	0.227
SF	0.200	0.227	0.154	0.500

Both the batter and pitcher want the best possible result. We set this up as a linear programming problem. Our decision variables are x_1, x_2, x_3, x_4 as the percentages to guess FB, CB, CH, SF respectively and V represents Howard's batting average.

Max V

Subject to

$$0.337x_1 + 0.283x_2 + 0.188x_3 + 0.200x_4 - V > 0$$
$$0.246x_1 + 0.571x_2 + 0.347x_3 + 0.227x_4 - V > 0$$
$$0.220x_1 + 0.339x_2 + 0.714x_3 + 0.154x_4 - V > 0$$
$$0.200x_1 + 0.303x_2 + 0.227x_3 + 0.500x_4 - V > 0$$
$$x_1 + x_2 + x_3 + x_4 = 1$$
$$x_1, x_2, x_3, x_4, V > 0$$

We solve this linear programming problem and find the optimal solution (strategy) is to guess the fastball (FB) 27.49%, guess the curveball (CB) 64.23%, never guess changeup (CH), and guess split-finger fastball (SF) 8.27% of the time to obtain a 0.291 batting average.

The pitcher then wants to also keep the batting average as low as possible. We can setup the linear program for the pitcher as follows.

Our decision variables are y_1, y_2, y_3, y_4 as the percentages to throw the FB, CB, CH, SF respectively and V represents Howard's batting average.

Min V

Subject to

$$0.337y_1 + 0.246y_2 + 0.220y_3 + 0.200y_4 - V < 0$$
$$0.283y_1 + 0.571y_2 + 0.339y_3 + 0.303y_4 - V < 0$$
$$0.188y_1 + 0.347y_2 + 0.714y_3 + 0.227y_4 - V < 0$$
$$0.200y_1 + 0.227y_2 + 0.154y_3 + 0.500y_4 - V < 0$$
$$y_1 + y_2 + y_3 + y_4 = 1$$
$$x_1, y_2, y_3, y_4, V > 0$$

We find the RHP should randomly throw 65.93% fastballs, no curveballs, 3.25% changeups, and 30.82% split-finger fastballs for Howard to keep only that 0.291 batting average.

Now, we also have statistics for Howard versus a left handed pitcher (LHP).

Howard/ LHP	FB	CB	CH	SF
FB	0.353	0.185	0.220	0.244
CB	0.143	0.333	0.333	0.253
CH	0.071	0.333	0.353	0.247
SF	0.300	0.240	0.254	0.450

We setup as before and solve as a linear programming problem.

$$Max \ V$$

Subject to

$$0.353x_1 + 0.143x_2 + 0.071x_3 + 0.300x_4 - V > 0$$
$$0.185x_1 + 0.333x_2 + 0.333x_3 + 0.240x_4 - V > 0$$
$$0.220x_1 + 0.333x_2 + 0.353x_3 + 0.254x_4 - V > 0$$
$$0.244x_1 + 0.253x_2 + 0.247x_3 + 0.450x_4 - V > 0$$
$$x_1 + x_2 + x_3 + x_4 = 1$$
$$x_1, x_2, x_3, x_4, \ V > 0$$

We find the optimal solution for Howard versus LHP. Howard should guess as follows: never guess fastball, guess curveball 50.82%, guess Change up 21.15% and guess split-finger 28.02% for a batting average of.307. For the LHP pitchers facing Howard we set the following linear program.

$$Min \ V$$

Subject to

$$0.353y_1 + 0.185y_2 + 0.220y_3 + 0.244y_4 - V < 0$$
$$0.143y_1 + 0.333y_2 + 0.333y_3 + 0.253y_4 - V < 0$$
$$0.071y_1 + 0.333y_2 + 0.353y_3 + 0.247y_4 - V < 0$$
$$0.300y_1 + 0.240y_2 + 0.254y_3 + 0.450y_4 - V < 0$$
$$y_1 + y_2 + y_3 + y_4 = 1$$
$$x_1, y_2, y_3, y_4, \ V > 0$$

The pitcher should randomly throw 0.33% fastballs, 67.88% curveballs, 0% change ups, and 31.78% split-finger so Howard batting average remains at.307.

So you are manager of the opposing team and you are in the middle of a close game. There are two outs and runners in scoring position with Ryan Howard coming to bat. Do you keep your LHP in the game or switch to a RHP. The percentages say to switch to the RHP since 0.291 <0.307. You tell the catcher and pitcher to randomly select the pitches to be thrown to Howard.

Did the NY Yankees conduct in-depth analysis in order to get an optimal strategy and type pitchers against Howard in the World Series? We have gathered some statistical data on Howard based upon facing a RHP and a LHP seeing just FB and CB. These are provided in Table 11.13.

Examination clearly shows that Howard batted poorly against RHP curveballs and against LHP. The NY Yankees took advantage of this by

TABLE 11.13

Howard versus right hand pitchers data for Example 11.13

| | | Howard versus RHP | | | | |
Year	Total FB	Percent FB	BA	Total CB	Percent CB	BA
2005	620	85.87	0.336	102	14.13	0.217
2006	880	82.47	0.399	187	17.53	0.297
2007	767	81.16	0.386	178	18.84	0.074
2008	875	81.10	0.286	205	18.90	0.27
2009	802	75.66	0.337	258	24.34	0.283
Versus LHP						
2005	117	78.52	0.172	32	21.48	0.250
2006	437	77.89	0.323	124	22.11	0.276
2007	517	83.11	0.264	105	16.89	0.133
2008	512	83.11	0.266	104	16.89	0.160
2009	465	72.77	0.225	174	27.23	0.143

throwing more LHP to pitch to Howard and more curveballs resulting in only a 0.174 batting average.

As Ryan Howard's manager you want to improve his batting ability against both curve ball and LHP. Only by improving against these strategies can he effect change.

11.7 Partial Conflict Games Illustrative Examples

In this section we present some example of partial conflict games and their solution techniques.

Example 11.14 Writers Guild Strike, 2007–2008 (Fox, 2008)

The 2007–2008 Writers Guild of America strike was a strike by the Writers Guild of America, East (WGAE) and the Writers Guild of America, West (WGAW) that started on November 5, 2007. The WGAE and WGAW were two labor unions representing film, television, and radio writers working in the United States. Over 12,000 writers joined the strike. These entities will be referred to in the model as the Writers Guild.

The strike was against the Alliance of Motion Picture and Television Producers (AMPTP), a trade organization representing the interests of 397 American film and television producers. The most influential of these are eight corporations: CBS Corporation, Metro-Goldwyn-Mayer, NBC Universal, News Corp/Fox, Paramount Pictures, Sony Pictures

Entertainment the Walt Disney Company, and Warner Brothers. We will refer to this group as Management.

The Writers Guild has indicated their industrial action would be a "marathon". AMPTP negotiator Nick Counter has indicated negotiations would not resume as long as strike action continues, stating, "We're not going to negotiate with a gun to our heads – that's just stupid."

The last such strike in 1988 lasted 21 weeks and 6 days, costing the American entertainment industry an estimated $500 million ($870 million in 2007 dollars). According to a report on the January 13, 2008 edition of *NBC Nightly News*, if one takes into account everyone affected by the current strike, the strike has cost the industry $1 billion so far; this is a combination of lost wages to cast and crew members of television and film productions and payments for services provided by janitorial services, caterers, prop and costume rental companies, and the like.

The TV and movie companies stockpiled "output" so that they could possibly outlast the strike rather than work to meet the demands of the writers and avoid the strike.

Build a model that presents how each side should progress.

Game Theory Approach

Let us begin by stating strategies for each side. Our two rational players will be the Writers Guild and the Management. We develop strategies for each player.

Strategies:
Writers Guild: Their strategies are to strike (S) or not to strike (NS).
Management: Salary Increase and revenue sharing (In) or status quo (SQ).

First, we rank order the outcomes for each side in order of preference. (These rank orderings are ordinal utilities.)

Writers' Alternatives and Rankings

Strike–Status Quo S SQ–writers' worst case (1)
No strike–Status Quo NS SQ–writers' next to worst case (2)
Strike–Salary increase and revenue sharing S IN – writers' next to best case (3)
No strike–Salary increase and revenue sharing NS IN–writers' best case (4)

Management's Alternatives and Rankings

Strike–Status Quo–management's next to best case (3)
No strike–Status Quo–management's best case (4)
Strike–Salary increase and revenue sharing – management's next to worst case (2)
No strike–Salary increase and revenue sharing–management's worst case (1)

		Management (Colin)	
		SQ	*IN*
Writer's (Rose)	*S*	(1,3)	(3,2)
	NS	(2,4)	(4,1)

FIGURE 11.27
Payoff matrix for Writers Guild strike.

		Management	
		SQ	*IN*
Writer's	*S*	(1,3)	(3,2)
	NS	(2,4)	(4,1)

FIGURE 11.28
Movement diagram for Writers Guild strike.

This provides us with a payoff matrix consisting of ordinal values, see Figure 11.27. We will refer to the Writers as Rose and the Management as Colin.

Payoff Matrix for Writers and Management

We use the movement diagram, see Figure 11.28, to find (2,4) as the likely outcome.

We notice that the movement arrows point towards (2,4) as the pure Nash equilibrium. We also note that this result is not satisfying to the Writer's Guild and that they would like to have a better outcome. Both (3,2) and (4,1) within the payoff matrix provide better outcomes to the Writers.

We can employ several options to try to secure a better outcome for the Writer's. We can first try Strategic Moves and if that fails to produce a better outcome then we can move on to Nash Arbitration. Both of these methods employ communications in the game. In strategic moves, we examine the game to see if "moving first" changes the outcome, if threatening our opponent changes the outcome, or if making promises to our opponent changes our outcome, or a combination of threats and promises in order to change the outcome.

11.8 Exercises

1. Resolve the penalty kick problem if the values are:

		Goalie	
		Dive Left	**Dive Right**
Kicker	Kick left	0.50,–0.50	0.95,–0.95
	Kick right	0.83,–0.83	0.75,–0.75

2. Resolve a pitcher-hitter duel if the values are

		Pitcher		
		Fast ball	**Change up**	**Curve ball**
Hitter	Fast ball	.325	.125	.200
	Change up	.100	.375	.263
	Curve ball	.202	.200	.280

3. Given the following partial conflict game, solve the game for the Nash equilibrium. Find the security levels and find the Nash arbitration point.

		Colin		
		C1	**C2**	**C3**
Rose	R1	(–2,–4)	(5,–2)	(1,4)
	R2	(–3,–3)	(2,1)	(3,4)
	R3	(2,3)	(1,1)	(3,–1)

11.9 References and Suggested Further Reading

Aiginger, K. (1999). *The Use of the Game Theoretic Models for Empirical Industrial Organizations*. Kluwer Academic Press, Dordrecht, FRG, pp. 253–277.

Aumann, R. (1987). *Game Theory. The New Palgrave: A Dictionary of Economics*, Palgrave Macmillan, London, UK.

Barron, E. N. (2013). *Game Theory: An Introduction*. John Wiley & Sons, New York.

Bazarra, M., H. Sherali, and C. Shetty. (2006). *Nonlinear Programming*, 3rd ed. John Wiley & Sons, New York.

Brams, S. (1994a). Theory of moves, *American Scientist*, 81: 562–570.

Brams, S. (1994b). *Theory of Moves*. Cambridge University Press. Cambridge, UK.

Camerer, C. (2003). Description and Introduction. In *Behavioral Game Theory: Experiments in Strategic Interaction*. Russell Sage Foundation, New York, 1–25.

Cantwell, G. (2003). Can two person zero sum game theory improve military decision-making course of action selection? Monograph. School of Advanced Military Studies, United States Army Command and General Staff College, Fort Leavenworth, KS.

Chatterjee, K. and W. Samuelson (2001). *Game Theory and Business Applications*. Kluwer Academic Press, Boston. MA.

Chiappori, A., S. Levitt, and T. Groseclose (2002). Testing mixed-strategy equilibria when players are heterogeneous: The case of penalty kicks in soccer, *American Economic Review*, 92(4): 1138–1151.

Crawford, V. (1974). Learning the optimal strategy in a zero-sum game. *Econometrica*, 42(5): 885–891.

Danzig, G. (1951). Maximization of a linear function of variables subject to linear inequalities. Chapter XXI in T. Koopman (ed.), *Activity Analysis of Production and Allocation Conference Proceeding*, John Wiley Publishers, pp. 339–347.

Danzig, G. (2002). Linear programming, *Operations Research*, 50(1): 42–47.

Daskalakis, C., P. W. Goldberg, and C. H. Papadimirtriou (2008). The complexity of computing a Nash equilibrium. *Communications of the ACM*, 50(2): 89–97.

Dixit, A. and Nalebuff, B. (1991). Thinking strategically: The competitive edge in business, politics, and everyday life. W. W. Norton Publisher, New York.

Dorfman, R. (1951). Application of the simplex method to a game theory problem. Chapter XXII in T. Koopman (ed.), *Activity Analysis of Production and Allocation Conference Proceeding*. John Wiley Publishers, pp. 348–358.

Dutta, P. (1999). *Strategies and Games: Theory and Practice*. MIT Press, Cambridge, MA.

Fox, W. P. (2008). Mathematical modeling of conflict and decision making "the Writers Guild strike 2007–2008", *Computers in Education Journal*, XVIII(3): 2–11.

Fox, W. P. (2010). Teaching the applications of optimization in game theory's zero-sum and non-zero sum games, *International Journal of Data Analysis Techniques and Strategies (IDATS)*, 2(3): 258–284.

Fox, W. P. (2012). *Mathematical Modeling with Maple*. Cengage Publishers, Boston, MA.

Fox, W. P. (2014). Finding alternate optimal solutions in a zero-sum game with MS-Excel, *Computers in Education Journal*, 6(3): 10–19.

Gale, D., H. Kuhn, and A. Tucker (1951). Linear programming and the theory of games. Chapter XIX in T. Koopman (ed.), *Activity Analysis of Production and Allocation Conference Proceeding*. John Wiley Publishers, pp. 317–329.

Game Theory. http://en.wikipedia.org/wiki/List_of_games_in_game_theory (accessed April 26, 2013).

Gillman, R. and D. Housman(2009). *Models of Conflict and Cooperation*. Providence: American Mathematical Society, pp.189–195.

Gintis, H. (2000). *Game Theory Evolving: A Problem-Centered Introduction to Modeling Strategic Behavior*. Princeton University Press, Princeton, NJ.

Giordano, F., W. Fox, and S. Horton (2014). *A First Course in Mathematical Modeling*. Cengage Publishers, Boston.

Harrington, J. (2008). *Games, Strategies, and Decision Making*. Worth Publishers, New York.

Isaacs, R. (1999). *Differential Games: A Mathematical Theory with Applications to Warfare and Pursuit, Control and Optimization*. Dover Publications, New York.

Klarrich, E. (2009). The mathematics of strategy. *Classics of the Scientific Literature*, October. (www.pnas.org/site/misc/classics5.shtml).

Kleinberg, J. and D. Easly (2010). *Networks Clouds, and Markets: Reasoning about a Highly Connected World*. Cambridge University Press, Cambridge, UK.

Koopman, T. (1951). *Activity Analysis of Production and Allocation Conference Proceeding*. John Wiley Publishers, New York.

Kuhn, H. W. and A. W. Tucker(1951). Nonlinear programming. In J. Newman (ed.), *Proceedings* Second *Berkeley Symposium on Mathematical Statistics and Probability*. University of California Press, Berkeley, CA.

Lay, D. (2003). *Linear Algebra and Its Applications*, 3rd ed. Pearson Publishing. Essex, UK.

Leyton-Brown, K. and Y. Shoham (2008). *Essentials of Game Theory*. Morgan & Claypool Publishers,San Rafael, CA.

Mansbridge, J. (2013). Game theory and government. www.hks.harvard.edu/news-events/publications/insights/democratic/jane-mansbridge (accessed March 7, 2013).

McCormick, G. and L. Fritz (2009).The logic of warlord politics. *Third World Quarterly*, 30(1): 81–112.

Miller, J. (2003). *Game Theory at Work: How to Use Game Theory to Outthink and Outmaneuver Your Competition*. McGraw-Hill, New York.

Myerson, R. (1991). *Game Theory: Analysis of Conflict*. Harvard University Press, Cambridge, MA.

Nash, J. (1950a). The bargaining problem, *Econometrica*, 18: 155–162.

Nash, J. (1950b). Equilibrium points in n-person games, *Proceedings of the National Academy of Sciences of the United States of America*, 36(1): 48–49.

Nash, J. (1951). Non-cooperative games, *Annals of Mathematics*, 54: 289–295.

Nash, J. (2009). Lecture at NPS. February 19, 2009.

Osborne, M. (2004). *An Introduction to Game Theory*. Oxford University Press, Oxford, UK.

Papayoanou, P. (2010). *Game Theory for Business, e-book*, Probabilistic Publishing, www.decisions-books.com/Links.html.

Rasmusen, E. (2006). *Games and Information: An Introduction to Game Theory*, 4th ed. Wiley-Blackwell, New York.

Shoham, Y. and K. Leyton-Brown (2009). *Multiagent Systems: Algorithmic, Game-Theoretic, and Logical Foundations*. Cambridge University Press, New York.

Smith, J. and G. Price (1973). The logic of animal conflict, *Nature*, 246(5427): 15–18.

Smith, M. (1982). *Evolution and the Theory of Games*. Cambridge University Press, Cambridge, UK.

Straffin, P. D. (1980). The prisoner's dilemma. *UMAP Journal*,1: 101–113.

Straffin, P. D. (1989). Game theory and nuclear deterrence, *UMAP Journal*, 10: 87–92.

Straffin, P. D. (2004). *Game Theory and Strategy*. Mathematical Association of America, Washington.

Von Neumann, J., and O. Morgenstern (2004). *Theory of Games and Economic Behavior* (60th Anniversary ed.).Princeton University Press, Princeton, NJ.

Webb, J. (2007). *Game Theory: Decisions, Interaction and Evolution*. Springer, London, UK.

Wiens, E. (2003). Game theory: Introduction, battle of the sexes, prisoner's dilemma, free rider problem, game of chicken, online two person zero sum game. www.egwald.ca/operationsresearch/gameintroduction.php (accessed April 26, 2013).

Williams, J. D. (1986). *The Compleat Strategyst*. New York: Dover Press (original edition by RAND Corporation, 1954).

Winston, W. L. (1995). *Introduction to Mathematical Programming*. 2nd ed. Belmont: Duxbury Press, Chapter 11.

Winston, W. L. (2003). *Introduction to Mathematical Programming*, 4th ed. Duxbury Press, Belmont, CA.

12

Appendix: Using R

Before We Start: Setting Up the Workspace

Before working in *R*, it is necessary to set up the "workspace": the virtual environment in which you can load, manipulate, and analyze data. The code below cleans the workspace, erasing any previous objects or functions; sets the working directory, from which we'll load the data to analyze, and also loads a set of "packages" of useful functions that make data cleaning and analysis easier and faster.

```
###########################
##
## Setting up workspace
##
###########################
## Clear previous workspace, if any
rm(list = ls())
## Set working directory
os_detect<- Sys.info()['sysname']
if (os_detect == 'Darwin'){
setwd('/Users/localadmin/Dropbox/Research/StatsChapter')
}
## Load packages for analysis
pacman::p_load(
data.table, tidyverse, ggplot2, stargazer, easynls, pscl, pander
)
###########################
##
## Reading in the data sets used in this chapter
##
###########################
## Read in spring data
spring_data<- read_csv('./Data/01_correlation.csv')
```

```
## Parsed with column specification:
   ## cols(
   ## x = col_integer(),
   ## y = col_double()
   ##)
```

sigacts_data<- read_csv('./Data/06_poisson_sigacts.csv')

```
## Parsed with column specification:
   ## cols(
   ## sigacts_2008 = col_integer(),
   ## ggi_2008 = col_double(),
   ## literacy = col_double(),
   ## poverty = col_double()
   ##)
```

recovery_data<- read_csv('./Data/02_exponential_decay.csv')

```
## Parsed with column specification:
   ## cols(
   ## T = col_integer(),
   ## Y = col_integer()
   ##)
```

shipping_data<- read_csv('./Data/03_sine_regression_shipping.csv')

```
## Parsed with column specification:
   ## cols(
   ## Month = col_integer(),
   ## UsageTons = col_integer()
   ##)
```

afghan_data<- read_csv('./Data/04_sine_regression_casualties.csv')

```
## Parsed with column specification:
   ## cols(
   ## Year = col_integer(),
   ## Month = col_integer(),
   ## Casualties = col_integer()
   ##)
```

war_data<- read_csv('./Data/05_bin_logit_conflict.csv')

```
## Parsed with column specification:
   ## cols(
   ## side_a = col_integer(),
   ## cd_pct = col_double()
   ##)
```

sigacts_data<- read_csv('./Data/06_poisson_sigacts.csv')

```
## Parsed with column specification:
   ## cols(
   ## sigacts_2008 = col_integer(),
   ## ggi_2008 = col_double(),
   ## literacy = col_double(),
   ## poverty = col_double()
   ##)
alliance_data<- read_csv('./Data/07_bin_logit_alliance.csv')
## Parsed with column specification:
   ## cols(
   ## statea = col_character(),
   ## stateb = col_character(),
   ## alliance_present = col_integer(),
   ## igo_overlap = col_integer()
   ##)
## Format and subset casualties data
   afghan_data<- mutate(
   afghan_data
   , Date = as.Date(paste0(Year, '-', Month, '-', '01'), format = '%Y-%m-%d')
   ) %>% filter(
    Date > = as.Date('2006-01-01')
   , Date < = as.Date('2008-12-01')
   ) %>% mutate(
   DateIndex = 1:36
   )
## Print data as a tibble
   print(spring_data)
## # A tibble: 11 × 2
   ##    x    y
   ##   <int>  <dbl>
   ## 1   50 0.1000
   ## 2  100 0.1875
   ## 3  150 0.2750
   ## 4  200 0.3250
   ## 5  250 0.4375
   ## 6  300 0.4875
   ## 7  350 0.5675
   ## 8  400 0.6500
   ## 9  450 0.7250
   ## 10  500 0.8000
   ## 11  550 0.8750
```

Correlation in R

Using the *cor()* command in *R* on the data table:

```
## Calculate and print correlation matrix
   print(cor(spring_data))
## x y
   ## x 1.0000000 0.9992718
   ## y 0.9992718 1.0000000
```

Plotting in R

```
## Generate a plot visualizing the data
   spring_cor_plot<- ggplot
   aes
   (
   x = x, y = y
   )


   , data = spring_data
   )
   +

   geom_point
   (
   )
   +


   annotate
   (

      'text'
      , x = 100
      , y = 0.75
      , label = 'Correlation coefficient:\n 0.999272'
      , hjust = 0) +
   ggtitle
   (
   'Spring data scatterplot'
   )
   +
```

```
theme_bw
(
)
```

```
## Print the plot to console
plot
(
spring_cor_plot
)
```

Fitting an Ordinary Least-Squares (OLS) Model with Form yx + ε to the Spring Data in R

```
## Fit OLS model to the data
spring_model<- lm(
    y ~ x
    , data = spring_data
    )
```

Correlation Matrix in R

```
## Calculate and print correlation matrix
print(cor(recovery_data))
## T Y
## T 1.0000000 –0.9410528
## Y –0.9410528 1.0000000
```

Quadratic Regression of Hospital Recovery Data

```
## Generate model
recovery_model2 <- lm(Y ~ T + I(T^2), data = recovery_data)
## # A tibble: 15 × 6
## T Y index predicted residuals pct_relative_error
##    <int><int><int>  <dbl>    <dbl>    <dbl>
## 1  2   54  1 52.460836  1.5391644   2.8503045
## 2  5   50  2 47.640993  2.3590072   4.7180144
## 3  7   45  3 44.575834  0.4241663   0.9425917
## 4  10  37  4 40.200199  –3.2001992  –8.6491871
## 5  14  35  5 34.780614  0.2193857   0.6268164
```

## 6	19	25	6 28.672445	−3.6724455	−14.6897820
## 7	26	20	7 21.364792	−1.3647924	−6.8239618
## 8	31	16	8 17.033457	−1.0334567	−6.4591042
## 9	34	18	9 14.790022	3.2099781	17.8332119
## 10	38	13	10 12.213370	0.7866302	6.0510012
## 11	45	8	11 8.844363	−0.8443634	−10.5545422
## 12	52	11	12 6.926437	4.0735627	37.0323886
## 13	53	8	13 6.770903	1.2290967	15.3637082
## 14	60	4	14 6.511355	−2.5113548	−62.7838691
## 15	65	6	15 7.214379	−1.2143795	−20.2396576

Prediction

```
## Create a set of hypothetical patient observations with days in the hospital
   from 1 to 120
patient_days = tibble(T = 1:120)

## Feed the new data to the model to generate predicted recovery index values
predicted_values = predict(
 recovery_model2
, newdata = patient_days
)
```

Nonlinear Regression: Exponential Decay Modeling of Hospital Recovery Data

```
## Fit NLS model to the data
## Generate model
recovery_model3 <- nls(
 Y ~ a * (exp(b * T))
, data = recovery_data
, start = c(
 a = 1
, b = 0.05
)
, trace = T
)
```

Sinusoidal Regression

The functional form for the sinusoidal model we use here can be written as:

$$Usage = a * sin(b * time + c) + d * time + e$$

This function can be expanded out trigonometrically as:

$$Usage = a * time + b * sin(c * time) + d * cos(c(time)) + e$$

```
## Generate model
  shipping_model2 <- nls(
  UsageTons ~ a * Month + b*sin(c*Month) + d*cos(c*Month) + e
  , data = shipping_data
  , start = c(
  a = 5
  , b = 10
  , c = 1
  , d = 1
  , e = 10
  )
  , trace = T
  )
```

Sinusoidal Regression of Afghanistan Casualties

Visualizing data on casualties in Afghanistan between 2006 and 2008 shows an increasing trend overall, and significant seasonal oscillation. Once again, we want to fit a nonlinear model that accounts for the oscillation present in the data. We use the same sinusoidal functional form

$$Casualties = a * sin(b * time + c) + d * time + e$$

which as before can be expressed as

$$Casualties = a * time + b * sin(c * time) + d * cos(c * time) + e$$

The logistic model in R is treated as one case of a broader range of generalized linear models (GLM), and can be accessed via the conveniently named *glm()* function. Note that because *glm()* implements a wide range of generalized linear models based on the inputs provided, it is necessary for the user to specify both the family of model (binomial) and the link function (logit).

```
## Generate model
  war_model <- glm(
  side_a ~ cd_pct
  , data = war_data
  , family = binomial(link = 'logit')
  )
```

Poisson Regression

Visualizing count data in a histogram is a useful way of assessing how the data are distributed.

Histogram Plot

   ## `stat_bin()` *using* `bins = 30`. *Pick better value with* `binwidth`.

Poisson regression in *R* is also treated as a special case of GLMs, similar to the logistic regression covered in the previous section. As such, it can be implemented using the same *glm()* function, but now specifying the model family as 'Poisson', which tells R to implement a Poisson model. The model we use here can be specified as

$$Y = e^{\beta_0 + \beta_1 GGI + \beta_2 Literacy + \beta_3 Poverty}$$

   ## *Generate model*
   sigacts_model<- glm(
   sigacts_2008 ~ ggi_2008 + literacy + poverty
   , data = sigacts_data
   , family = poisson
)

Index

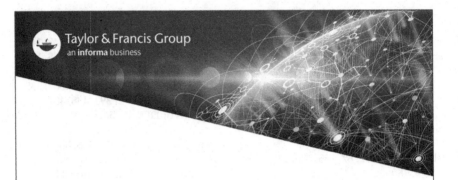

Printed in the United States
by Baker & Taylor Publisher Services

Printed in the United States
by Baker & Taylor Publisher Services